RADAR AND AIS
FOR WATCHKEEPING OFFICERS

Radar and AIS for Watchkeeping Officers

BY

DAIRE BRUNICARDI
Lt.Cdr. Irish Navy (retired), BSc(Hons), FNI, Master Mariner

GLASGOW
BROWN, SON & FERGUSON, LTD. NAUTICAL PRINTERS
4-10 DARNLEY STREET

First Edition   *2012*

ISBN 978-1-84927-027-4
© 2012 BROWN, SON & FERGUSON, LTD., GLASGOW, G41 2SD
Printed and Made in Great Britain

FOREWORD

Daire Brunicardi was one of the first people I met when I joined the navy in 1973. I have always remembered his kindness and empathy as we both joined having spent some time in merchant shipping. He left the navy in 1983 and joined the Nautical Department of the Cork Institute of Technology as a lecturer. Much of his work until his retirement in 2008 has been in radar training both on the college's training ship and in simulators. When the Nautical Department became the National Maritime College of Ireland he was the senior lecturer involved in the development and installation of the comprehensive simulator suite and associated equipment at the NMCI and lead instructor on the full mission simulators. Professional mariners both serving and in the future are fortunate that he has used his vast experience to write this book, experience which included seatime in a range of merchant and naval ships where he learned techniques unique to both services.

The book is designed to be user friendly emphasising the practical use of radar/AIS for navigation and collision avoidance in modern ships. Theory is referred to so that the OOW had an understanding of the basic technology and thus an appreciation of the limitations of the equipment.

Manual radar plotting related to the presentations on ARPA/ATA screens is included in a new way to develop in the OOW's mind a real appreciation of relative motion, which is a sense vital for young officers to develop.

Collision avoidance is dealt with in detail, but with particular emphasis on data presented on ARPA, traffic in multiple target situations and traffic lane situations. Ship handling is included for the purpose of collision avoidance and avoiding 'close quarters' situations, as well as navigating ships it confined waters.

The use of radar for coastal navigation including parallel indexing is comprehensively dealt with. The author enlisted the help of Paul Chapman FNI, Senior Pilot at Maritime Safety Queensland, Australia in developing the section on concentric indexing, i.e. the use of radar in monitoring controlled turns.

FOREWORD

There can be no doubt of the central role of radar on the bridge of a modern ship. Modern technology has allowed the radar picture to be presented in many versatile ways, such as integrating it with the electronic chart display, on overhead screens, on multiple displays such as bridge wing manoeuvring stations. The 'radar picture' forms the basis for the anti-collision plot (or interception or fleet manoeuvring plot in naval/coast guard use).

The author consulted widely on this project and was assisted by the Irish Navy, the Commissioners of Irish Lights, TRANSAS and the National Maritime College of Ireland. However, what makes this book special is the application of the author's vast experience in the subject. He has produced a valuable text book for use in training establishments, but it is of equal use as a reference book for mariners of long or little experience.

Captain James Robinson DSM FNI Irish Navy (Retd)
President
The Nautical Institute

March 2012

CONTENTS

INTRODUCTION

Since radar first appeared on the bridge of merchant ships in the 1950s, it has undergone considerable change, both in its technology and its functionality. It has changed from being an occasionally used aid, brought into play in unusual and difficult navigational circumstances to a situation where it is a central and always available tool in the watchkeeper's console. The sea has not changed in those intervening years, nor the basic duties of the OOW on the bridge in navigating the ship safely and maintaining what is now called situational awareness. But ships have changed as have navigational practices, the regulation of marine traffic, and many other aspects of the OOW's job. The benefits or otherwise of modern technology, particularly radio and computer technology have impacted immensely and to such a degree, and in such a short space of time that the OOW of thirty years ago would be lost in a modern ship's bridge.

Radar has undergone such radical change, and continues to do so. Currently some of the technology is much the same as it was fifty years ago, but this too is now undergoing change. Much of the new technology over the past decades has been in solid state systems and much improved reliability. The presentation of the radar information to the watchkeeper has undergone considerable modernisation and computerisation. Recent new performance standards produced by the IMO have recognised this changing technology and are worded in such a way as to allow for new developments while maintaining or improving the information and its presentation to the watchkeeper.

This book is designed for the modern ship's watchkeeping officer, or more correctly the officer on the bridge of a modern ship with up to date radar equipment. While some aspects of older technology are touched upon where necessary to explain where modern developments or practices have originated, the emphasis is on modern equipment and current practices. Likewise some notes on developments currently taking place are mentioned so that the reader can anticipate new practices or equipment.

The book has students at nautical schools and colleges in mind, at all levels from cadet or trainee to those studying for ship master, or various nautical degrees or other qualifications but also as a reference book to be carried to sea. The requirements of STCW, and of the relevant IMO Model Courses are particularly taken into account.

This book is considerably influenced by the work of William Burger and his seminal work in the early days of radar training for the mercantile marine. His influence in training on radar all over the world is still felt today. He defined the balance of technical knowledge required by the watchkeeper so that he was fully aware of what he was seeing on a radar screen, the correct setting up, adjustment and use of the equipment and the interpretation of the data in assessing the navigational risks to the ship are still the underlying factors. In other works on the subject, in lecturers' notes in nautical colleges one can recognise William Burger's influence, sometimes even where the author of the notes never heard of William Burger!

In the modern context, radar cannot be considered in isolation (could it ever?). It is integrated to a greater or lesser degree with other navigational aids. In the most basic form of integration there is directional input (gyro or transmitting magnetic compass) and usually speed (Doppler or electromagnetic logs) and position (GPS). In the incoming IMO regulations AIS data will be required to be shown on a radar display; it already is on many makes of radar sets. Data from and to the electronic chart system is common, the most usual form is the radar picture overlaid on the electronic chart, but tracks and navigation lines from the chart can be shown on the radar, and ARPA acquired targets can be shown on the ENC (Electronic Navigation Chart).

This book looks at the fundamental technology behind the display on the bridge. An understanding of this is essential so that the watchkeeping officer realises what exactly the return presented on the display is. He/she must thoroughly understand the limitations of the equipment. In the age of precise navigational systems, that can give the ship's position to within a matter of metres, it is vital that the user realises that radar is a blunt instrument in comparison. An understanding of the technology will help explain why that is, and be at the back of the watchkeepers mind when analysing the information that the equipment is providing.

A certain amount of the book is devoted to radar plotting. Surely this is 'old hat' and disputes what has been said above about a book for the modern ship's officer. Automatic plotting and other plotting aids are fairly universal, so why the emphasis on plotting? There are two reasons for this. The first is that in extreme circumstances the professional ship's officer should be able to plot and determined risk of collision when the automatic or other aids are not available. The second reason is to build up an appreciation of relative motion, so that when looking at the data being presented by ARPA, the data 'looks' right; that in the event of a malfunction of the equipment, or an error in a setting, a 'warning bell' goes off in the OOW's head.

While training in the use of radar and its underlying technology is a comprehensive part of modern nautical education, this must, of its nature be generic. There are so many different types of radar set on the market that it is impossible for any educational or training institution to have even a small sample of the different types. This also applies to books such as this. This then puts the onus on the individual ship's officer, the owners and master of the ship to ensure adequate training in the particular equipment on the ship's bridge. While there was some attempt to standardise symbols for the different controls, which has fallen into disuse, and there is ongoing discussion of 'S-mode' for a standard display of navigational data, currently there is no standard format for ships' radar displays other than the most basic stipulation as to how the radar information is to be presented, and the functioning of some basic controls. Therefore it cannot be emphasised enough that the ship's officers must familiarise themselves with the equipment in their own ship. This takes time and should not be done while on watch. Too often an officer joins a ship and within a matter of hours, or even less, finds him/herself on watch on an unfamiliar bridge, with a complex piece of radar equipment which s/he has never seen before and even if the situation of the ship is not too demanding, exploring the functions of the radar is a distraction to good watchkeeping. In a busy traffic situation, to be trying out the controls of the radar is positively dangerous. Also, without structured training, too often ships' officers will work out the few functions which they feel they need and will never get familiar with many other useful or essential functions of the equipment.

With the international nature of the modern mercantile marine in mind, and many ship's officers not having English as a native tongue, the language in this book has been kept simple. Also taken into account in this regard is its use by new entrant to the nautical profession. Jargon has been avoided and obscure acronyms. Traditional nautical terms have, of course been retained.

Chapter 1

Radar Basic Principles

This chapter explains the basic functioning of a marine radar display. It examines:

- the functioning of the various components and their relationship with each other,
- the primary display modes,
- the measurement or range and bearing and
- the processing of radar data for presentation on the display screen.

Radar is a name that is derived from 'Radio Direction and Ranging'. It is the process of using radio transmissions to establish the range (distance) of an object from the transmitter in a particular direction. Before the Second World War the principle was well understood and experiments were being conducted in several countries with varying success. It was the invention of the magnetron during that war however which allowed the construction of a compact piece of equipment which could be conveniently fitted in ships and aircraft. This valve, still in use today, although new technology is producing alternatives, is small in size but can produce the extremely high radio frequency which allows a usable degree of precision for range measurement.

Two radio frequency bands are used for commercial shipborne radar. The more common is in the 3 cm band which offers probably a more advantageous compromise than the alternative 10 cm band. The advantages and disadvantages will be discussed later.

Apart from producing the required frequency the radar equipment must have the ability to transmit a very short high powered transmission and then change to receiving mode to receive the echoes or reflections of its own transmissions off objects within range. This requires another very sophisticated electronic device, commonly called the TR (Transmit/Receive) cell which directs the transmission to the scanner (antenna) while shutting off the receiver, and then opening the receiver to the scanner to allow incoming echoes access to it.

The aerial, antenna or scanner (different names for the same thing) is the device which establishes the direction of a detected object. Most people are familiar with the rotating scanners on ships and at airports. It is the direction in which the scanner is pointing which establishes the direction of a detected object. It is obvious then that the siting of the

1

scanner, its construction, the control of its rotation are all vital in providing accurate data to the equipment and ultimately to the navigation and watchkeeping team on the bridge.

Range measurement
    For all practical purposes we can use the speed of light for the speed of the transmission of radar electromagnetic energy. The speed of radar energy may be considered constant at

300,000,000........metres per second.

That is *161,987*.....nautical miles per second.
    As we normally consider time in microseconds (ms, sometimes written as μsec.) when dealing with the speed of radar energy and functioning of the radar equipment, this converts the above figures into

300.....metres per ms (or μsec.), and
0.162...nautical miles per ms.

The magnetron produces a short burst (pulse) of electromagnetic energy. This flies out of the scanner in the direction in which the scanner is pointing at that particular time. Once it has left the scanner, the TR cell now opens the receiver. When the pulse strikes a reflecting object, some of it is reflected (echoed) back to the scanner where it is passed to the receiver, amplified and presented on the display in the bridge. Measuring the time taken between the pulse leaving the scanner and the echo returning, and knowing the speed of the pulse and the echo (which travels at the same speed as the pulse), the distance of the object can be calculated.

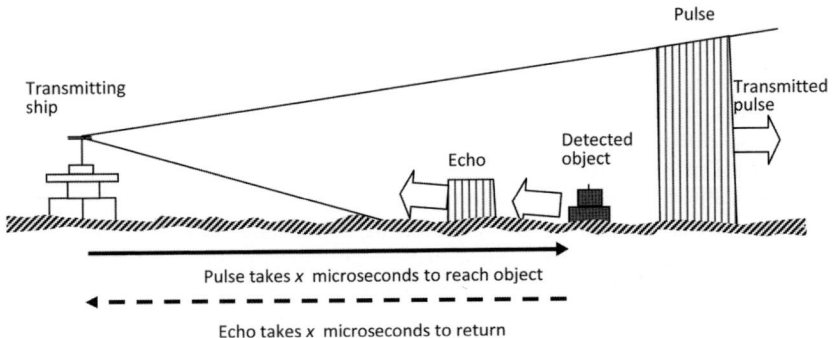

FIGURE 1.1   Pulse leaving the scanner and echo returning.

As a small exercise in understanding this principle calculate the time taken for a pulse to return from a target whose range is

   a. 50 metres.
   b. 12 nautical miles.
   c. 8 nautical miles.

Calculate the range where the time taken between transmission of a pulse and the reception of the echo is:

   a. 50 ms.
   b. 120 ms.
   c. 0.5 ms.

> *Note: One nautical mile is represented by elapsed time of 12.35 µs. Also 8 nautical miles gives a useful approximation. Work it out!*

It can be seen from these examples that the intervals to be measured are very short and the accuracy required is far beyond that of which ordinary time measuring devices are capable. The necessity of actually measuring time can be overcome by the use of a cathode ray tube. The earliest use of such was an oscilloscope. This is used to slow down or 'freeze' repetitious changes in electric current so that the wave pattern can be seen and measured. The use of this device produced the 'A-scan' which showed the timebase, a sample time period, in microseconds, which can also represent distance. The received voltages from the repetitious echoes are then shown and their time of arrival (or distance) (see Figure 1.2).

In early radars the measurement of time/range was overcome by the use of another cathode ray tube. This has a stream of electrons fired down the length of the tube to impinge on the inside surface of a specially coated screen which glowed in response to the electron stream. This electron stream was drawn out to the edge of the screen in time with the pulse leaving the scanner. The spot, the point of the electron stream being drawn out to the edge of the screen, moved at half the scale speed of the transmitted pulse. This had the effect that a returning echo, which, after amplification would cause the spot to glow brighter, would show it at the correct distance (range) from the centre. The speed of the spot, moving at a scale speed of 150 m/sec. would appear as a bright line on the screen (the timebase). This was made to rotate in coordination with the scanner. This then gave the correct bearing of the bright spot indicating a detected target. This form of presentation may still be found on older radar sets still in use (see Figures 1.3a and 1.3b).

Modern radar sets digitise the data received from the scanner. In

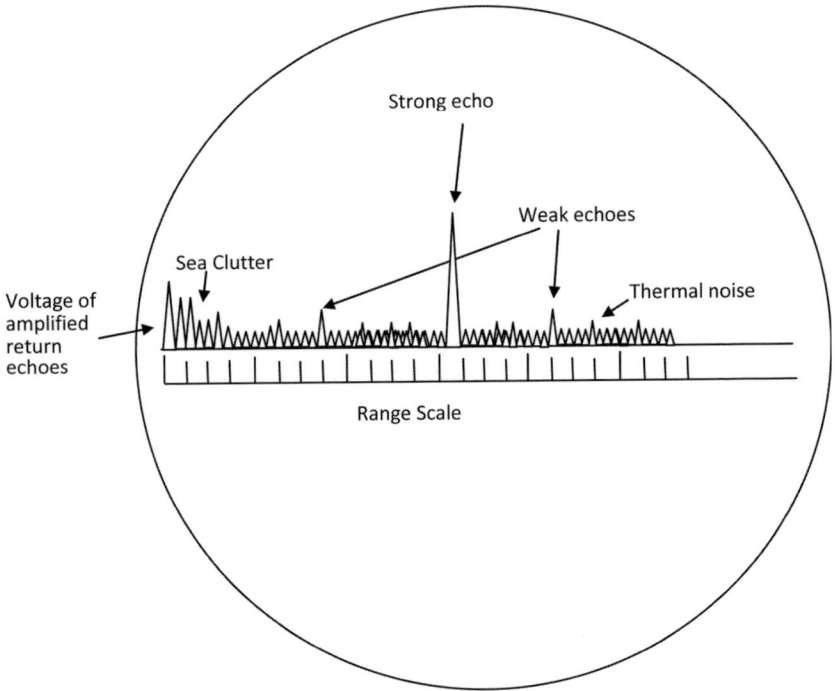

Figure 1.2 The A-Scan.

transitional technology, this digitised data was presented in the traditional rotating timebase described above. To the operator it appeared little different from a traditional display. However, this was soon replaced by raster technology which had the following benefits:

a. It allowed 'off the shelf' computer components to be used, thus reducing cost of production.
b. It reduced moving parts, i.e. the rotating coils required to cause the timebase to rotate in time with the scanner.
c. It permitted an increased use of 'soft' controls and adjustments (i.e. using roller-ball and 'click') thus reducing the necessity of elaborate keyboards and arrays of buttons.
d. More computer style graphics could be used for navigation and plotting tools and for displays of information.
e. It allowed colour to be use in the display.
f. Probably the best advantage is that it allowed daylight viewing.

Pulse leaving scanner

Echo returning to scanner

Scanner

Rotating trace, rotating with scanner

Spot brightens when Echo is received

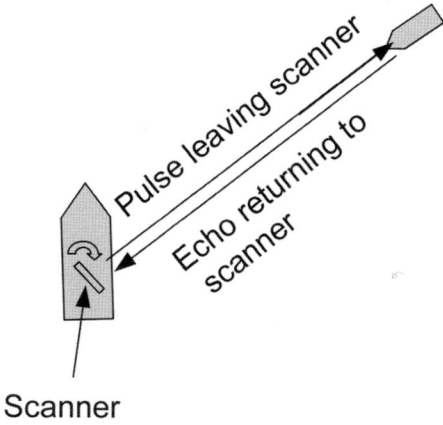

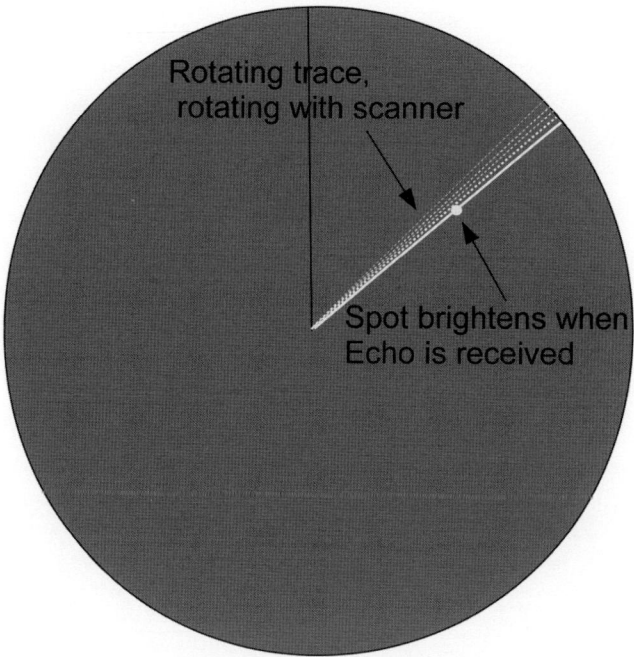

FIGURE 1.3a Rotating trace on CRT.

Scanner has rotated to
this angle from heading

Range of target

Scanner

Rotating trace,
Has rotated to this angle

Range of target

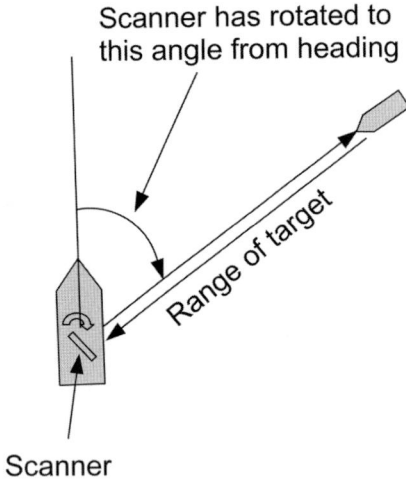

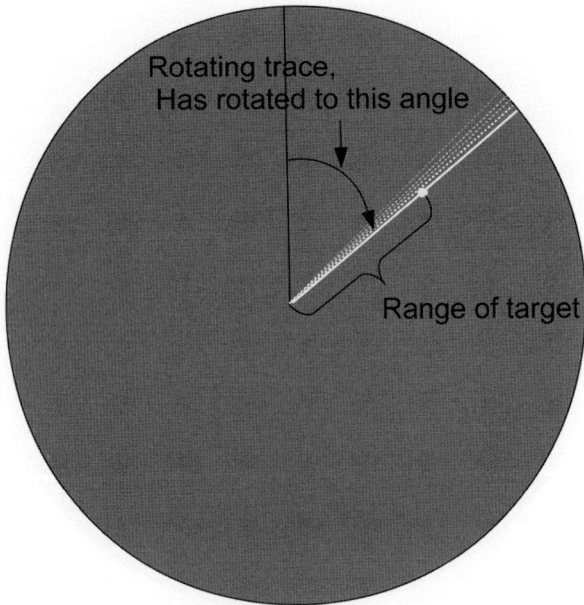

FIGURE 1.3b　Rotating trace on CRT.

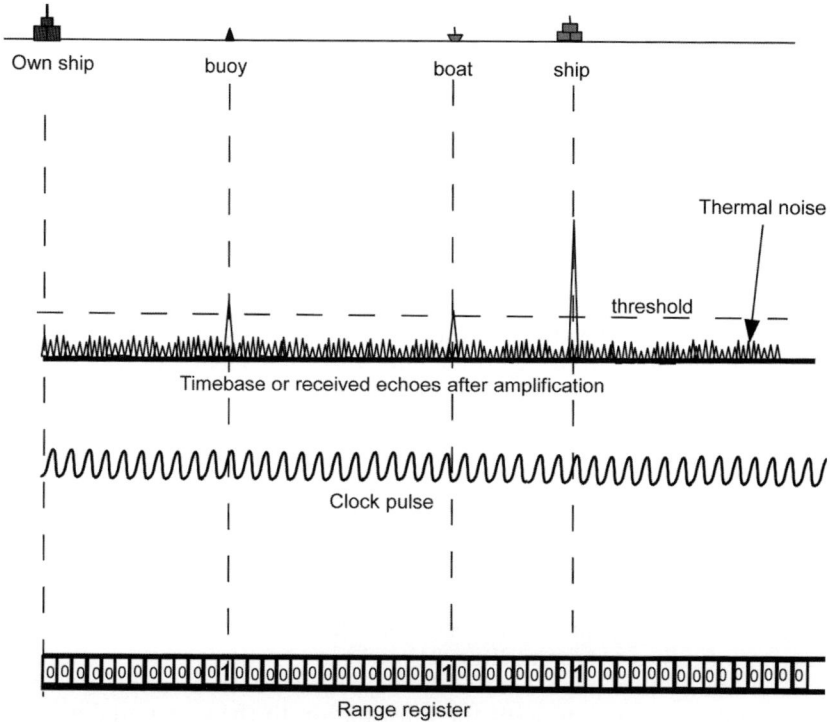

FIGURE 1.4   Digitised range data.

Returning to the display of target range, the incoming return of echoes from a pulse is amplified and then digitised by combining the voltages of the returned echoes against a clock pulse.

The clock pulse decides the number of cells for a particular range scale. When the voltage of a returning echo is above the threshold value it causes the associated clock pulse to record a value of '1' in the relevant cell, otherwise '0' is recorded in all the other cells.

Bearing Measurement

Having established the range of a target, the next requirement is to establish the direction. This is established by the direction in which the scanner is pointing at the same time as the range of the target is established.

Firstly a reference direction is established by carefully aligning a 'heading' contact switch in the scanner assembly with the ship's fore and aft line. This can be quite difficult, particularly when, in some cases the radar is not mounted on the centre-line in the ship. Even after the scanner is mounted, there is always an adjustment which can finely align the heading contact. When the scanner is pointing in the direction of the heading contact it causes a heading line or marker to appear on the display in the bridge, and all directions are subsequently referred to it. (Note: It is important that ship's officers check the heading marker occasionally. This should be done by observing visually a small object passing ahead of the ship and checking that when it is dead ahead that its echo is on the heading marker of the radar. Any error should be noted and applied to all bearings until the adjustment can be made to eliminate the error.) (See Figure 1.5.)

In older sets there was a direct linkage between the rotation of the scanner and the rotating timebase or trace on the radar screen in the bridge. In modern sets the rotation and direction of the scanner is digitised, usually by what is termed a 'shaft encoder'. This has multiple concentric rings, each with different numbers of segments. As the scanner rotates an electrical contact rotates across all the rings and making contact or otherwise with the segments. This writes a series of digital 'words' or binary numbers like the range measurement into a register of cells, containing either '1' or '0'.

Combining the range data with the bearing data a simple polar to rectangular conversion sends the information to the screen and a small group of pixels on the raster screen shows the relative position of the target (see Figure 1.6).

Range and bearing Precision

The concept of a 'point' target is a theoretical notion to help describe the most precise functioning of the equipment in displaying the smallest possible target. It is supposedly a target with no dimension, but with good radar reflective properties. Obviously there can be no such thing but a narrow dimension steel navigational perch might be the nearest thing to it in real life.

Two factors govern the display of our 'point' target. One is the 'size' of the pulse, determined by beam width and pulse length which is discussed in the chapter on the scanner. The other is the pixel dimension on the display screen. A 'pixel' is a 'picture cell' on a computer screen. The radar (or computer) raster screen is constructed of thousands of tiny squares in rows and columns, each of which can be activated or not to produce a 'Spot' image. In monochrome display nothing can be

Echo of target is just coming right ahead

Radar screen

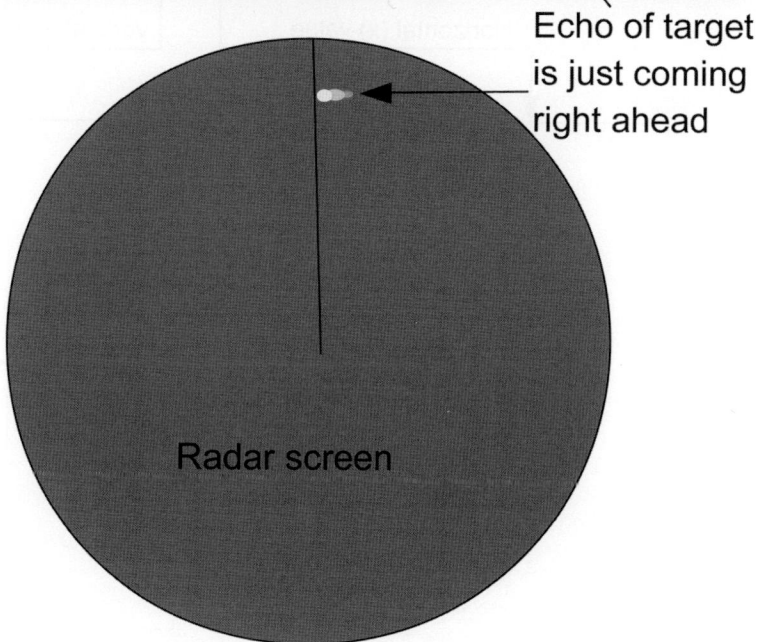

FIGURE 1.5 Checking HM.

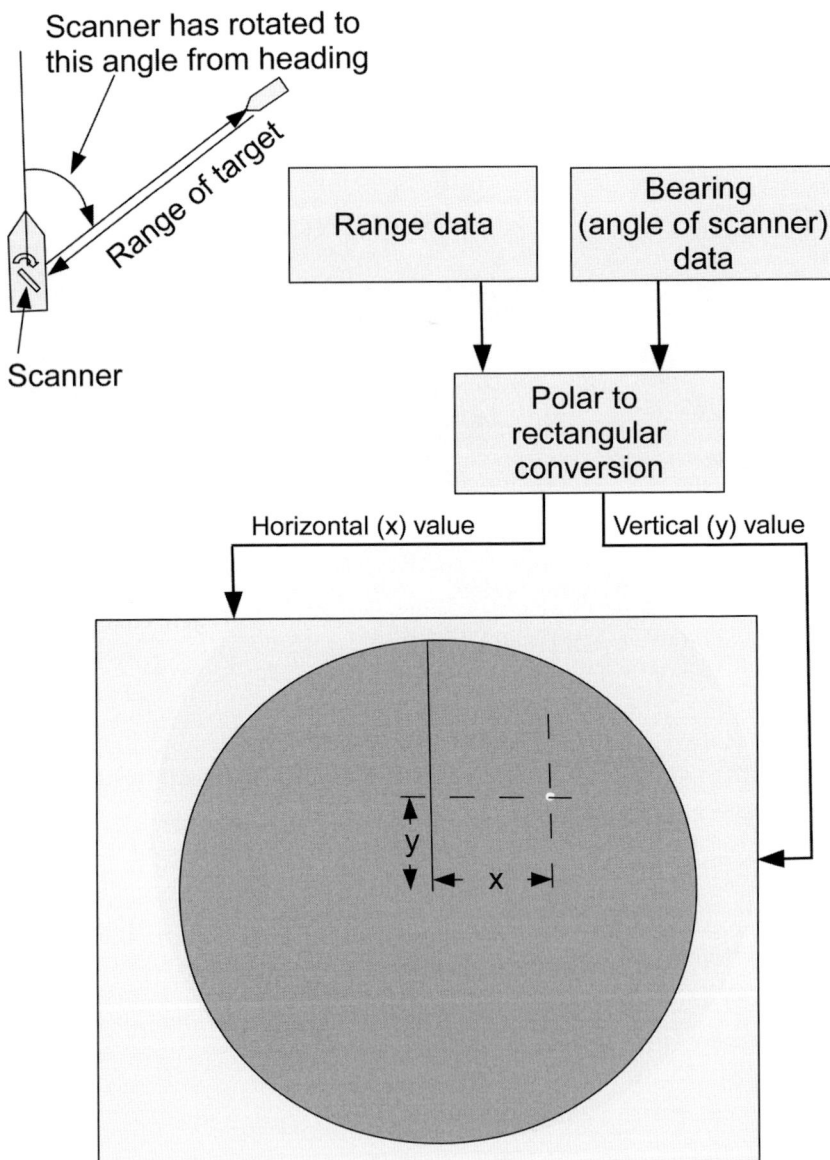

FIGURE 1.6   Range/bearing digital conversion.

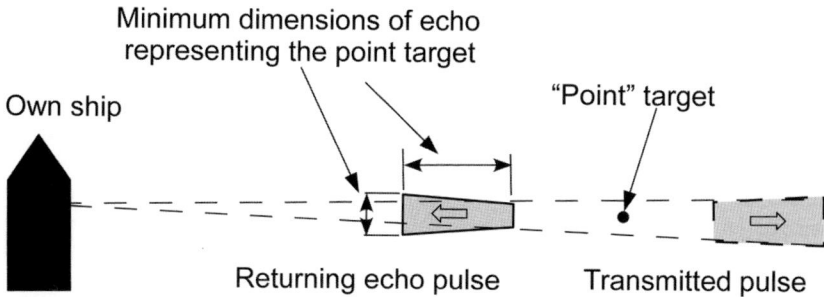

FIGURE 1.7 Echo size.

displayed smaller than a pixel. In a colour display there is a compromise in so far as the pixels are grouped in threes (called triads) for each colour, blue, green and red, so nothing can be displayed smaller than a three pixel group (see the chapter on the display). The varying degree of colour in each of the three pixels can represent a vast number of colours and shades of colour to the human eye.

A target displayed on the screen, no matter what its actual size, cannot be displayed smaller than a pixel, or group of three pixels on a colour display. This is not usually of concern to the observer on the bridge as the governing factor in determining the smallest possible return to be displayed is the dimension of the pulse. The image therefore of a point target is made up of a group of pixels. Let us take the radar set on the 1.5 mile scale. If we take a pulse length of 0.1 ms and a beamwidth of $1°$ and a 'point' target at 1 mile from own ship. The actual size of the pulse echo returning to the scanner will be length $300 \times 0.1 = 30$ m and width at one mile $\sin 1° \times 1852$ metres $= 32.3$ m. If the radius of the display is 170 mm, this represents 1.5 miles, or $1.5 \times 1852 = 2778$ m. Let us suppose that the number of pixels in a radius is 300. Then one pixel represents $2778/300 = 9.26$ m $= 85.75$ square metres. Our 'point' target has a minimum size of 30×32.4 metres $= 972$ square metres, therefore it is represented by $972/85.75 = 11.34$ pixels which is about 3×3 pixels. From this it can be seen that even the smallest target image on the screen is represented by a number of pixels.

Carrying out a similar exercise at a longer range, the 12 mile range scale, using of 0.5 ms pulse length and a 'point' target at 10 miles; the pulse length will then be $300 \times 0.5 = 150$ m and width at 10 miles, $\sin 1° \times 1852 \times 10 = 323$ m. The number of pixels in the radius is still 300. The radius of the display now represents 12 miles or $12 \times 1852 =$

22,224 metres. One pixel then represents $22,224/300 = 74.08 = 5487.8$ square metres. Our point target has a minimum size of $150 \times 323 = 48,450$ square metres so it is now represented by $48,450/5487.8 = 8.82$ pixels which is 3×3 pixels.

It can be seen from this that the smallest possible target detected by the equipment is represented by a number of pixels. It can also be seen that the governing factor in the size of a target's representation on the screen is pulse length and beam width and that the screen resolution (the number of pixels) is not a limiting factor. In fact most modern sets have a greater screen resolution than our examples above.

The Build-up of the Picture

The 'picture' on the radar screen is constructed from a combination of pixels representing the targets and other contacts (echoes) which are returning echoes to the scanner with sufficient power to be amplified and processed. We will take a simple example of a ship off a coast in the diagram below. As the scanner rotates it receives echoes from the objects which fall within its beam. For each echo received in response to each pulse transmitted a series of pixels will be printed on the screen. As the scanner rotates, the pixels printed in response to the received echoes will build up a series of images on the screen representing the coastline and the ship in their correct bearing and distance relative to the scanner. Thus, by the continuous rotation of the scanner a complete 'picture' is built up of the radar responsive contacts around own ship (see Figure 1.8).

Orientation

The Heading marker is produced by a pair of contacts which close the instant the radar beam crosses the ship's fore and aft line. The whole picture as it appears on the screen can be rotated so that the heading marker coincides with either the 000° graduation on the bearing scale around the screen, or with that graduation which corresponds with the ship's course (heading). These are the traditional orientations. These two ways of orientating the display are called '.Head Up.'(also called 'Ship's head Up' or 'SHU') and '..North Up...' (also called 'True North Up' or 'TNU'), respectively. In most modern sets the bearing scale itself can be rotated so that course of the ship is uppermost. This orientation is called 'Course Up'. There are further variations of 'Course Up' and 'Head Up' which are discussed below.

A 'Head Up' display is usually without compass stabilisation. This means that when the ship alters course the change of course is not

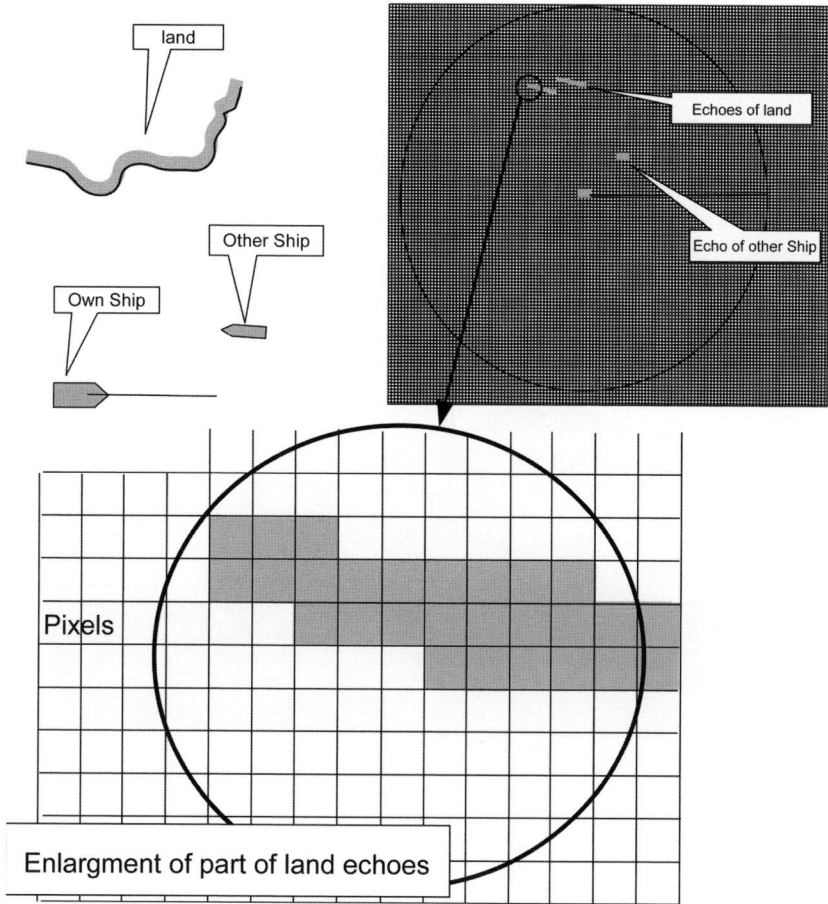

land

Echoes of land

Other Ship

Echo of other Ship

Own Ship

Pixels

Enlargment of part of land echoes

FIGURE 1.8 Build up of the picture.

applied to the heading marker. The heading marker remains at 000°
while everything on the screen rotates. It should be noted however, that
some modern displays will give a 'Head Up' display, but the bearing
scale will rotate in accordance with the ship's heading; in other words,
the heading marker will remain stationary, pointing up the screen, but
the bearing scale around the edge will show the ship's heading in line
with the heading marker, and bearings taken will be 'true'. If the ship
alters course the heading marker will remain where it is, but the bearing

scale will rotate to the new heading. This, of course requires gyro or compass input.

A 'North up' display is usually used with compass stabilisation. The bearing graduation remains stationary with 000° uppermost. This means that when the ship alters course the change of course is applied to the heading marker. The heading marker rotates to the new course on the bearing graduations and the rest of the picture remains stationary.

A 'Course up' display is used with compass stabilisation. This means that when the ship alters course the heading marker rotates to the new course and is no longer vertically up the screen. It then must be reset for the new course (see Figures 1.9a, 1.9b and 1.9c).

FIGURE 1.9a   Orientation. View before alteration of course.

FIGURE 1.9b   Orientation. View after alteration of course.

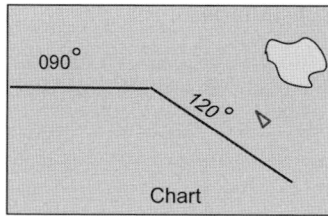

Chart

Head Up   North Up   Course Up

Before alteration of course

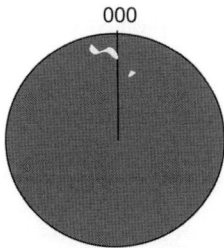

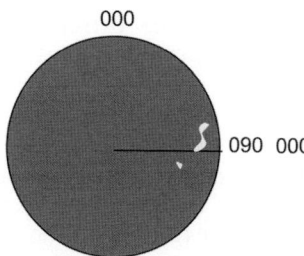

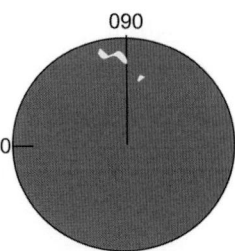

After alteration of course

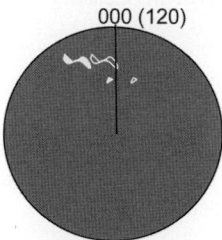

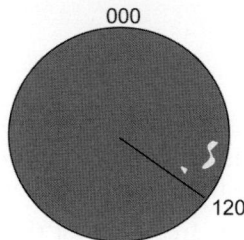

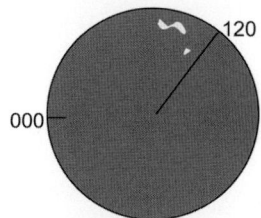

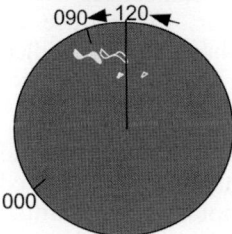

After alteration of course,
"Course Up" mode
requires reset
To bring heading
marker "up" again

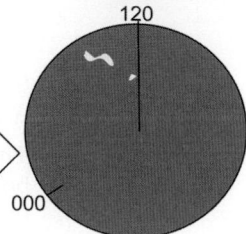

Head Up with rotating bearing scale

FIGURE 1.9c   Orientation. Chart and radar screens.

Main component of a radar installation

Marine Radar consists mainly of four elements – the transmitter, the scanner (also called aerial or antenna), the receiver and the display.

THE TRANSMITTER

Trigger Unit	Times the pulse (sets PRF). This changes with the range scale in use.
Modulator	Sets the pulse length and shape and powers the magnetron.
Magnetron	Sends out the extremely high frequency pulse in response to the modulator.

THE SCANNER

This concentrates or focuses the pulse into a narrow beam of radio magnetic energy. It also then acts as a receiver to pick up the returning echoes of its own pulses. Its continuous rotation through 360° is carefully regulated and its directional data (scanner rotation signal) is sent to the central processing unit (CPU).

THE RECEIVER

TR Cell	Shuts off the amplifier when the pulse is being transmitted and otherwise open the amplifier to the scanner to receive any incoming echoes.
Mixer	Mixes the frequency of the very weak echoes with a frequency from a local oscillator, to produce an 'Intermediate Frequency' (IF) that can be handled by the next stage of the receiver. The Tuning control works on the oscillator frequency.
IF Amplifier	This accepts the echoes produced by the mixer and boosts them to a level at which they can be detected by the next stage, the video amplifier. The manual gain control operates on the IF Amplifier. This will amplify not only the wanted echoes, but also the thermal noise, clutter and interference.

Video Amplifier This consists of a 'detector' which extracts the rectangular pulses of echoes from the amplified signal from the IF amplifier so that it can be used as input to computer memory for storage and processing. The 'amplifier' boosts these signals to the level required for producing an image on a cathode ray tube or LCD.

THE DISPLAY

Cathode Ray Tube [CRT] or Liquid Crystal display [LCD]
The CRT is a process whereby a beam or stream of electrons is projected onto the inside of a glass screen. This, by a process of directing the stream across the screen, forms the radar 'picture'. The intensity of the electron stream is governed by the signal from the video amplifier. The LCD produces the same effect by electrical impulses across different layers of conducting and non conducting materials.

Shift Register This is a process whereby the data is from the video amplifier is stored before being sent to the CRT or LCD display. There is usually a process of 'correlation' to help eliminate unwanted echoes.

Central processing Unit (computer)
This carries out the function of combining the data from different sources such as correlated range data, bearing data, and data from other sources such as graphics (range rings etc.) for display on the screen.

Scanner Rotation Signal
This is digitised data from the shaft encoder on the scanner to the CPU, which provided the bearing data.

Figure 1.10 shows the different components in their related connections.

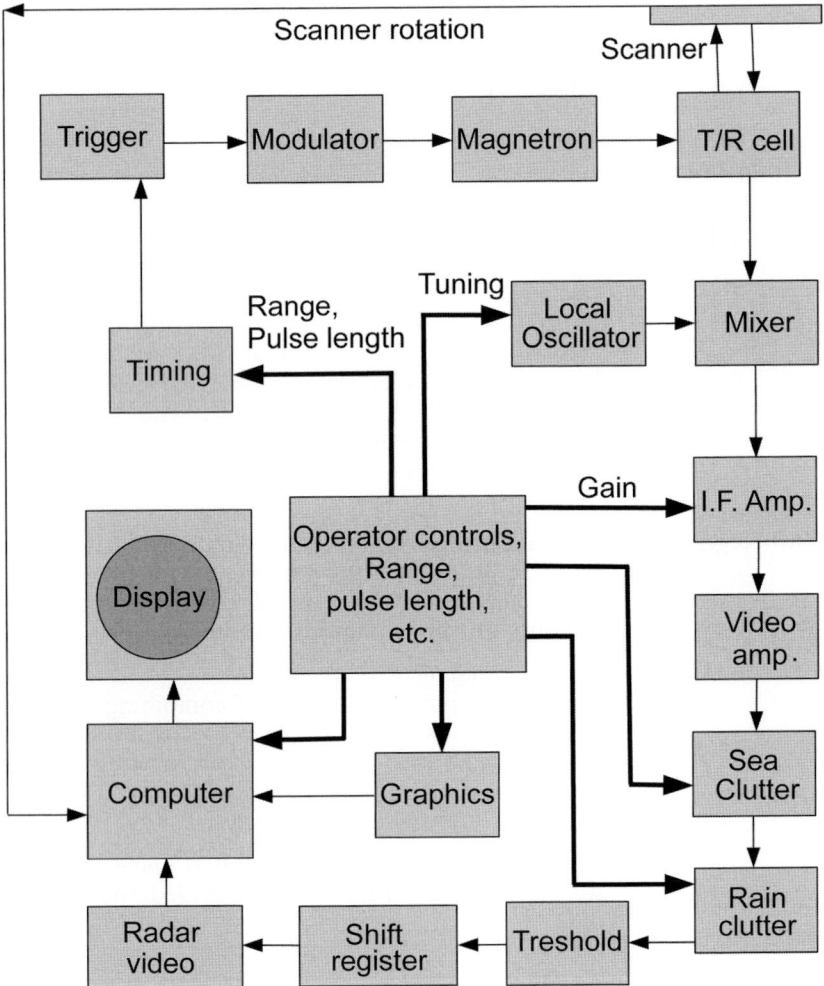

FIGURE 1.10   Basic block diagram of components.

Pulse Length and Pulse Repetition Frequency

In order to produce good range discrimination, that is to distinguish between two objects which are close together, and to give good range measuring accuracy, the pulses transmitted need to be short, and 'square'. That is they need to achieve maximum amplitude almost immediately. We can take it that the magnetron produces the necessary 'squareness'. The pulse length is set by the modulator timing. The term 'pulse length' may be a bit misleading as it is usually quoted in time measurement, that is, in microseconds. This, of course can be easily converted into a physical dimension for length. If a pulse length is quoted as being 1 ms, and the speed of the pulse is 300 metres per microsecond, then the actual length of the pulse is 300 metres.

There is usually an operator input to pulse-length, most sets offering a choice of 'Short' or 'Long' for each range scale, but this choice of pulse-lengths will be a function of the range-scale in use at the time. Pulse-lengths vary from 0.025 ms, to 1.3 ms. Different makes of radar sets offer different selections of pulse lengths. For example in the short range scales (0.5 to 3 miles) the operator might have a choice of 'Short Pulse' or 'Long Pulse' and this might give pulse lengths of 0.05 ms and 0.1 ms, respectively. Whereas on longer range scales 'Short Pulse' might be 0.5 ms and 'Long Pulse' 1.0 ms. Some sets will have intermediate pulse lengths corresponding to operators settings for 'Long' and 'Short' pulse-lengths for intermediate range scales. The manual for the radar set will list the actual pulse lengths for the different range scales.

What is not available to change by the operator is the Pulse Repetition Frequency (PRF) sometimes called the Pulse Repetition Rate (PRR). This is the number of pulses transmitted per second. The PRF is set by the manufacturer and is dependent on the selected range scale. Like other aspects or radar design, the pulse length and the PRF are a compromise. The same scanner is used for both transmission of the pulses and the reception of the returning echoes. The probability of detecting a target, particularly a weak one at long range will depend, among other factors, upon the pulse length. This cannot be too long,

a. for range discrimination, and
b. to allow the long period between pulses for echo reception.

An example of the various PRF and pulse lengths on a typical modern radar are set out in the table below. Note that this set gives the operator a choice of three pulse lengths in several of the intermediate range scales:

3cm radar:	Pulse length (in micro seconds) and pulse repetition frequency (PRF)									
Range Scale	0.125	0.25	0.5	0.75	1.5	3	6	12	24	48
Short pulse	0.06	0.06	0.06	0.06	0.06	0.15	0.3	0.3	1	1
PRF	3200	3200	3200	3200	3200	3200	1600	1600	800	800
Medium Pulse	0.06	0.06	0.06	0.15	0.15	0.3	0.5	0.5	1	1
PRF	3200	3200	3200	3200	3200	1600	1600	1600	800	800
Long Pulse	0.15	0.15	0.15	0.3	0.3	0.5	1	1	1	1
PRF	3200	3200	3200	1600	1600	1600	800	800	800	800

FIGURE 1.11   Table of pulse length and PRF 3 cm radar.

The time between pulses is extremely long compared to the pulse length. This is to allow the echoes from detected objects to be received. The time between one pulse and the next is the Pulse Repetition Period (PRP) in micro seconds. Typically for a pulse of 0.5 ms, there could be an interval of 1,000 ms. So it is obvious that for most of the time the set is in receiving mode. The shorter the PRP, the greater the PRF, in other words, with a shorter interval between pulses, the greater the number of pulses per second. At short range scales, the PRF is comparatively high, the PRP is less than in the longer range scales. This is because the time needed for echoes to return at the shorter ranges is less than that for the longer ranges. Also, the number of short pulses needed to get an acceptable return from weak targets is increased by the higher PRF.

Scanner Rotation and Pulse Repetition Frequency

The more 'strikes' a 'point' target receives, the greater the return for detection. As the scanner is rotating constantly, the number of pulses which strike a target will be function of the PRF, the horizontal beamwidth (HBW) and scanner rotation rate. Take, for example the scanner rotating at 30 revolutions per minute, the horizontal beamwidth is 2° and the PRF is 1000 Hz (pulses per second).

The scanner takes 60/30 seconds to make one rotation of 360°, that is 2 seconds.

This means that in that 2 seconds, 1000 × 2, or 2000 pulses are transmitted.

Therefore for each 1° of the 360° of the rotation, 2000/360 pulses are transmitted, which gives 5.5 pulses per degree. If the HBW is 2°, the number of pulses to strike a point target will be 5.5 × 2 which is 11.

Range discrimination

As the pulse has a physical length, this governs its ability to distinguish between two small or 'point' targets along the same line of bearing. In our example above we mentioned a pulse-length of 1 ms. This gave a pulse-length of 300 metres. If we use a shorter pulse length, of say 0.05 ms, this will give a length in metres of 15 metres. At any range scale, the theoretical range discrimination will be half the pulse-length. So, in our two pulse lengths mentioned here, at longer ranges targets must be separated by at least 150 metres to be distinguishable, at short ranges by 7.5 metres. This theoretical discrimination will be achievable in the shorter range scales, but in the longer ones, the actual pixel size on the display screen will be the deciding factor in distinguishing two point targets on the same line of bearing.

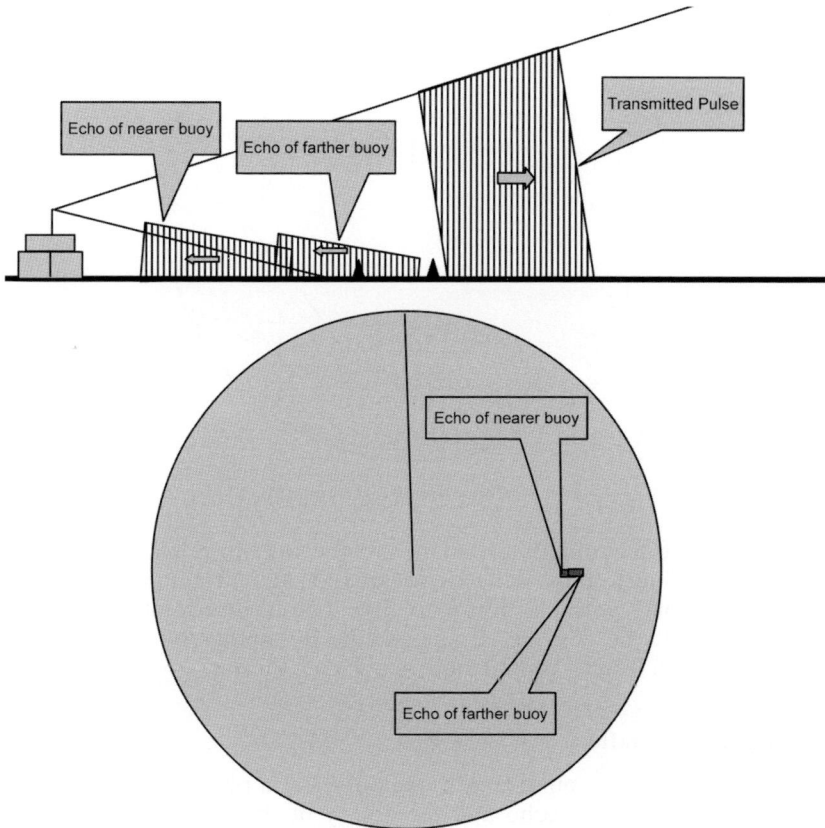

FIGURE 1.12a   Range discrimination 1.

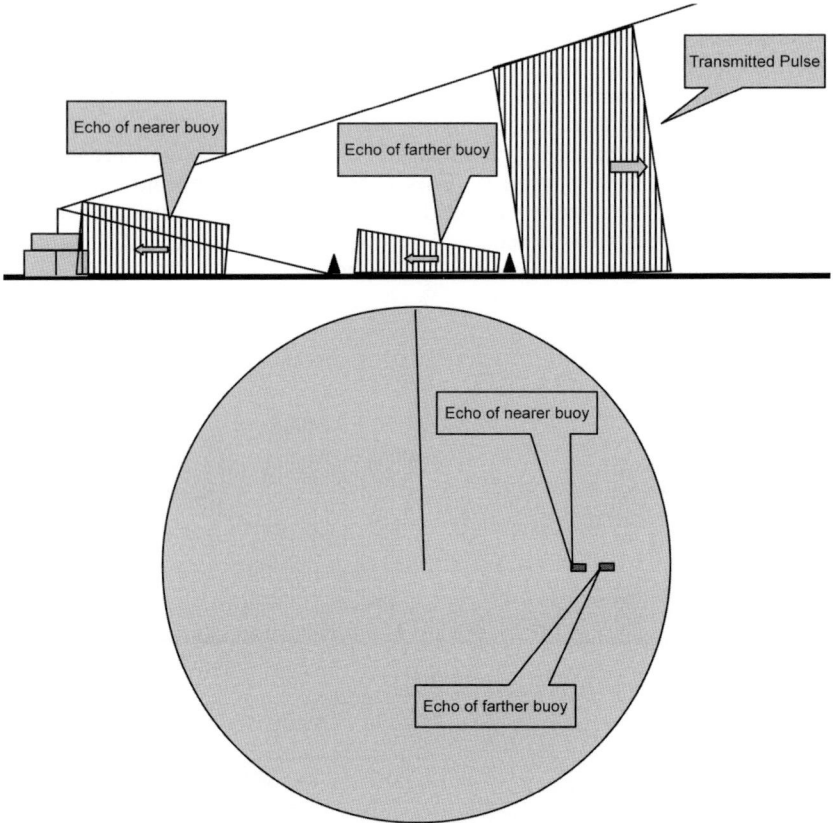

FIGURE 1.12b   Range discrimination 2.

It can be seen from this, that where the operator has a choice of 'Long' and 'Short' pulse lengths for any particular range scale, that selecting 'Short' pulse is going to give better discrimination. This, however has to be considered against the possibility of weak targets being lost when 'Short' pulse length is selected.

Bearing discrimination

In a similar way to pulse length, the pulse has a physical width, which determined the size of echo and the ability of the radar to distinguish between two small targets which are close together. This beam-width is decided by the scanner design. Unlike the pulse length, this beam-width

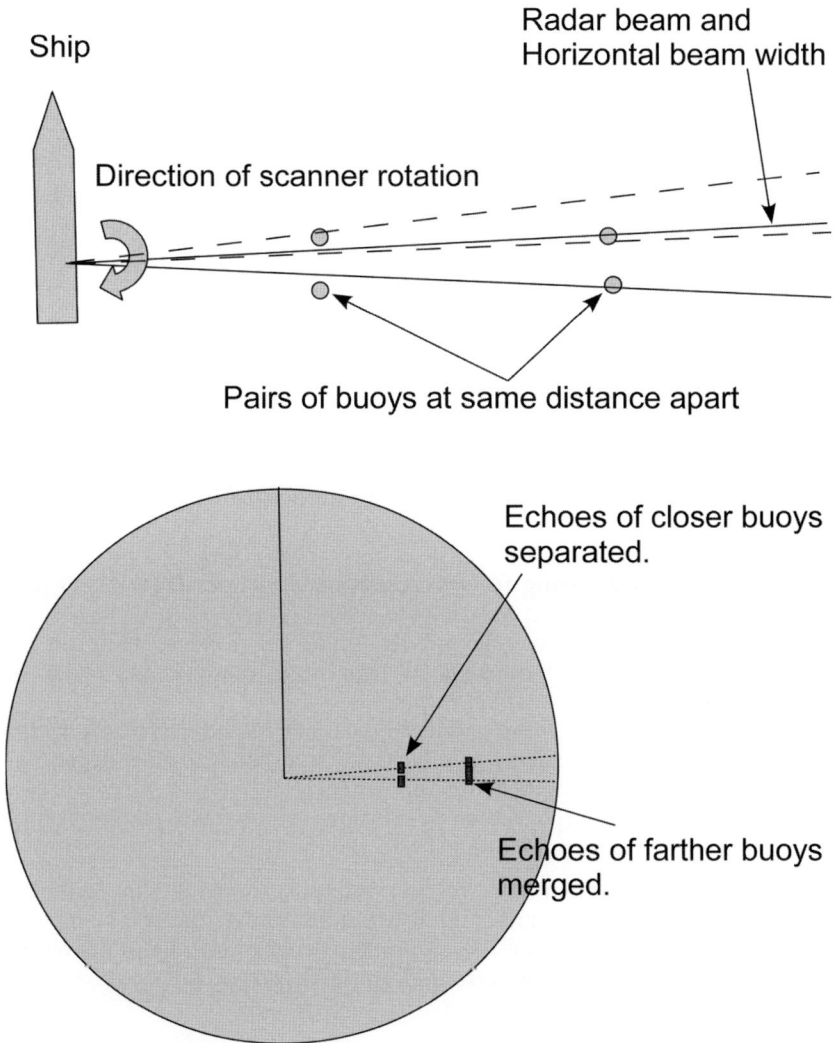

Ship

Radar beam and
Horizontal beam width

Direction of scanner rotation

Pairs of buoys at same distance apart

Echoes of closer buoys
separated.

Echoes of farther buoys
merged.

FIGURE 1.13   Bearing discrimination.

does not depend on range scale or any selection by the operator, but it fans out in width, so that two small targets which are distinguishable in azimuth at closer range will merge at a greater range (see Figure 1.13).

Picture Distortion

These factors of the size and shape of the pulse have implications in the distortion of the radar picture, particularly in distinguishing and indentifying features of land. The accompanying diagrams show how headlands can appear more or less pronounced when compared to the chart, narrow inlets might not appear at all, and offshore islands could appear joined to the adjacent land. This is also discussed later in the chapter on 'Radar for Navigation' (see Figure 1.14).

Conclusions

In this chapter we have looked at the radar in its most basic. Summarising we see that:

- The radar pulse is extremely short and moves at close to the speed of light requiring time measurement in millionths of a second (micro seconds).
- The time transmitting the pulse is very short compared to the time receiving, thus allowing for the reception of echoes from short and long distances,
- The number of pulses per second depend on the range scale in use,
- The radar equipment consists of four main components, each of which is discussed in further chapters,
- The size and shape of the radar pulse affect the ability of the radar to distinguish between targets near each other and the degree of precision in measuring bearing and range.
- In modern radars the screen resolution (number of pixels) is not normally a factors in the precision of the radar.

In the following chapters we will look at some of the technical features in more detail, sufficient for watchkeeping officers to have a basic understanding of how the equipment works and of the degree of reliability of the data being presented on the display.

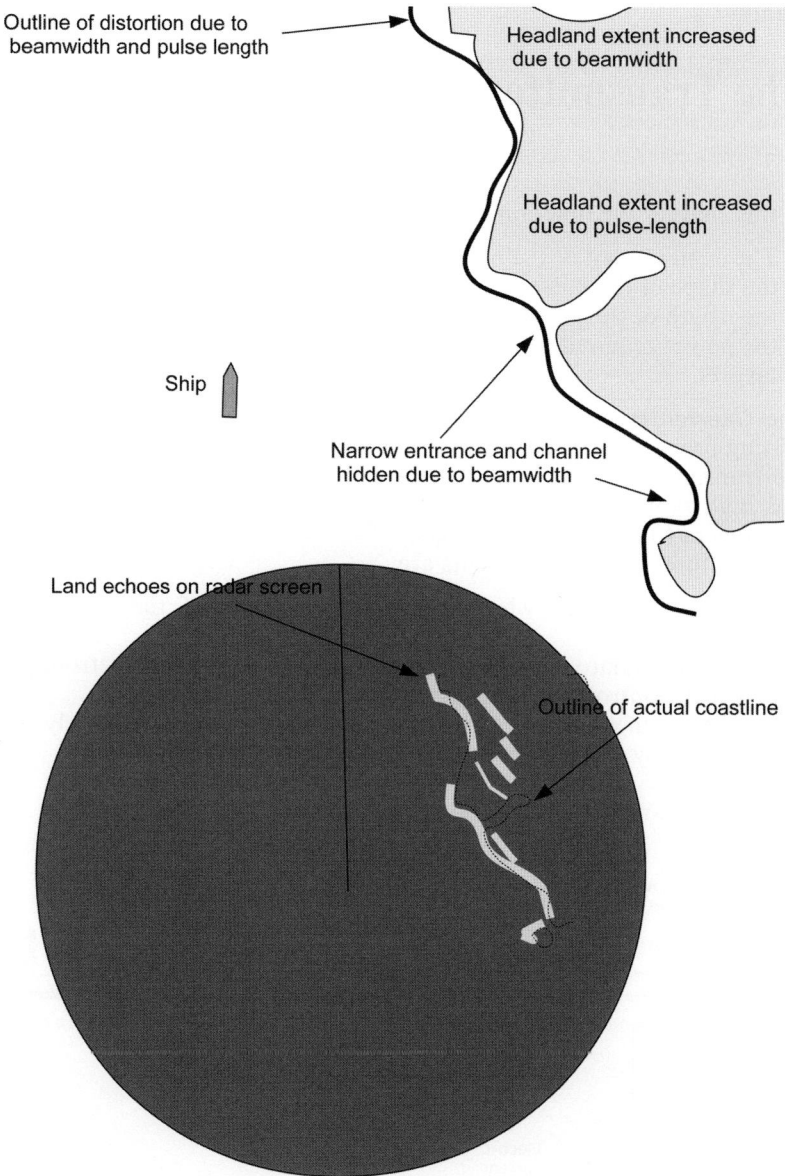

Outline of distortion due to beamwidth and pulse length

Headland extent increased due to beamwidth

Headland extent increased due to pulse-length

Ship

Narrow entrance and channel hidden due to beamwidth

Land echoes on radar screen

Outline of actual coastline

FIGURE 1.14   Showing distortion of land echoes.

Chapter 2

The Transmitter

In this chapter the function of the transmitter are discussed:

- The achievement of the number of pulses per second,
- the shape of the pulse,
- the length of the pulse and
- the power required to receive detectable echoes from small or distant targets.

The transmitter has to achieve three things; it has to start the transmitting process at a precise time interval, it has to define the duration of the transmission and it has to generate the extremely high frequency and high power pulse to be sent to the scanner. It consists of three component to achieve these things.

These three main components are:

1. The trigger unit which controls the timing of the transmitted pulses.
2. The modulator which provides power to the magnetron, and which controls the length, power and shape of the pulses.
3. The magnetron which sends out the pulses at the extremely high frequency and power on being activated by the modulator.

The Transmitter

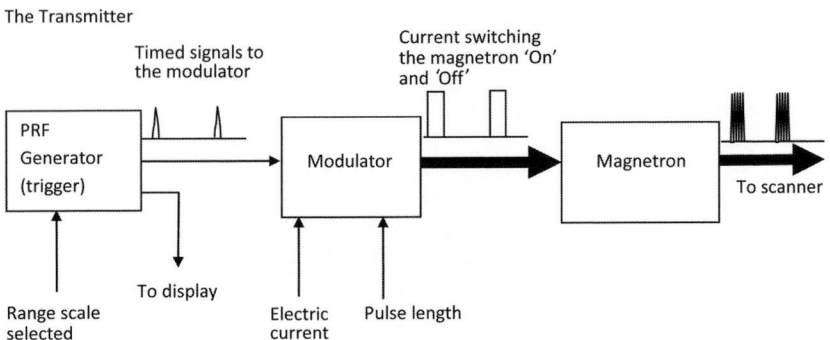

FIGURE 2.1   Components of the transmitter.

26

The Trigger

The first component is the Trigger, also called the PRF (pulse repetition frequency) generator. This incorporates a free running oscillator which generates a series of precisely timed low voltage pulses. The trigger can also be provided with a timing sequence from the central processing unit (computer) clock. The frequency generated by the oscillator is usually the highest frequency required by the system, and this can be subdivided to provide lower frequencies for longer ranges. The trigger fires the modulator which is discussed below. It also provides the timing to the receiver, as the timing of the pulse leaving the scanner provides the datum by which the time of returning echoes is measured. This timing is governed by the pulse repetition frequency (PRF) which is decided by the range scale selected by the operator. The trigger unit also sets the modulator charging function to charge up the capacitors in the pulse forming network (see below) of the modulator.

The Modulator

The modulator has two functions; it accumulates power over the comparatively long periods between transmitted pulses and it provides the short duration high power to activate the magnetron. The pulse length is decided by the range scale which has been selected, and the pulse length option within each range scale. These options, to be selected by the operator, are usually 'Long' or 'Short' for the particular range scale. As discussed under range discrimination, the different range scales will give different choices of pulse length.

In order to achieve the square pulse necessary for accurate range measurement, the power from the modulator to the magnetron must rise almost instantaneously from zero to several magnitudes of kilo volts (e.g. 10 kV or 20 kV). The 'Pulse Forming Network' (PFN) is an accumulator of electrical energy which is charged up in the period between pulses and is discharged suddenly when triggered. The trigger signal actives a very fast operating switch, usually called a controlled silicon rectifier which prevents current flowing to the magnetron until activated by the trigger. The PFN is a series of capacitive and inductive elements, which accumulates the electrical energy and then allows the constant rate of discharge. Normally when a capacitor discharges, the rate of discharge decreases gradually. This would be unsatisfactory for radar where a 'square' pulse is required; that is where the voltage increases suddenly from zero to the required level, remains at that for the duration of the pulse, and then drops immediately to zero again. The group of elements in the PFN, acting together achieve this. The

The Modulator

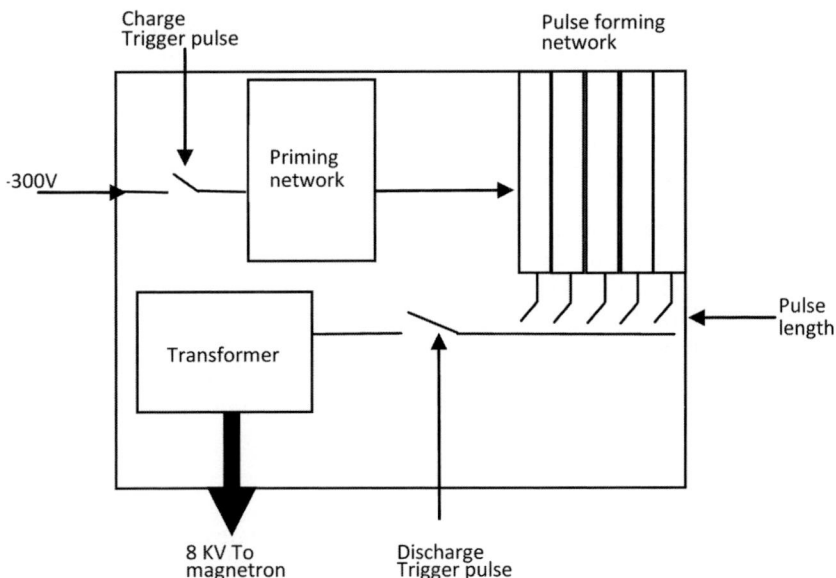

FIGURE 2.2   Diagram of the modulator.

cut-off, where the voltage drops to zero, is decided by the full discharge of the elements in the PFN. The pulse length is determined by the number of elements which are activated which in turn is determined by the range scale and pulse length decided by the operator.

The magnetron

The third component of the transmitter is the magnetron. This has been the core of marine surface radar for generations, and while there have been improvements in the design and reliability, the essential principle and construction has remained unchanged. There are modern developments which replace the magnetron and provide a more stable and reliable method of achieving the required transmissions, but the days of the magnetron are not yet over and it is still the essential component of most current shipborne radars.

The magnetron is a multi-cavity device with several resonant cavities, usually eight, all operating in the same frequency. These cavities are drilled around the inner edge of a cylindrical copper block which constitutes the anode. The cathode is a small cylinder placed centrally in the copper block (see Figure 2.3). The slots which open the cavities to the central chamber constitute the effective capacitor of the cavity while the cavity itself constitutes the inductor. The copper cylinder is placed between a powerful permanent magnet so that the flow of electrons between the cathode and anode is across the magnetic field.

The voltage from the modulator is applied to the cathode. The flow of electrons thus produced would under normal circumstances go directly to the anode, the copper block containing the cavities. However, the magnetic field of the permanent magnet affects the flow of the electrons as does the size and shape of the cavities, and their slots into the main chamber serve to produce the micro-wave frequency and

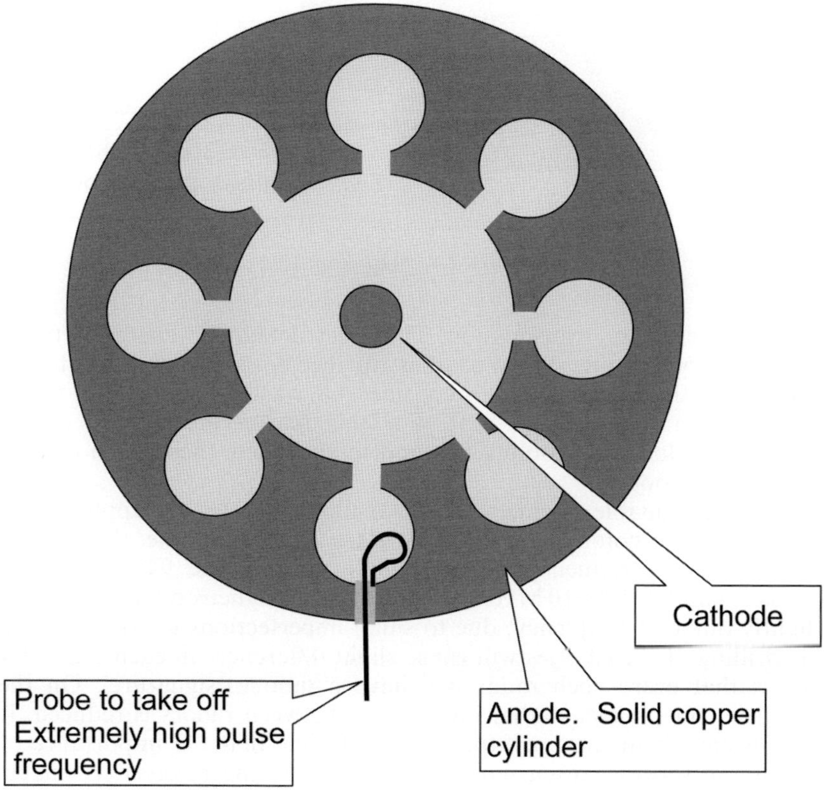

Cathode

Probe to take off Extremely high pulse frequency

Anode. Solid copper cylinder

FIGURE 2.3   Magnetron 1.

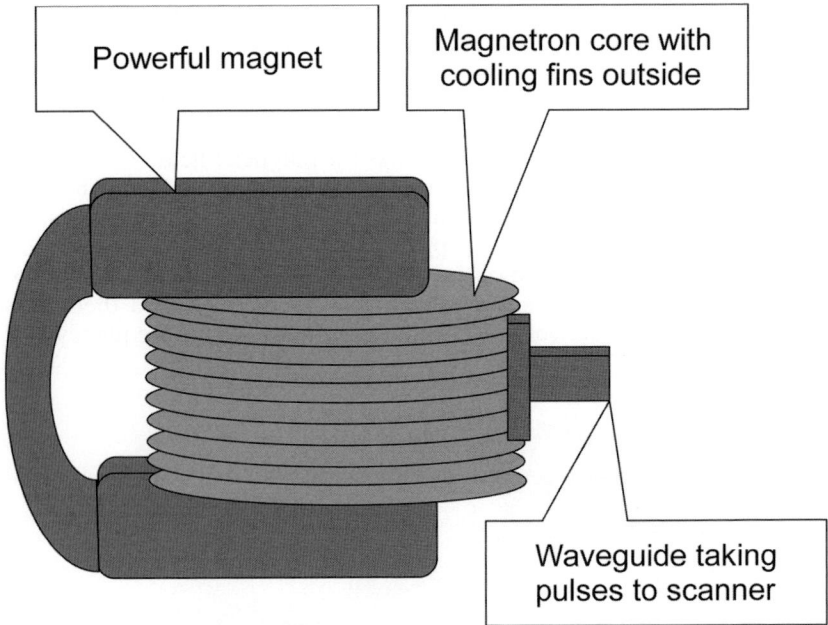

FIGURE 2.4   Magnetron 2.

power required for transmission. This electromagnetic energy is taken off by a probe in one of the cavities for feed to the wave-guide to the scanner. In this process there is considerable waste of energy and this is dissipated in the form of heat. For this reason the outside of the magnetron block has cooling fins and in some sets there is an electric fan to assist cooling.

Although tunable magnetrons are available for certain applications, those made for commercial marine radars are not tunable, that is they work at a set frequency. These radars work in the $9410 \pm 30\,\text{MHz}$ (X-band) and $3050 \pm 10\,\text{MHz}$ (S-band). Each magnetron will work at a slightly different frequency, due to small imperfections in construction; the drilling of the cavities will cause slight differences in each one. This means that every such radar will have a unique 'signature'. On the beneficial side, it means that interference between radars is reduced. It also means, from the operator's point of view that the importance of the tuning process, discussed in the section on the receiver, is under-stood. Also, as the magnetron heats up when started from cold, there will be a slight change in transmitted frequency which necessitates tuning, as discussed later, and as the equipment ages there will also be some slight change in frequency.

In the pressure for assigning bandwidth internationally, the commercial radar frequencies are being pressed to 'tidy up' the two internationally agreed bands, the X-band and the S-band. The magnetrons produce a certain amount of out-of-band transmissions, and these can interfere with other systems using frequencies close to the commercial radar frequencies. Also, magnetrons generally are not very efficient in the sense that they dissipate more energy, as heat, than they actually output. The new technology radars have much tighter bandwidth (the amount that a certain radio frequency is allowed each side of a central or assigned frequency) than conventional radars. Such new technology radars are called 'coherent' radars as opposed to the magnetron driven sets which are 'incoherent'. This term applies not only to the bandwidth but to other aspects of radar technology which are discussed elsewhere.

As stated earlier modern magnetrons have improved in so far as the life expectancy has been increased from about 2000 hours to over 10,000 and also have allowed for a reduction in the 'Peak Power' of transmissions from about 75 kW to 25 and even 10 kW.

The Duty Cycle

The Process of operation of the transmitter is called the 'Duty cycle', and is commonly referred to in works and manuals provided with the radar equipment. This is the relationship between the pulse length and the Pulse Repetition Period (PRP), the length of time between the start of one pulse and the start of the succeeding one. The pulse is only transmitted for a very short time compared to the pulse repetition period, say 0.5 ms, in a PRP of 1000 ms. So for 999.5 ms the transmitter is idle. (Well not quite idle as the Pulse Forming Network (PFN) in the modulator is accumulating energy for the next pulse). The duty cycle is the time the magnetron is active as a proportion, or a percentage of the PRP. For example, a pulse of 0.5 ms, in a PRP of 1000 ms is $0.5/1000 = 0.0005$ or 0.05%.

The 'Peak Power' is not, as is sometimes assumed the maximum instantaneous power of the sine wave carrier, the actual radio frequency, but the average power over one full carrier frequency cycle at the maximum value of the pulse output. It is approximately half the maximum instantaneous power (see Figure 2.5).

The 'Average Power' is the average power used during the entire PRP. If the peak power is, say 25 kW in the PRP mentioned above of 1000 ms and a pulse of 0.5 ms, then the average power is

Radar Pulse

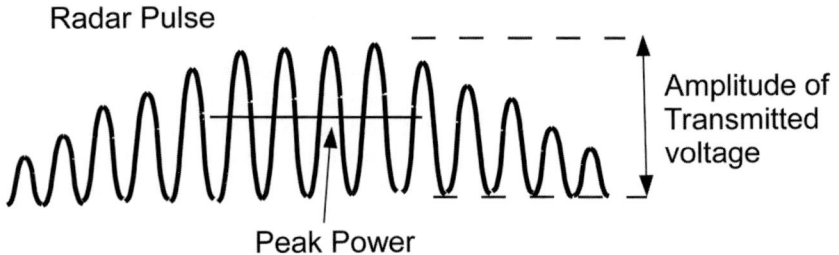

FIGURE 2.5   Peak power.

$$P_{avg} = (0.5/1000) \times 75\,kW$$
$$= 37.5\ watts.$$

The duty cycle parameters of a magnetron is important and while the operator normally need not concern him/herself with it, it is useful to see how changing range scales can influence the work and hence the life of the magnetron. In a set with a nominal peak output power of 25 kW and a magnetron with an average power capability of 80 W:

At short ranges the pulse length is 0.06 ms and the PRF is 3600 pulses per second.

At medium ranges the pulse length is 0.5 ms and the PRF is 1800 pps,

At long ranges the pulse length is 1.0 ms and the PRF is 900 pps.

In the first instance the PRP is 1,000,000/3600 ms (3600 pulses in 1 million ms) = 277.778 ms

$$P_{avg} = (0.06/277.778) \times 25\,kW$$
$$= 0.0054\,kW = 5.4\,W.$$

In the second instance above the PRP is 1,000,000/1800 = 555.556 ms

$$P_{avg} = (0.5/555.556) \times 25\,kW$$
$$= 0.0225\,kW = 22.5\,W.$$

And similarly the third instance for long ranges the average power is 22.5 W.

In all these instances the operation is well within the capabilities of the magnetron.

Conclusions

This chapter has explained some of the theory of a conventional radar transmitter. It gives the operator/OOW some idea of what is going on in this part of the radar equipment. It has, of course to be read with the chapters on the other components of the equipment and their interrelationship understood. It should also be read in conjunction with the chapter on New Technology(NT) radar which has fundamental differences in the transmission of radar energy.

Chapter 3

The Scanner

Introduction

In this chapter the following details are discussed:

- The construction of the most common type of marine radar scanner (also called aerial or antenna),
- The radar 'beam',
- The siting of the scanner aboard ship, and
- The false echoes produced as a result of the scanner's position in relation to the ship's structure.

The scanner, aerial or antenna, is one of the most fundamental components of the radar installation. The transmission of the pulses and the reception of the incoming echoes pass through it. Its design therefore is crucial to achieve four things:

1. The effective and efficient transmission and reception of pulse and corresponding echo.
2. A narrow horizontal beamwidth to achieve good bearing discrimination and accuracy.
3. A wide vertical beamwidth to allow for the ship's rolling and pitching movement.
4. A steady and consistent rate of rotation to achieve an accurate 'map' of the surrounding area and targets, and this must be achieved even in situations of high winds and violent motion.

The position of the scanner is also the effective position of the radar itself. It is the position from which ranges and bearings are measured and while its horizontal distance from the conning position is not usually of concern in many ships, in an age of very precise navigational data, and very large ships, such a remove from the central conning position may need to be taken into account. This is called the Consistent Common Reference System (CCRS) and the position where all navigational measurements are referred to is the Consistent Common Reference Point (CCRP).

Scanner Construction and radar beam

The most common early marine scanners took the form of a parabolic reflector. This is quite a simple concept where the end of the waveguide from the transmitter/receiver is at the focal point of a curved reflecting surface. The shape of the reflecting surface concentrates the transmission into a narrow horizontal beam, and a wide vertical one. In the reverse function, any incoming echo signals caught by the reflecting surface are focused into the end of the waveguide and thus to the receiver.

Such scanners have not been in general marine use for many years although they may still be seen in aviation, military and naval applications. The most common modern form of scanner is the end-fed slotted waveguide. This is a long rectangular-section metal box with a series of not quite vertical slots cut in one face of it. The waveguide from the transmitter/receiver is led through one end of the box and the pulse is transmitted through the slots. The spacing and the alignment of the slots are carefully adjusted to achieve a narrow horizontal beam.

Parabolic reflector scanner

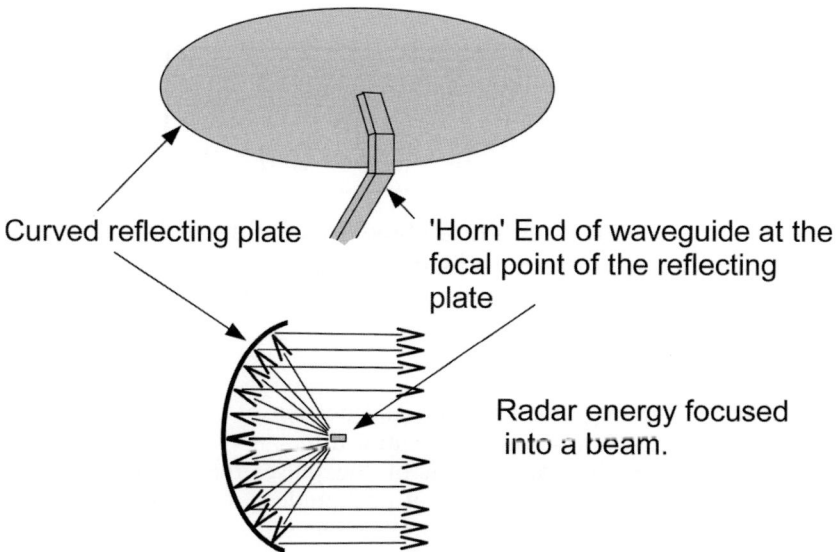

Curved reflecting plate

'Horn' End of waveguide at the focal point of the reflecting plate

Radar energy focused into a beam.

FIGURE 3.1 Parabolic reflector.

Section through antenna

Front view of antenna

Part of antenna showing slots angled
To control radiated energy

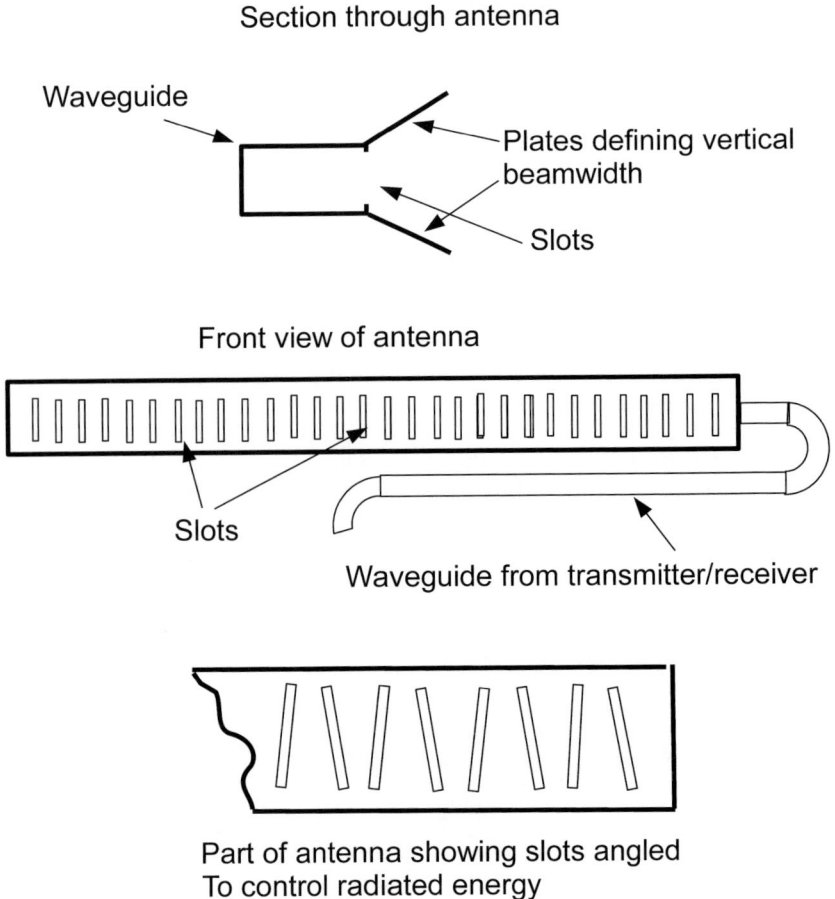

FIGURE 3.2   Slotted waveguide Antenna (scanner).

Metal plates running along the top and bottom of the side containing the slots achieve the wide vertical beam width.

In considering the transmission of the pulse, the scanner box, as described above is in fact a section of waveguide, and the conduct of the electromagnetic energy around its walls is the same as in the waveguide connecting the transmitter/receiver to the scanner. The slots, each of which is half a wavelength in length, and half a wavelength apart, act as a series of dipoles transmitting in phase. The vertical slots produce a horizontally polarised transmission. This has been found to be better than vertically polarised for marine use, as most marine targets are wider than they are high and thus respond better to horizontal polarisation.

One dipole will transmit an omnidirectional signal, in other word it will transmit all around equally and will have no directional properties.

A series of them in a line and separated by half a wavelength, transmitting simultaneously in phase, will produce a strengthened signal along an axis at right angles to the line of the array. This is caused by the individual transmissions from each of the dipoles falling into and out of phase with all the others depending on the distances that the wave patterns have travelled. It will also produce lesser concentrations of electromagnetic energy at divergent angles from the main concentration, called side lobes.

As can be seen from Fig. 3.3, the main lobes are in two directions, on each side of the line of the array of dipoles. The slotted waveguide scanner eliminates one of these by having a metal reflector at the back of the slotted waveguide, which also achieves the wide vertical beamwidth required.

The slots in the waveguide are cut at various angles to the line of the waveguide. As the current flows around the walls of the waveguide, a vertical slot will radiate horizontally polarised EM energy and the angle of the slot will control the amount of energy being radiated. Using this principle, the fact that the energy is being fed in from one end of the waveguide, and therefore the energy would tend to be radiated more strongly from the slots at that end than from those furthest away, the amount of energy being radiated from each slot can be controlled and equalised. Thus the slots nearest the incoming waveguide are nearly vertical, and thus tending to restrict the amount of energy being radiated, and those at the furthest end have a considerable angle of several degrees. By this method an equal amount of radiation is transmitted from each slot.

In order to avoid the beam being skewed by the angles of the slots, the slots are paired as to their angular characteristics, so if the first slot is at x° away from the feed end of the scanner, the next one will be x° towards it, and so on for the length of the scanner. Thus, in each pair of slots, the angular skew effect of one is cancelled by the other.

The end of the scanner poses a problem if the residual energy which has not been radiated resonates or echoes off the end away from the feed end, thus interfering with the radiations from the slots. This is dealt with by several methods. One way is an absorbent attenuator which absorbs the residual energy and dissipates it as heat. Another is a reflecting short circuit at the end, a quarter wavelength from the last

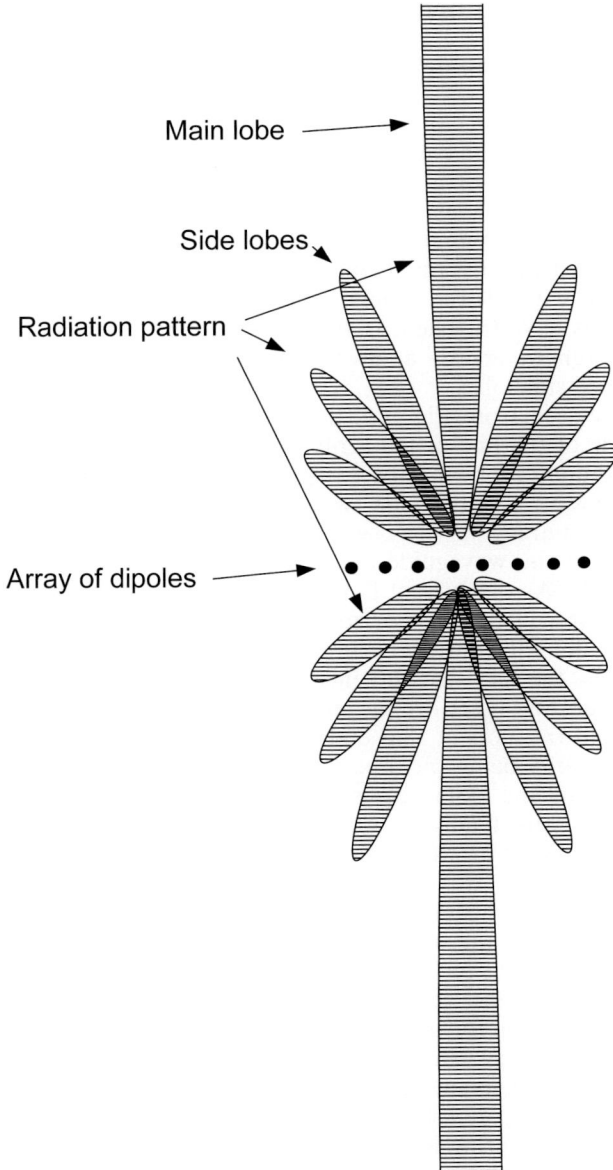

Main lobe

Side lobes

Radiation pattern

Array of dipoles

FIGURE 3.3   Array of dipoles and resulting polar diagram.

slot. This will resonate at the required phase to reinforce the radiation from the slots.

The horizontal radar beam thus produced will have several characteristics. The main focus of radiated energy is along the central axis at right angles to the scanner. On either side of this line, the radiated energy falls away rapidly. On each side of the main axis, points called the 'Half power points' are where as the name suggests the radiated power is half that in the main axis. It is the angle subtended by the half power points which defines the effective radar beam, and the degree of precision of the beam for measuring the bearing of detected objects (see IMO specifications).

On each side of the beam, and at large angles to it are lesser beams of radiated energy called side-lobes. The number and extent of these depends on scanner design. They are generally a nuisance but only come into effect for targets at very close range (see also Chapter 7 on false echoes). Their effect is to either produce a series of lesser echoes each side of the primary one or merging or 'smearing' of it in a curved pattern.

Because of minor unavoidable differences between the transmitted frequency and the dimensions of the slots, the axis of the main beam can emanate from a position slightly off centre from the scanner and slightly off right angles to it. This is called 'squint'. It has serious implications for bearings but can be, and is, offset by adjustment of the heading marker contact so that the correct bearings are determined by the radar beam and not the actual direction of the scanner.

Odd as it may seem, the wider the scanner, the narrower the beamwidth. To achieve a narrow beamwith, such as an effective one of 1° as required by IMO regulations, the width of the scanner is a function of the radar frequency. A typical 4m 'X' band (3cm) scanner can have an effective horizontal beamwidth of 0.65°, whereas one of 3m would have 0.85°.

The power, or strength of the main lobe, or radar beam diminishes with distance from the central axis of the beam, much like the way the light from a powerful torch or search-light is concentrated centrally in the beam. Where this energy diminishes to half the power of the central axis is considered the point at which the beam is non-effective and thus defines the width of the beam. This is effective for targets away from the ship, but with very strong echoes close in, the echo will be widened because energy from outside the half power points is being reflected. Such widening can then merge with the echoes produced by the side-lobes, mentioned above.

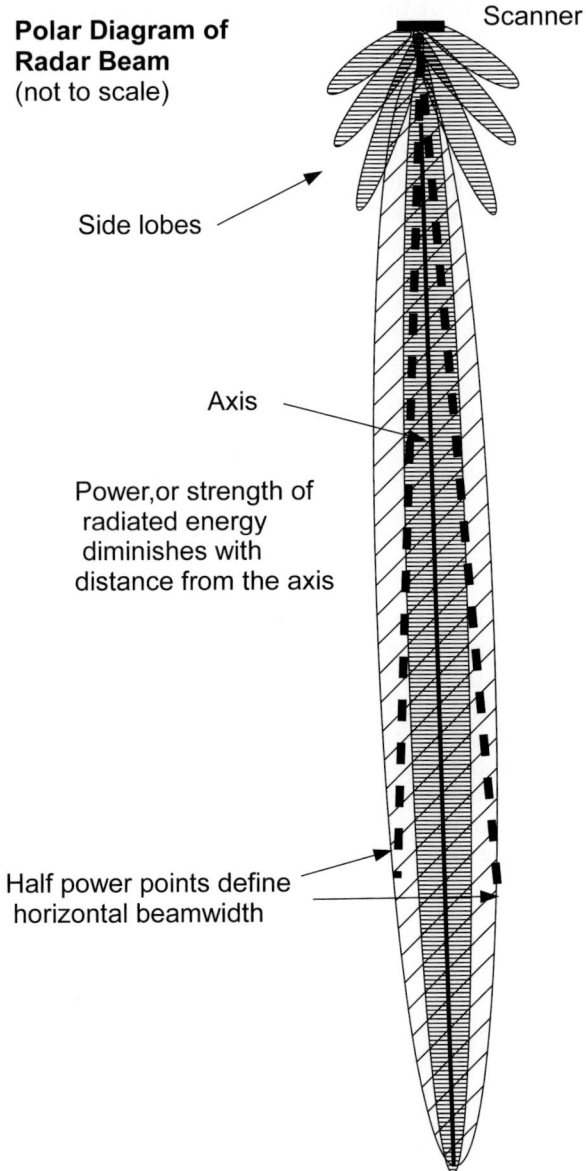

FIGURE 3.4   Polar diagram of radar beam.

Vertical beamwidth

The radar beam can be considered to be like a large fan rotating round the ship, very narrow looking down on it in plan view, but very wide in elevation, looking at it sideways. If the ship were suspended in space this fan shape would be an even 'teardrop' outline. But the ship is floating in the sea, and the surface of the sea, which is an excellent reflector of radar energy has a profound effect on the vertical shape of the radar beam.

The part of the radar beam which strikes the surface of the sea reflects, and changes phase by 180° as it reflects. This reflected energy interacts with that part coming directly from the scanner. This produces areas of increased effect and areas of diminished effect. These areas of increased effect are called vertical lobes. They are stronger than the radar energy which would be produced if the ship were suspended in space, thus giving enhanced detection power and range. As the first of

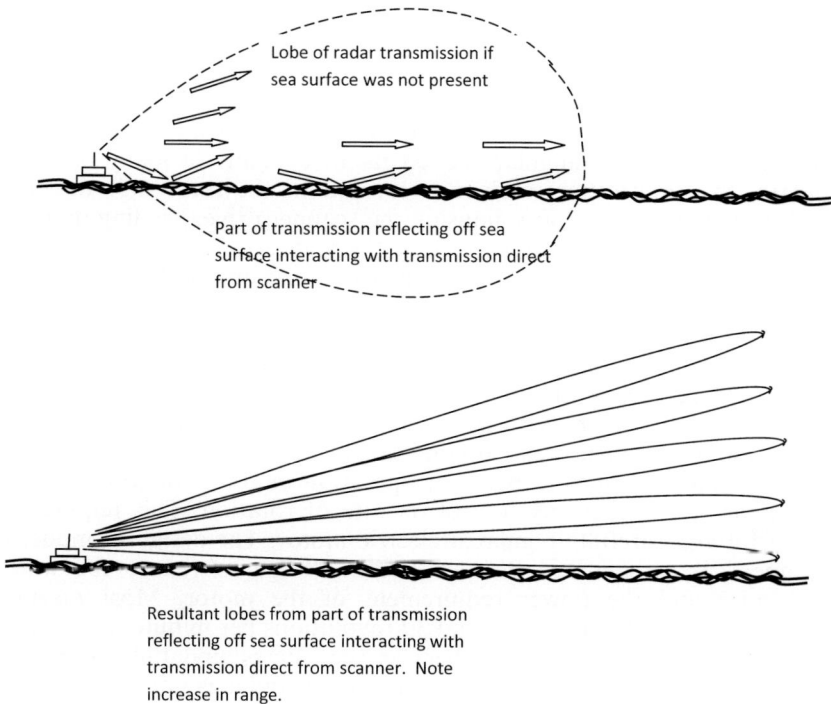

Lobe of radar transmission if sea surface was not present

Part of transmission reflecting off sea surface interacting with transmission direct from scanner

Resultant lobes from part of transmission reflecting off sea surface interacting with transmission direct from scanner. Note increase in range.

FIGURE 3.5 Vertical beam and lobes.

these lobes is along the surface of the sea, the fact that there are areas between the lobes of diminished radar energy is not of great concern to the merchant ship's officers. The one effect that is, is the fact that this bottom lobe lifts gradually off the surface of the sea thus lessening the effect for detecting low targets at a distance from the ship. Oddly enough, the movement of the ship, rolling and pitching does not affect the disposition of the vertical lobes.

One other effects of the reflective nature of the sea surface is to produce circular areas of reduced effect around the ship called 'Fresnel Zones'. The radar energy reflected off the sea surface and changed in phase by 180° causes a diminished effect close to the sea surface. This effect is usually not very high, about a metre or two, and the rings or circles, concentric with the scanner, are usually about at about 2 miles intervals and about a cable or two in 'thickness'. These do not usually cause any difficulty for normal ship borne navigational radar, but it does mean that small or low target may be temporarily lost. It could conceivably cause a problem if a guard zone or ring happened to coincide with one of the Fresnel zone rings, but practically the guard zone rings are much wider that the Fresnal rings and should allow the target to be picked up while within the guard zone, all other factors being allowed for.

Connecting the scanner to the transmitter receiver is the waveguide. This is a hollow rectangular section metal conduit. In many modern radars, the transmitter/receiver is located in the mast unit directly under the scanner, and also housing the scanner drive, heading marker switch and scanner rotation encoder. Such an arrangement eliminates long runs of waveguide, which are prone to various problems such as damage, dampness, vibration and the necessity for joints and other weak points. It also allows more versatility in the location of the scanner, such as sometimes seen, in the bows of certain ship types. The down-side of this is that it can make repair and maintenance more difficult when it may have to be carried out high on a mast, with limited room and in inclement weather conditions.

The scanner must be capable of operation in 100 knot winds (IMO requirement) without any loss of its rate of rotation. This requires a powerful and carefully regulated electric motor. The design of modern scanners, in streamlined Perspex housing helps to reduce the wind resistance and the power requirement of the motor. Most marine scanners rotate at between 20 and 30 revolutions per minute. Fast craft often have faster rotating scanners as the rate of updating the radar video (the picture on the screen) in a normal radar is not fast enough for safe navigation of such craft.

An essential part of the scanner assembly is the directional data to be sent to the display. The returning echo from a target indicates the range of the target, the direction of the scanner at that instant indicates the direction of the target.

The first item in establishing direction is the heading marker switch. This is quite a simple contact switch that send a signal to the display each time the scanner is pointing ahead (see the note on 'squint' above). This must be carefully aligned when the scanner is being installed. Even after installation there is usually scope for fine adjustment which can be made subsequently, if necessary. Such alignment can be quite difficult, particularly if the scanner is not placed on the ship's centreline, or if there is no fore-part of the ship clearly visible from the scanner position. As stated in the sections on operation and setting up, watch-keeping officers should use the opportunities of checking the heading marker when small targets pass ahead of the ship (see Chapter 1, 'Radar Basic Principles' and Figure 1.5).

The rotation of the scanner is usually taken from a device called a 'shaft encoder'. This consists of a series of concentric rings, each with a varying number of segments. As the scanner rotates contact is made and broken with these segments sending a binary series of '1s' and '0s' to the CPU computer. These binary series encode the direction of the scanner for any instant. Obviously, the greater the number of rings and segments the greater the precision of bearing data, but all that needs to be achieved in this regard is a precision some degree more precise than the effective radar beamwidth.

Before leaving the subject of the scanner, it is worth looking at possible future developments. In certain other radar applications, mostly military, the radar beam can be rotated electronically. In this, there are as it were, an array of fixed scanners disposed around a central mast, or post. It is possible to change the direction of the beam across the face of each of the fixed scanners in turn thus achieving a 360° coverage. Such a system, called 'phased array' allows a more 'intelligent' use of the radar beam such as having it 'slow down' to ensure more pulses strike a weak target, but mainly it dispenses with the rotating mechanism, requiring the powerful electric motor, the eventual 'wear and tear' and the vulnerability to damage. Such a scanner or antenna system needs the refinement of 'coherent' radar (see Chapter 18 on New Technology (NT) radar) and so is not likely to appear until more such radars appear on the market and meet IMO type approval.

Scanner Siting

It will not often be the case where a ship's officer is involved in where a radar scanner is sited or positioned; usually one joins a ship and there they are. But the position of the scanner in relation to the ship and its structure are important for several reasons that the ship's watchkeeping officers should be aware of. In positioning the scanner/s, very often it is a matter of convenience rather than the optimum from a radar detection point of view. Some modern feeder container ships have a tower of a bridge structure perched on the stern of the ship, the main purpose of which is that the bridge will be high enough to see over the stack of containers. The radars then will be on a short mast on top of the bridge. The naval architect designing a cruise liner is more concerned with the attractive lines of the ship than the position of the radar scanners; they will be mounted, but in a streamlined 'mast', curved and shaped to match the overall design and not appear ugly and out of place.

The height of the scanner is the first consideration. The radar's view is almost 'line-of-sight', in other words, the visible horizon is almost that of the radar. This would suggest that the higher the scanner the longer the detection range of the radar, which is correct. However, on the negative side, the higher the scanner, the greater the extent of sea clutter for any given situation. This has implications for the detection and tracking of weak targets close to the ship.

The imposition of parts of the ships structure into the radar beam have several implications. In the first instance they cause shadow and blind sectors.

Shadow sectors are those where the horizontal radar beam is wider than the obstruction. This can reduce the effect of the beam, but not fully obstruct it. The response of targets passing into a shadow sector can be reduced, with the effect that strong targets appear less strong and weak ones can disappear altogether, and if being tracked by ARPA/ATA, may be 'lost'.

In a blind sector the obstruction is almost as wide as or wider than

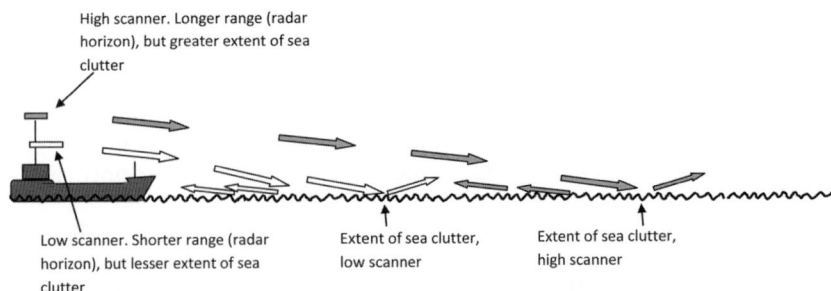

High scanner. Longer range (radar horizon), but greater extent of sea clutter

Low scanner. Shorter range (radar horizon), but lesser extent of sea clutter

Extent of sea clutter, low scanner

Extent of sea clutter, high scanner

FIGURE 3.6   High scanner v low scanner.

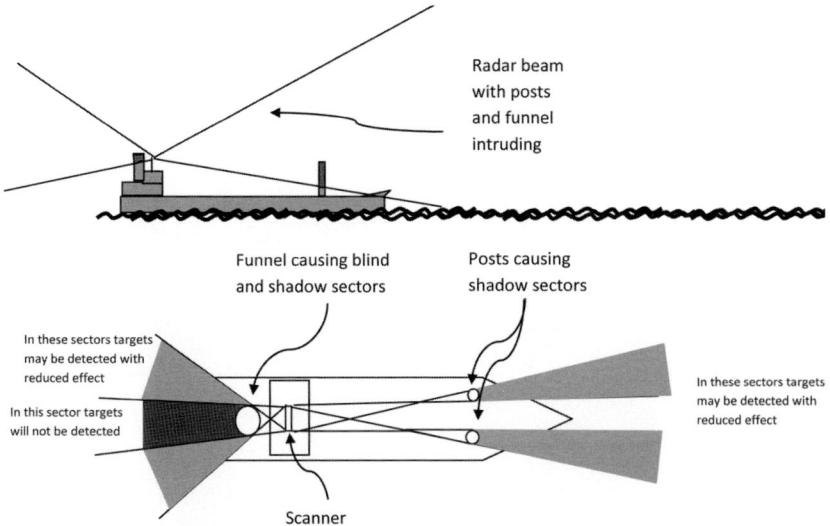

Radar beam
with posts
and funnel
intruding

Funnel causing blind
and shadow sectors

Posts causing
shadow sectors

In these sectors targets
may be detected with
reduced effect

In this sector targets
will not be detected

In these sectors targets
may be detected with
reduced effect

Scanner

FIGURE 3.7   Radar blind and shadow sectors.

the horizontal radar beam. In this situation, the beam is completely obstructed and targets in the sector cannot be detected. A target echo passing into such a sector will be lost, in every sense of the word although if the sector is not very wide the automatic tracking, if it is an 'acquired' target, may pick it up again as it comes clear of the sector.

At times of heavy sea clutter the blind and shadow sectors can usually be seen. The shadow sectors show as sectors of reduced sea clutter return; that of blind sectors show as sectors of no sea clutter return. It is also usual to see a sector of 'shadow' on each side of a blind sector. This is an area of reducing radar effect as the beam is progressively obscured by the obstruction, e.g. the funnel. The ship's officers should be aware of these sectors, both 'blind' and 'shadow', and where they exist there should be a diagram, near the radar displays showing them. Often, but not always, these can be astern, such as that from a funnel, but cranes, masts, and other high structural features can cause them in more important sectors (see Figure 3.8).

The fore part of the ship can also cause blind areas ahead. These will depend on many factors in relation to the size and design of the ship and the location of the scanner. The trim of the ship can also be a variable factor governing a blind area ahead. For example, a VLCC in ballast and trimmed by the stern, with a comparatively low scanner

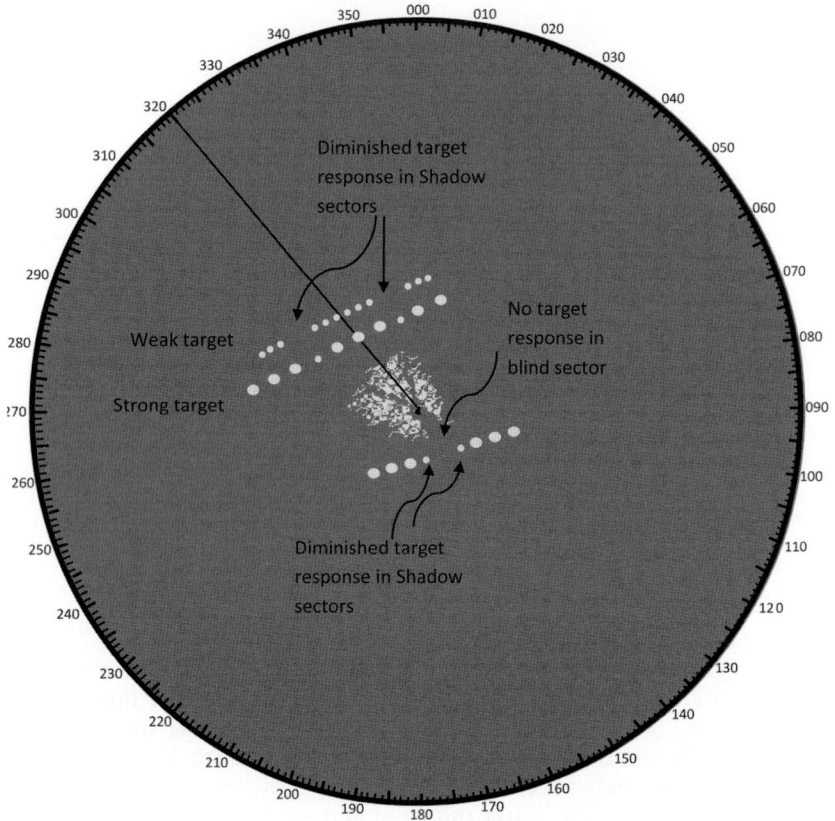

FIGURE 3.8   Radar screen showing shadow sectors in sea clutter and targets.

can have a blind area under the bows for a considerable distance ahead (see Figure 3.9).

False Echoes related to scanner siting

This section should be read in conjunction with Chapter 7 False Echoes.

There are two main generators of false echoes related to the scanner and parts of the ship's structure which obtrude into the radar beam. The first of these is the side-lobes. The power of these is considerably less than that of the main lobe, but a good reflecting target close to own ship will return an echo, not only from the main beam, but also from

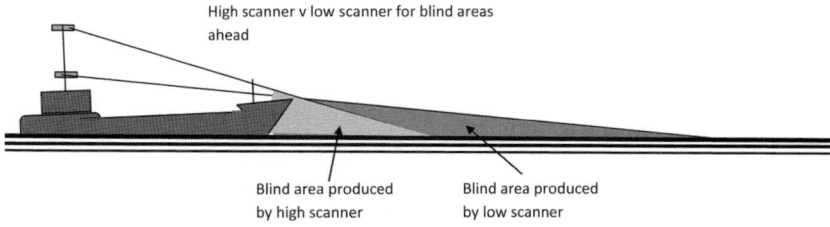

High scanner v low scanner for blind areas ahead

Blind area produced by high scanner

Blind area produced by low scanner

FIGURE 3.9 Blind areas ahead.

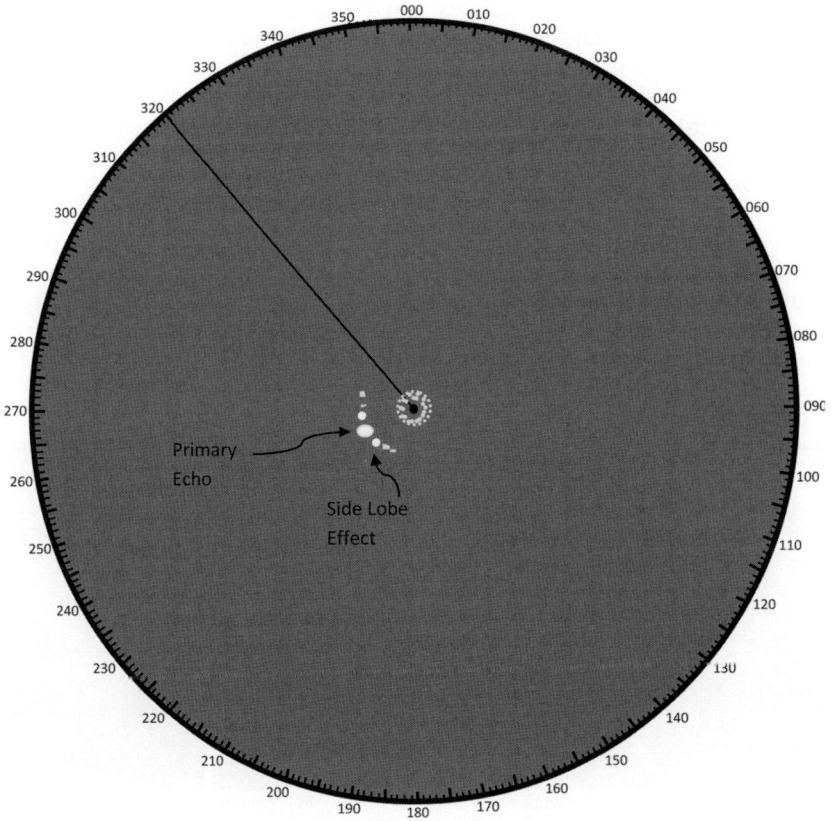

Primary Echo

Side Lobe Effect

FIGURE 3.10 Radar screen with sidelobe effect.

the side lobes (see above). This usually only occurs when the target is very close to one's own ship. It has the appearance of elongating the main target response echo in an arc both sides of the main echo, or else a series of smaller echoes in an arc each side of it and diminishing in size with distance from it (see Figure 3.10).

What is of more serious concern is the situation where part of the ship's structure re-radiates the beam and the echoes, in other words part of the ship's structure acts as a reflector, both of the transmitted pulse and also of the returning echo. These are called indirect echoes and can cause quite firm and strong echoes in positions where no target exists. Very often these are in blind or shadow sectors, and if these have been noted, then the appearance of an echo in these sectors can be suspected as being indirect or false echoes.

However, such indirect echoes can be caused by features which do not form blind or shadow sectors. Items which may be quite low on the ship but fall within the broad vertical beam of the radar can cause such a reflecting surface; for example it has been noted that the aft facing part of containers stowed on deck can provide it. They can also be caused by quite small structural features which are too narrow to form blind or shadow sectors and would tend to be ignored, but, with targets with strong echo response they can cause indirect echoes.

Less serious, but still of concern is the false echoes of land and other similar multiple targets, reflected off such structural features which can clutter the screen and obscure other wanted targets. This can be a real nuisance and while land causes most of it, other sources such as concentrations of shipping and fishing vessels can have similar effects.

Structure causing re-reflection   Re-reflection of land echoes

FIGURE 3.11   Photos showing structure causing indirect echoes.

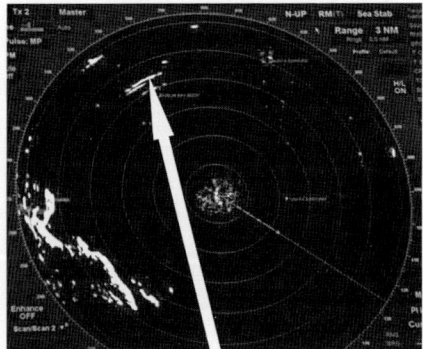

One source of such interference is the proliferation of off-shore wind farms. While they themselves show strong returns and are identifiable by their charted positions and usually regular layout, they do cause the sort of strong reflection off ship's structural features being discussed here and cause false and unwanted echoes in various part of the screen. It is rarely the case that the false echoes being discussed here, such as from land or wind farms would be mistaken for the real thing, but it is of course possible that such an error could be made.

As stated at the start of this section, it is rare that a ship's officer will have any input to the positioning of the scanner; but it may happen, if for example the officer is standing by a new build, or a new radar set is being fitted and therefore the officer should have some knowledge of the problems and factors. What is more likely is that the officer is presented with a situation that exists and therefore has to work with that situation. He/she must then look at the positioning of the scanner, and observe the screen, and assess the radar screen presentation in the light of the scanner position and ship's structure which impinges in the radar beam.

Conclusions

The design and construction of the scanner is the main factor in focussing the radar beam, both in its horizontal and vertical aspects and achieving the required beam-width in both. The scanner position is essentially the position of the radar; it is the position from which bearings and ranges are measured. The construction of the scanner must meet stringent specifications laid down by the IMO so that it can operate in the most extreme conditions. The height of the scanner above sea level and above the deck has implications for the range at which targets may be detected, the range of sea clutter interference and the possible blind area ahead of the ship. The surrounding ship's structure in relation to the position of the scanner may have serious implications for detecting and tracking targets and for producing false echoes. Ships' officers should be fully aware of these structural features and the effect they have on the display.

Chapter 4

The Receiver

In this chapter the components and the role they play in producing usable data to the display are discussed. These are:

- the TR Cell and switching between transmitter and receiver,
- the various stages of amplification, and
- the reduction of clutter and unwanted echoes.

In many ways the receiver can be considered the refining part of the system. The transmitter sends out a fairly crude and powerful pulse. The return echoes are only a tiny fraction of the power transmitted and it is the function of the receiver to detect these tiny fractions, to distinguish them from 'noise' and clutter and to amplify them in a way that they can be presented as useful images on the display.

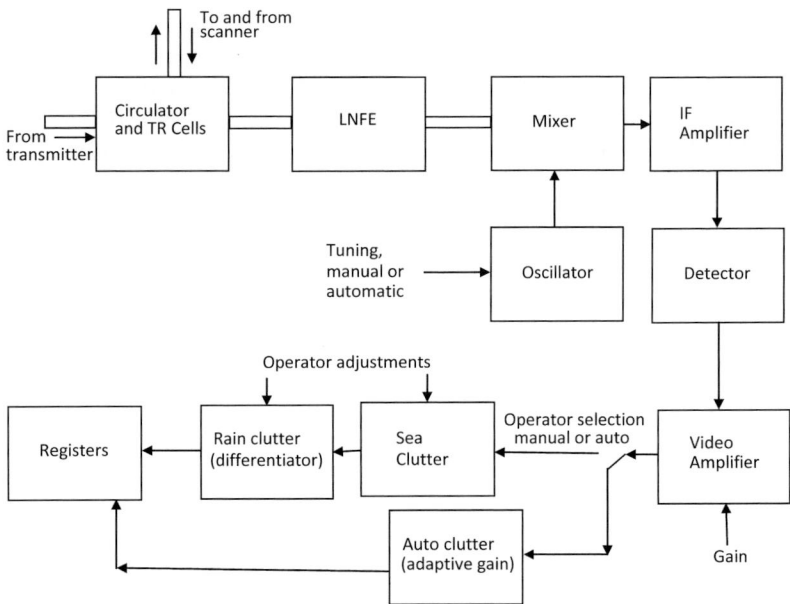

FIGURE 4.1   Receiver block diagram.

TR (Transmit/receive) Cell

The first item to be considered in the receiver is really shared between the receiver and the transmitter. Both use the same scanner, so there needs to be a form of switch that switches the scanner between transmit and receive modes. In addition to this, the receiver must be protected from the very powerful transmitted pulse, considering its function is to detect and amplify extremely weak signals. The switching process is called duplexing, where one device switches between two functions. The component that fulfils this function is traditionally called the 'T-R cell' (T-R for transmit/receive).

The T-R cell must be an extremely fast acting device; any delay in switching over to receive, after the transmission pulse, means the possibility of close-by targets not being detected. Its action must match the PRF, which could be as high as 4000 times per second. For this reason a mechanical device would be incapable of meeting this requirement.

In outgoing technology in the T-R cell a spark gap was used to block the branch of the wave-guide to the receiver during transmission. Two metal cones are placed across the narrower side of the waveguide. When the powerful transmission is taking place, the cell ionises causing a spark across the waveguide which short circuits (blocks) the waveguide to the receiver. When the transmission ceases, the cell deionises and the waveguide short circuit is removed, thus allowing incoming voltages from the scanner to the receiver. In order to ensure rapid action, and particularly to ensure the immediate ionising process when the transmission pulse commences, a 'keep alive' electrode is provided which keeps an auxiliary spark between transmissions. This is not sufficient to ionise and thus block the receiver from the scanner.

Most modern radar installations have a device called a circulator as the first stage in the duplexing process. This is a device which can have several branches into a central core, but the route of RF energy is always in the same direction around the central core. Its effect can be considered to be like a roundabout at a road junction except that the traffic must take the next available exit. In the case of the radar there are three branches in the circulator; the scanner, the transmitter and the receiver. The pulses from the transmitter enter the circulator and take the next exit which is to the scanner. The incoming echoes from the scanner take the next exit from the circulator which is to the receiver.

While the circulator ensures the route taken by the transmission pulses and by the echo pulses, and that the transmission pulses do not go directly to the receiver, it is usual to have added protection to block any transmission energy getting to the receiver. For this reason, even

Circulator

To and from
scanner

From
transmitter
(magnetron)

To receiver

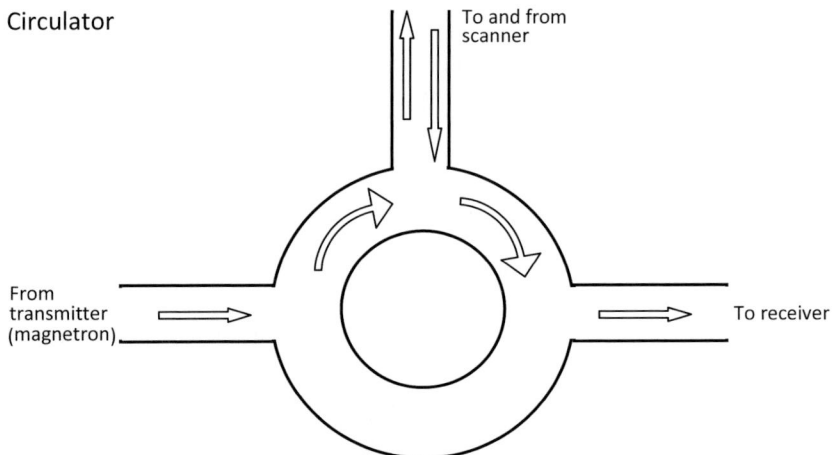

FIGURE 4.2   The circulator.

with a circulator as described, there is usually a T-R cell in the wave guide to the receiver.

LNFE (Low noise front end) Amplifier

In modern sets the first part of the receiver is called pre-amplification or Low Noise Front End (LNFE). This is a process to distinguish the extremely weak echo returns from the surrounding 'noise'. Noise is a term from audio radio and it is the background hiss or crackle that can be heard on unsophisticated radio systems. In various radio systems, including navigational systems, the 'Signal to Noise Ratio' (SNR) is often quoted. This is a ratio demonstrating how much the signal projects above the average level of the surrounding noise, and gives an indication of how well or not the particular receiver will be able to pick up and process the signals. In radar terms noise is atmospheric radio frequencies close to that of the radar frequency and it can hide the echo signals if their SNR is too low, that is, if their amplitude is not great enough to project above the noise. The LNFE process filters the incoming signals for consistency, the noise being of a random nature and the echoes showing a more consistent frequency. It does not amplify the echoes, nor does it entirely eliminate noise or unwanted echoes. Sea clutter for example will show a consistency which will allow it past the LNFE.

Mixer

The next stage of the receiver is called the mixer. This is a process called superheterodyning. As the extremely high frequency of the incoming pulses is very difficult to amplify, it is combined with a slightly lower frequency to produce a lower frequency, called an 'Intermediate Frequency' (IF) which the various amplifier stages can handle.

The frequency fed into the mixer is supplied by a local oscillator. This frequency usually is 60 mHz below the magnetron (and hence the incoming echo pulse) frequency. The local oscillator, usually a device called a Gunn diode, is tuned to the magnetron frequency ±60 mHz (usually -) to produce the 60 mHz which the amplifier can receive, but it has a fine tuning process which allows for the small variations in the frequency transmitted by the magnetron. This is the tuning process which the operator carries out at the display. In older technology the combining of the frequencies is carried out by a crystal diode which produces the 60 mHz of IF. Modern radars have a 'balanced' mixer which has two Gunn or crystal diodes spaced half a wave length apart. This makes the noise appear out of phase and helps to eliminate it and other components which do not fall within the acceptable bandwidth of the amplifier.

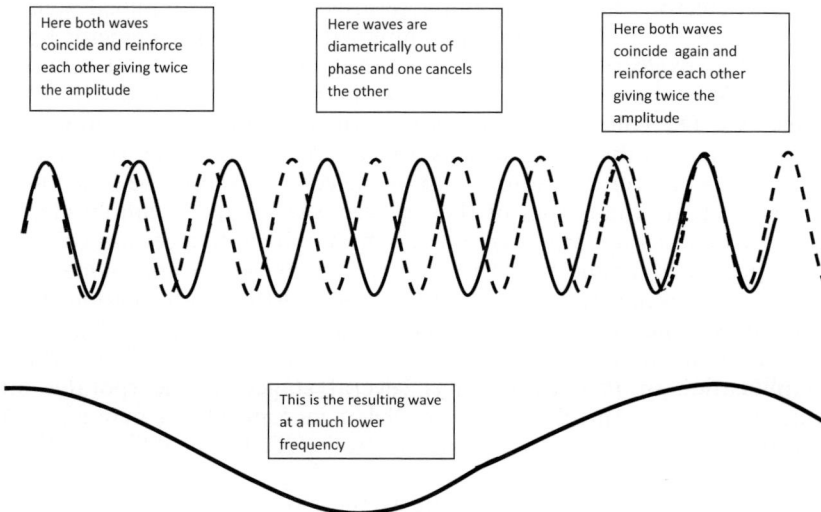

Here both waves coincide and reinforce each other giving twice the amplitude

Here waves are diametrically out of phase and one cancels the other

Here both waves coincide again and reinforce each other giving twice the amplitude

This is the resulting wave at a much lower frequency

FIGURE 4.3   Mixing two frequencies.

Amplifier

The amplifier is usually in several stages. There are two techniques used in marine radar. The first stage is a called a linear amplifier which amplifies incoming signals by a small amount. A linear amplifier amplifies all signals by an equal amount, say by a multiple of ten times their amplitude. This will substantially increase the amplitude of weak echoes, but very strong echo will reach a saturation level for the amplifier, and will not be amplified above this level; this is called the limiting output. Each stage is so limited by the amount it can amplify a signal and it is for this reason that there are the multiplicity of stages. The second technique is called a logarithmic amplifier. This gradually reduces the degree of amplification in relation to the strength of the incoming signals, hence weaker signals are amplified more than stronger ones.

The several stages of a logarithmic amplifier, the most common in modern radars, although some may have a linear amplifier circuit in parallel to the logarithmic one, will each have an input to the detector stage. The inputs from the various stages are combined to give a unified output to the detector.

The detector

The detector 'detects' the pulse returns from the amplified signal. The echo pulse at this stage is an amplified 60 mHz. The detector finds the 'pulse envelope' around the alternating current signal and converts it to a DC voltage suitable for use in the next stage, called the video stage. The DC voltage must reflect both the duration and the amplitude of the pulse, so that weak targets appear weak and strong ones large. It must also reflect the pulse length of the echo. It should be noted that in older radars the output of the receiver was limited in amplitude, in other words, strong targets would only be amplified to a certain voltage. This prevented undue brightness on the radar screen and damage to the screen coating. It is usual that there is a further amplification stage here, called the video amplifier. This increases the pulse voltage to a degree that they register for storage in memory. This video amplification in modern sets has several stages. It incorporates an automatic gain control, but in parallel it incorporates a manual gain control, and also 'sensitivity time control' (STC) and 'differentiation' (Rain Clutter).

The automatic gain control adjusts the gain of the video amplifier based on the average strength of received echo signals. Sometimes called 'adaptive gain control', it can increase and decrease the amount of amplification for the returns of each transmitted pulse so that it can deal with different amounts of echo response from different areas around the ship.

Echo register

The incoming stream of echoes coming from the video amplifier is a rapidly varying positive voltage; stronger targets showing as large voltages and weaker ones showing lesser values. This stream of varying voltage is mixed with an alternating current of the computer 'clock'. Where an echo voltage above a certain value coincides with a positive voltage of the clock, a binary 'one' is recorded in the register, otherwise, for each positive clock voltage where no echo voltage coincides a 'zero' is recorded. In this way, all the echoes over a certain strength are recorded in the register as a series of 'ones' in a binary 'word' of usually several thousand 'cells' (see Chapter 1, 'Radar Basic Principles' and Figure 1.4).

The register described above is in fact a series of registers or series of stages. In the first stage, all echoes up to a certain voltage are registers. This is a low voltage, and catches the weaker echoes. The next stage or register will be set a higher threshold voltage, and only those echoes which reach or exceed this voltage will be recorded. There may be one or several more stages working on the same principle. The results of all these stages are then combined for display on the radar screen as images of different intensity, often, as mentioned above, as different colours.

Modern sets have, at this stage, several processes designed first of all to filter out unwanted echoes, and secondly to give some indication of the echo strength. In the old technology of 'raw' radar, the brightness of the spot on the screen gave some indication of echo strength. This is an important feature as it gives an indication of the possibility that the echo is a small vessel, a buoy, an ice floe, or a large ship or land echo. In the simplest form of single stage digitised register, all echoes coming through the receiver, would be displayed at equal strength and while the size of an echo might give some idea of the size of the actual target, the intensity of the image on the screen would give no variation, something which was a feature of the old type radars.

Correlation

Let us look at the process of rejection of unwanted echoes first. There are two processes used in modern sets; one is pulse to pulse correlation. In this, the echoes from one pulse are compared with those of the next. If an echo is not received in the same place in both or several pulses, it is assumed that it is a spurious echo, and is rejected.

The next stage is scan to scan correlation. The echoes for one rotation of the scanner are stored and compared with those from a second rotation of the scanner. Once again, echoes which do not appear in the same position for each scan are rejected. If it is still there after one rotation it is shown on the screen, often initially as a 'weak' target. Then, as it appears for successive rotations it is shown as an increasingly stronger target. In one make of radar the initial echo indication is coloured brown, then orange and finally a bright yellow. In this way strong, good radar contact echoes are shown as bright yellow, while weaker and intermittent targets are shown as orange or brown.

The received video data, effectively the range of each echo for each pulse sent from the rotating scanner, with some indication of the echo strength is ready to be sent to the CPU (central processing unit or computer) of the display. All that is now needed is the direction, or bearing that each echo was when it was received. This is achieved by data from the scanner rotation to the central processing unit (see Chapter 3 on the scanner).

Sea Clutter

Sea clutter is the random echo returns from waves close in to the vessel. The sea surface is an excellent reflector of radar energy, and, as has been discussed in the previous chapter on the scanner and radar beam, in flat calm conditions, the part of the radar beam striking the sea surface will be reflected away from the scanner. However, as soon as there is any wind, waves are formed, and the sides of the waves facing the scanner now can present a reflecting surface which reflects some energy back to the scanner.

Various factors affect the formation and size and steepness of the waves, the main ones being wind strength, duration and fetch. The depth of water and the direction and strength of current or tidal stream in relation to the wind are others. Whatever the cause, the size and steepness of the waves are going to affect radar response, and if the waves are breaking then there will be further material in the form of spray and spume which will cause reflection, scattering and attenuation of radar energy. Generally speaking, as the wind increases, so does the height of the waves and so larger targets are formed. This will also extend the range at which sea clutter is apparent. The responses from the larger waves close to the ship may saturate the receiver. Sea clutter thus becomes a serious problem in rough seas and under conditions where visibility may be compromised.

Another factor, discussed in the section on scanner siting, is that the height of the antenna extends the range of sea clutter. As mentioned above the sea clutter echo returns may saturate the receiver and even in circumstances where it does not, it may mask any close-in targets.

A target, which does not return a stronger echo than the surrounding sea clutter may not be detected. Watchkeeping officers must always bear this in mind.

Additionally with ARPA and ATA, it may cause target swop where the tracking system loses the acquisition of a weak target to the echo from the waves.

Heavy sea clutter will be seen on the radar screen as a large circular area of close spaced random echoes around own ship. In Figure 4.4 below it can be seen that the ship target (a) close to own ship is obscured by the surrounding sea echoes. Distant targets (b) are not affected by sea clutter and for the average vessels in rough weather sea clutter will not affect target detection or acquisition past 1.5 nautical miles or so.

Sea clutter can be suppressed by modifying the amplification of the echo returns in the IF amplifier, that is, by reducing the 'Gain'. The gain would then be raised slowly to allow targets received with echo strength which exceeds the sea clutter to be detected. Such a process of course will reduce all echoes. For this reason, radars have 'Sea Clutter' controls which deal with the suppression of targets close to the ship.

Sea clutter suppression can be manually or automatically controlled. The level of initial amplitude suppression, also referred to as swept gain, is adjusted to compensate for the level of sea clutter echo return. The swept gain function of the receiver protects the receiver from saturation, The IF amplifier is shut down while the magnetron is transmitting, to protect it from any leakage of energy through the T-R cell system, it is then gradually but fairly rapidly restored to full amplification under normal conditions. The adjustment for sea clutter delays this restoration to full amplification (see Figures 4.5a and 4.5b).

Sea clutter while random in nature has a predicted echo strength which lessens as the range from the ship increases. Automatic control, called adaptive gain, samples the immediate echo returned power and adjusts the swept gain amplification so that it is less close to the ship and increases gradually with range, the degree to which this suppression is applied depends on the average strength of the echo returns in the area close to the ship, the stronger the average echoes, the greater the suppression. This is done in relation to each timebase, that is, the echo return from each pulse transmission. This results in different degrees of

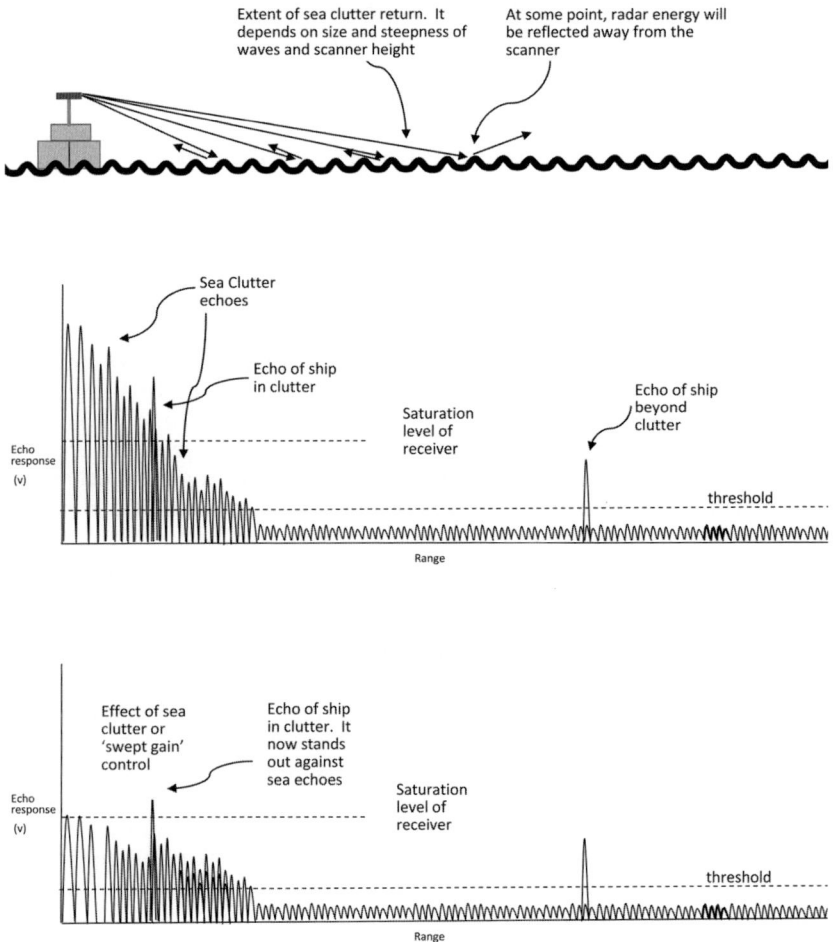

FIGURE 4.4   Sea clutter.

suppression in different areas of clutter, such as to windward, something which manual adjustment would not be capable of. It should be noted that while most radars will work using automatic control, manual control, which is an IMO requirement, should be used for close-in working as automatic control can masks smaller wanted targets.

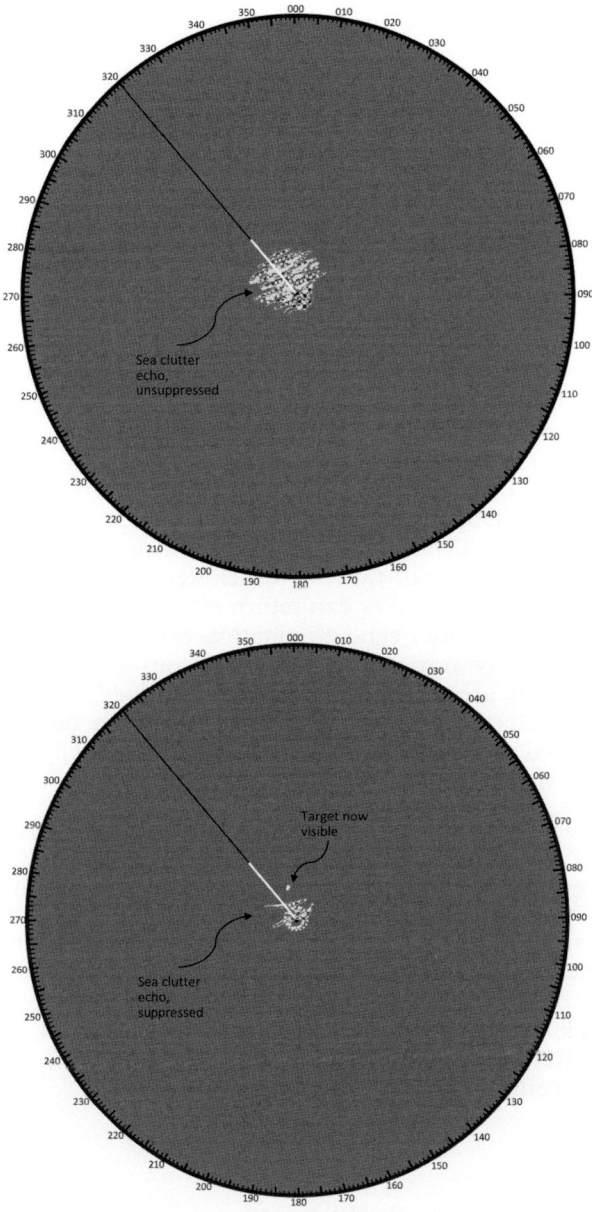

FIGURE 4.5   Effect of applying sea clutter suppression.

In addition to the above swept gain method of eliminating unwanted sea clutter, and increasing the probability of detecting weak targets, most modern radars incorporate the digitised comparison process which identifies consistency. This process, called correlation, is described above in the section on video processing. In relation to clutter control and the sophistication of this elimination of random echoes, modern sets are able to apply this process, in different degrees in different parts of the screen so that the distinguishing of weak echoes is applied where needed and not where there is a lesser degree of clutter or interference.

New Technology (NT) radar, which works on slightly different principles from conventional radar (see the chapter on NT radar) also uses Doppler shift, in addition to the correlation process, to detect consistent echoes. A solid target amid the equally strong or stronger echoes from the surrounding waves will probably have a different movement from those waves. The process in the receiver will detect this more consistent Doppler shift and show the target while eliminating the less consistent Doppler shift echoes of the waves.

In spite of the excellence of modern radars in suppressing sea clutter, and in distinguishing weak targets, sea clutter remains as a problem. In heavy sea conditions waves can return strong and consistent echoes which escape the random echo elimination processes. Also, the weak wanted targets may return a degree of inconsistency in their echo response that they are eliminated as "random echoes". For this reason, the automatic clutter control, or adaptive gain can be by-passed or switched off and the sea clutter control adjusted manually. This should be done from time to time for the following reasons:

1. In clear weather when there are weak targets nearby, to test the system and reassure the watchkeepers of the effectiveness of the clutter control, both manual and automatic. For example, does a buoy, normally hidden in clutter, that can be distinguished by manual adjustment not appear when the auto clutter control is switched on. Can it be distinguished or otherwise by either method.
2. In thick weather where weak targets are expected or suspected, a sea-clutter search should be carried out. This process is described under operation of the set.

Rain Clutter

Precipitation, in the form of rain, sleet, snow, mist and fog, and all forms of water in the atmosphere can cause problems for target detection by radar. Precipitation can cause unwanted echo returns which are sometimes very strong and can cause wanted targets to be obscured or fail to return a receivable echo. Such interference is generally called "Rain Clutter", although it may be caused by other forms of precipitation, and even moisture in the atmosphere which is not falling such as from low clouds.

The interference, or clutter thus formed is quite different from sea clutter and the means of suppressing it is different from the process of dealing with sea clutter. Rain (or other forms of precipitation) can be considered as millions of individual droplets. The size and material of the droplets in relation to the radar pulse wavelength and polarization produces different degrees of reflection and attenuation, and generally 10 cm (S-band) radar is less affected by rain clutter than 3 cm (X-band).

Each droplet can be considered as providing material for a good radar target but its shape is conducive to reflecting back only a fraction of the radar energy which strikes it; the rest will be scattered in all directions. Therefore two things happen, some of the energy is reflected back to the scanner, but more is scattered and lost thus reducing the effectiveness of the pulse, in other words attenuating or weakening it. The rain can therefore obscure targets by the echo return from the rain itself, and it also reduces the radar's effect and the echo return from wanted targets.

Obviously the size and intensity of the droplets are significant. But size of the droplets must be considered in relation to the wavelength. When the droplets are small in relation to it, any increase in diameter increases the scattering effect in proportion to the square in increase in diameter. So doubling the droplet size will increase the scattering effect fourfold. Considering a given particle size, the opposite hold true in relation to wavelength; the scattering effect is inversely proportional to the square of the transmitted wavelength. For this reason there will be less interference presented by precipitation on an S-band radar than on an X-band, for any given situation.

The intensity of the precipitation, the number of droplets in a given volume of atmosphere is obviously going to affect the amount of interference, both the echo return and the attenuation. The heavier the rain, the worse the effect on the radar.

It should also be noted that to radar a large rain or storm cloud that has not yet released its rain can appear as a band of rain as the vertical beam width is wide enough to catch these clouds. These clouds contain a great deal of moisture which to the bandwidth of radar frequencies are much the same as rain.

Rain clutter control is either automatic or manual. In automatic control the radar system measures the depth of the rain front and

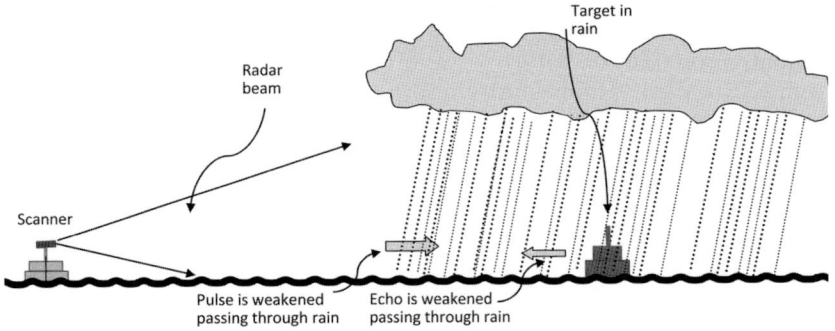

Target in rain

Radar beam

Scanner

Pulse is weakened passing through rain

Echo is weakened passing through rain

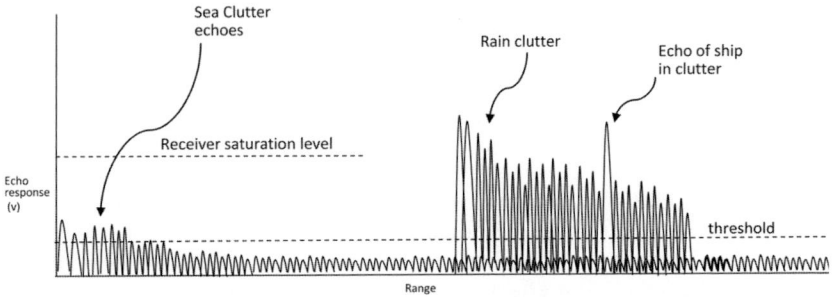

Sea Clutter echoes

Rain clutter

Echo of ship in clutter

Receiver saturation level

Echo response (v)

threshold

Range

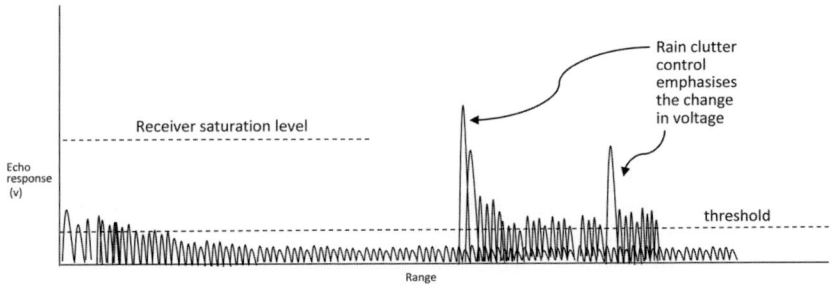

Rain clutter control emphasises the change in voltage

Receiver saturation level

Echo response (v)

threshold

Range

FIGURE 4.6   Rain clutter.

gradually increases the rain clutter control to compensate. If there are no targets detectable within the rain system then the radar will constantly adjust the rain clutter control, if targets are available then it will adjust the rain clutter control for the detection of these targets.

The manual control of rain clutter is much the same, when adjusting the rain clutter controls the operator is adjusting the threshold and differentiation response of the fast time constant circuit (detecting the change in echo strength). In lighter precipitation using little rain clutter control, the depth or size of target is still discernable. Using heavy anti rain clutter will just leave the leading edge of the target and the size of target response will be lost.

To counteract the effects of precipitation we can use:

Gain control

Sometimes adapting the gain control to the conditions will produce targets.

Short pulse length

Using short pulse length will return less power and give better discrimination. Reducing the range may also reduce the pulse length.

Anti sea clutter control

If the rain is near the origin then using anti sea clutter may help to detect the target.

Rain clutter control

The anti rain clutter circuit which is also called a fast time constant (FTC) circuit or differentiator. This circuit basically differentiates the incoming signal. A differentiation circuit looks for a change in signal amplitude; it shows the change in amplitude of the signal but not duration of the signal. In the radar fast time constant circuit only a rising change of amplitude is detected. As can be seen in Figure 4.6 only the leading edge is differentiated (see also Figure 4.7).

As discussed above, the transmitted pulse passing through an area of precipitation has much of its energy dissipated or attenuated by the drops of moisture. This will effect the detection of targets which are not in the area of precipitation, but beyond it. For these targets operating the rain clutter control, either manually or automatically will have no effect. The strength of the pulse striking them is attenuated by its passage through the area of precipitation and the echoes of such targets is also attenuated by their passage back to the scanner. This means that targets detected beyond precipitation will be reduced in strength.

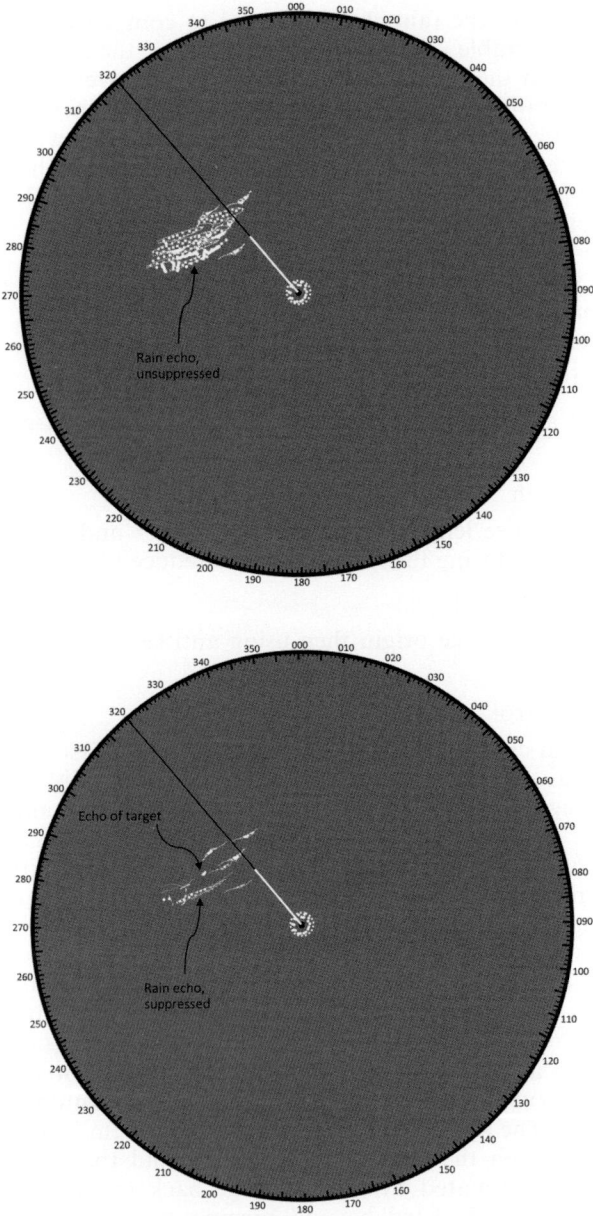

FIGURE 4.7   Effect of applying rain clutter suppression.

The only action possible to aid the detection of such targets is:

a. to increase the pulse length in order to increase the power of the transmitted pulse and also to increase the power of the returning echo.
b. Turn up the gain to amplify any returning echo as much as possible.

It should be noted that taking such action is a very temporary measure as it interferes with the rest of the radar picture, and once a 'search' in the area beyond the rain has been carried out, normal settings should be restored.

Automatic Clutter Control

This uses a process of correlation, as discussed above, in conjunction with a process of sampling the returns from echoes. When the sampler detects strong echoes close to the ship, combined with a certain amount of a random character of such strong echoes it automatically reduces the gain to a degree depending on the strength of the echoes. It also can effect this to varying degrees around the ship, so that if there is more clutter in one area, this will be suppressed more than where there is less. Often one can see a greater extent of sea clutter to windward. Such automatic clutter control works not only close to the ship but over all the area of the screen. Generally speaking, such automatic systems are very good in suppression clutter and still detecting small targets, but there are still situations where the operation of the manual controls will detect small targets which do not appear when the auto-clutter in on.

Usually when auto-clutter is switched on the manual sea and rain clutter controls are inoperative.

The following images give an idea of the effect of the various clutter suppressions. The first image (Fig. No. 4.8) shows the situation with no clutter suppression applied. The second one (Fig. No. 4.9) has a small amount of sea clutter suppression applies and small targets close to own ship are visible. The third image (Fig. No. 4.10) shows full rain clutter suppression applies in addition to the small amount of sea clutter suppression. In this, the leading edges of targets can be picked out including echoes which previously were obscured by the saturation of land echoes. The fourth image (Fig. No. 4.11) shows auto clutter applied. It is interesting to note what is suppressed; the land echoes close to the south of own ship, for example.

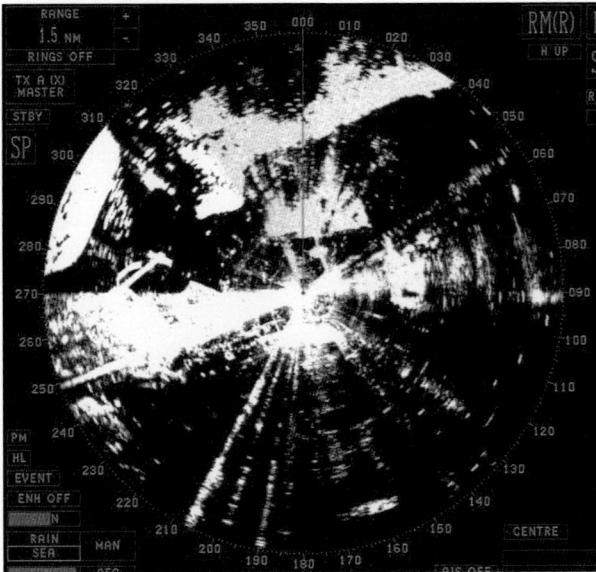

FIGURE 4.8 Radar screen, no clutter suppression applied.

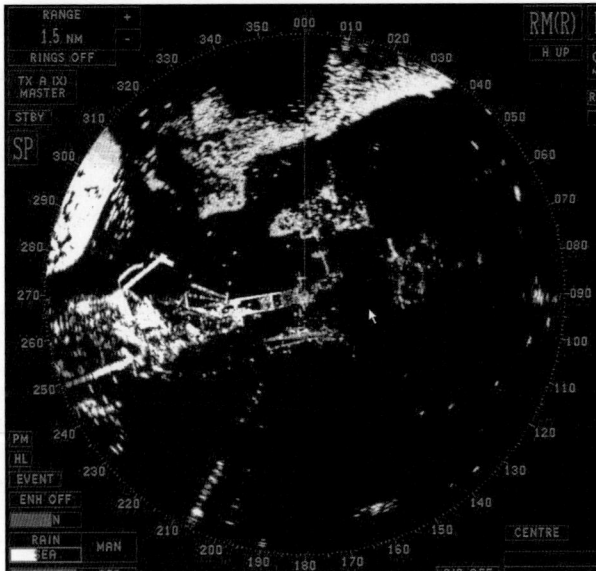

FIGURE 4.9 Radar screen with sea clutter suppression.

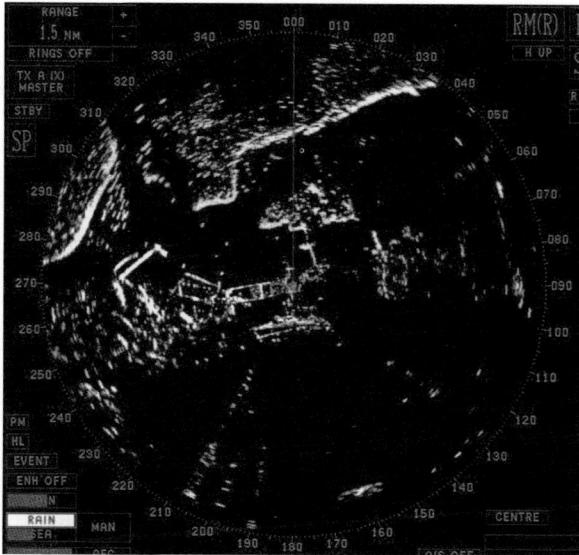

FIGURE 4.10 Radar screen with sea clutter suppression and full rain clutter suppression.

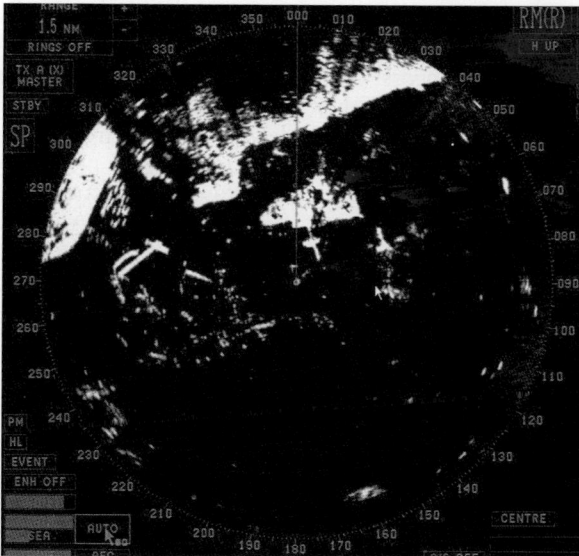

FIGURE 4.11 Radar screen with auto clutter suppression.

Other unwanted echoes

As can be seen from the above series of radar screen images, the adjustment of the clutter control has an effect on all echoes. This includes much of the echoes from land and re-reflected echoes from structures near the scanner (some of which are recognisable by the heavy and not so heavy series of echoes in definite lines radiating from the origin). Many of these echoes are 'unwanted' and are false (see the chapter on false echoes and on the scanner) and constitute 'clutter' in themselves. The adjustment of the clutter controls, both 'sea' and 'rain' tidies up the picture and makes the 'wanted' targets visible.

Conclusions

- The receiver is isolated and protected from the powerful transmitter by the circulator and T-R cell configuration. This also allows the same scanner to be used for transmission and receiving.
- The incoming very weak echoes have first of all to be converted, by the mixer to a voltage which can be handled by the amplifier and then amplified to a usable level for video.
- Amplification can be linear or logarithmic or more usually a combination of both.
- The detector converts the received echo pulses to a direct current value for use in video.
- The pulses are stored digitally in a series of registers
- The correlation process, comparing the received echoes in one register with that of another serves to help eliminate unwanted echoes and also to indicate the received echo strength.
- Sea and rain clutter controls operate on different principles. The rain clutter process detects changes in voltage (echo strength), while the sea clutter process is a form of gain control. Automatic clutter control can be very effective, but there are occasions when manual adjustment of clutter suppression is essential.

Chapter 5

The Display

The display consists of:

- The radar screen, sometimes called the Plan Position Indicator (PPI) which shows the radar picture.
- The data display, surrounding the radar screen, which provides the data on such things as the range scale, target data, radar status (tuning etc.), ship's head (gyro), ship's speed (ground or water track) and much other information.
- The control panel, the buttons, knobs and switches by which the operator controls the radar,
- The graphics generator, which provides nav. Lines, maps, EBL, VRM etc.
- The radar computer which carries out all the computations, converts the operator commands and presents the results to the radar screen and data display.

The form of display layout and the functions which are provided at the display are influenced by the provisions of IMO Resolution MSC192(79), 'Performance Standards for radar equipment for new ships constructed after 1 July 2008' and IMO Resolution MSC191(79), 'Performance Standards for the presentation of navigation-related information on shipborne navigational displays for new ships constructed after 2008'. These Resolutions may be accessed online at www.imo.org. Many of the provisions listed in these documents were current practice on modern radars and ARPAs and most sets which an OOW is likely to find on the bridge of a ship comply with them

The original radar display consisted of a cathode ray tube, a long 'Y' shaped glass tube through which a stream of electrons was projected onto the broad, almost flat end of the 'Y'. This stream of electrons was controlled by deflection coils which made it move from the centre out to the edge corresponding to each transmitted pulse from the scanner. The rate at which the point of this electron stream, called the 'spot' moved out from the centre was at half the scale rate of the actual pulse, so that an echo received from the pulse would show the spot at the correct range. When such an echo was received the spot glowed brighter.

The stream of electrons moving out from the centre to the edge of the screen appeared as a line to the observer stretching from the centre

Cathode Ray Tube

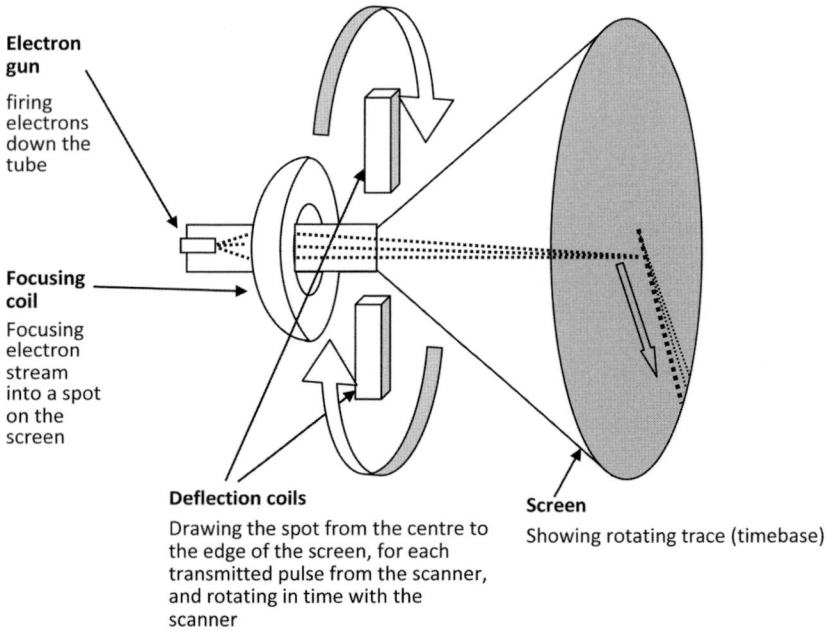

Electron gun

firing electrons down the tube

Focusing coil

Focusing electron stream into a spot on the screen

Deflection coils

Drawing the spot from the centre to the edge of the screen, for each transmitted pulse from the scanner, and rotating in time with the scanner

Screen

Showing rotating trace (timebase)

FIGURE 5.1   Cathode Ray tube display.

to the edge. This line was then made to rotate in time with the scanner. Thus echoes received were shown in their correct position relative to the ship (see Figure 5.1).

In more recent years, television and computer display technology was adopted. This still used a cathode ray tube, but in this technology the point of the electron stream moves very rapidly back and forth across the screen, staring at the upper left-hand corner and progressing line by line to the bottom right-hand corner and then starting at the upper left-hand corner again. By this method a series of closely spaced lines are formed. This forms a grid called a 'raster' of a series of tiny squares called 'pixels' an abbreviation for 'picture elements'.

In the raster and subsequent technology there is no relationship between the movement of the spot on the screen and the direction of the scanner or transmission of the pulse. The data of the scanner direction and the echo responses from targets has to be digitised before presentation to the display. This digitised data is sent from the receiver

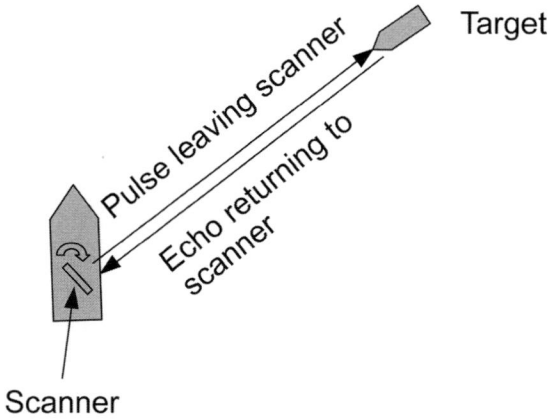

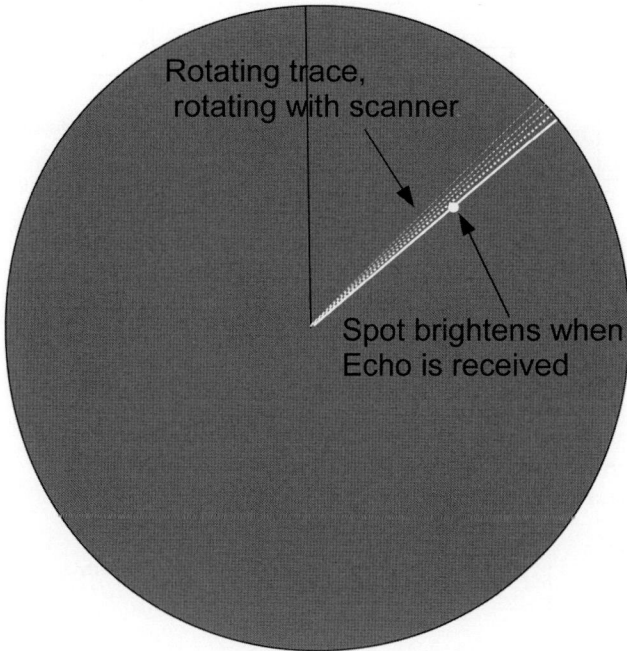

FIGURE 5.2   Rotating trace on CRT.

The Display

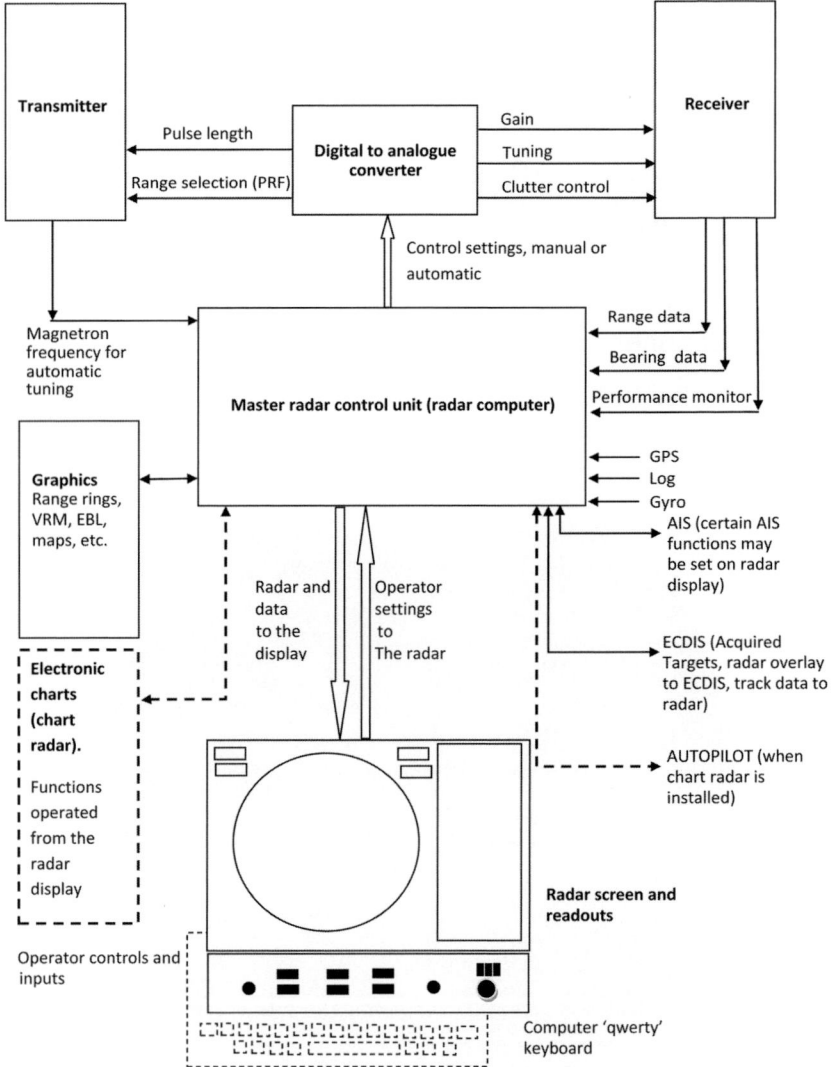

Figure 5.3   Block diagram of the display.

and causes the raster to display a series of illuminated pixels either brighter than the background illumination or in different colours.

In modern radars LCD (Liquid Crystal Display) is used in the display. This is the technology which is used in modern television and computer screens. Liquid crystals, as the name suggests have the characteristics of a liquid, but also have the characteristic of a solid in that the molecules can be lines up in different orientations. The orientation of the molecules is controlled by having a layer of liquid crystal between two layers of polarised glass, and layers of tiny transistors (thin film transistors,TFT) which divide up the screen into pixels. The alignment of the molecules allows the amount of light which can pass through the LCD panel (the screen). This alignment is controlled by the current which is passed through the transistors. The varying alignment of the molecules allows for a 'greyscale' to be produced, which is another way of describing the intensity or shade of the light, or the degree from white to black. A LCD screen can display 256 levels.

When colour is used, each pixel is subdivided into three sub-pixels, each red, green and blue. By varying the intensity or shade of each of the three colours, an almost infinite (over 1 million) variety of colours and shades of colour can be produced for each pixel.

Liquid Crystal Display

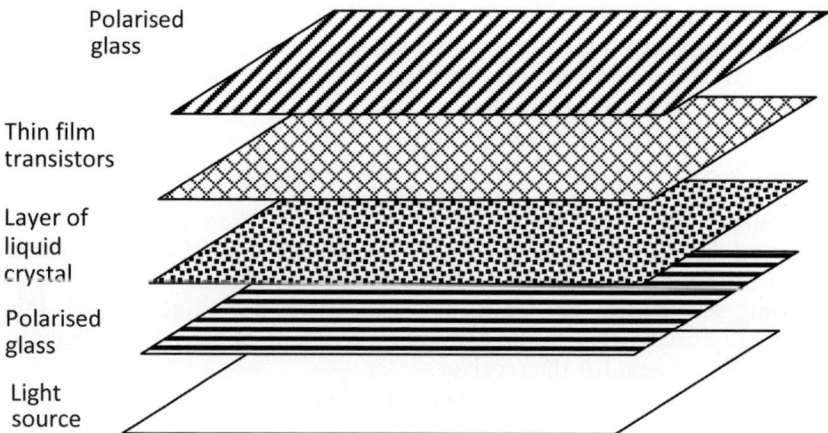

Polarised glass

Thin film transistors

Layer of liquid crystal

Polarised glass

Light source

FIGURE 5.4   Liquid Crystal Display.

A typical modern radar set will have a 1900×1200 pixel resolution or better. This describes the number of pixels horizontally and vertically on the screen. At normal viewing distance, of no closer than say 30 or 40 centimetres, the individual pixels are invisible and the picture appears smooth and normal; a page of text appears like a page printed on paper and it is only by very close scrutiny can the 'pixelated' form of the characters be seen. Pixelation is most noticeable, even in good resolution displays in curved graphics such as range rings and variable range markers (VRMs).

The technology in computer and television displays is constantly evolving. Developments in this area will affect radar displays (and ECDIS and other data presentations on the bridge of a ship). The ability of the radar manufacturers to adapt such 'off the shelf' technology helps to reduce cost, apart from benefiting from the improvements of such developing technology. This compares well to the historic situation where a radar cathode ray tube could only be used as such and each manufacturer had to design and produce its own tubes.

The use of modern raster and LCD technology allows considerable scope and flexibility for the way the radar information is presented. There are certain aspects of radar display which are controlled by the specifications set out in the IMO resolutions. Basically a radar display must look like a radar display and certain functions must be available to the OOW. How these functions are activated and controlled has changed over the years, again due to the developments in computer technology. Many modern sets have a minimal keyboard; many of the functions are activated from 'drop-down menus' or 'drag and drop' functions controlled by the roller ball. First generation ARPAs had a bewildering panel of switches and knobs; a modern set may have a roller-ball, three buttons, and maybe a panel with two rotating knobs controlling VRM and EBL, and range up and down buttons and one or two more and possibly not even these.

Again, unlike older technology, none of the functions or settings is controlled directly by the operator, they are all controlled by the radar computer. The controls and settings must be then converted from digitised data to analogue settings for many functions. Even in the case where rotating knobs control certain functions, such as may be found in some sets where the 'Gain' or 'Sea clutter' are so controlled, this gives a digitised signal to the computer, which must be changed back to an analogue signal for the receiver.

Controls and functions can be considered in four categories:

1. Controls and functions of immediate use. These are the functions that the operator needs to manipulate quickly and easily to take

FIGURE 5.5   An example of a modern control panel.

ranges and bearings, change range scale, acquire targets, change display modes, change vector modes and so on.

2. Controls and functions of lesser immediacy. These are functions like navigation or parallel index lines, changing orientation, vector time, clutter control adjustments, activating or deactivating maps or chart underlay, displaying or hiding routes.

3. Settings and adjustments. These are settings or adjustments that once set do not need to be re-set or adjusted very often. These would include functions like setting the alarm limits for CPA and TCPA, setting guard-zones, drawing maps

4. Maintenance and fundamental adjustments. These are functions that rarely need adjustments, and should only be done under full knowledge of the consequences. Many of them should only be accessed by a qualified technician. Usually a password is required for these functions. These includes the filtering rate, ship data such as length, beam, scanner position relative to CCRP (Consistent Common Reference Point), radar blind sectors.

The following list of controls is merely an example of the type of controls on modern radar sets and does not claim to be a complete list, there are too many different makes of type approved radars. It should be remembered that even a basic radar installation is a complex computerised system. There is no substitute for ships' officers getting their hands on the sets, in conjunction with the manuals, and getting familiar with all the functions and controls which they will normally need.

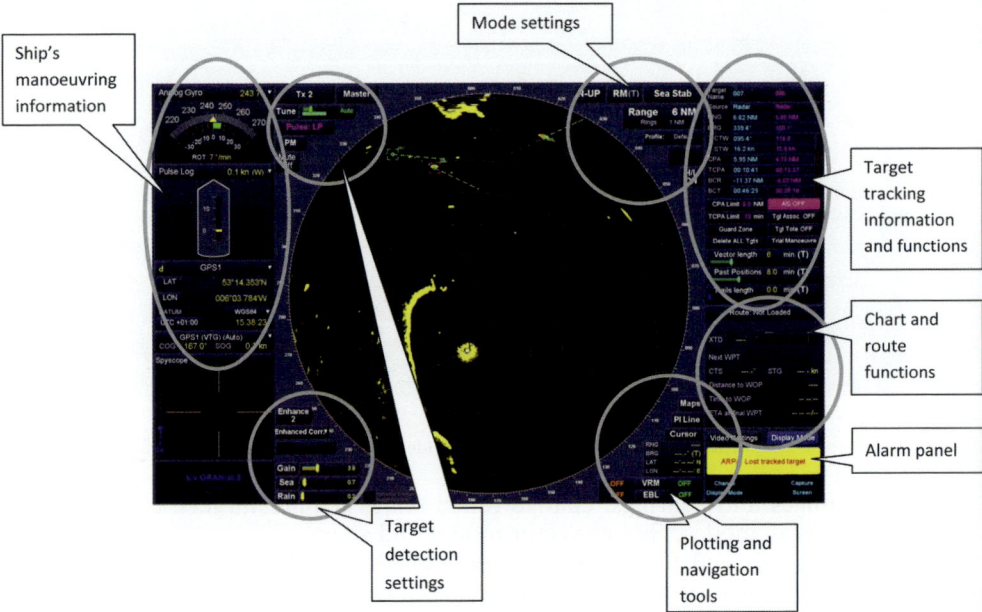

Ship's manoeuvring information

Mode settings

Target tracking information and functions

Chart and route functions

Alarm panel

Target detection settings

Plotting and navigation tools

FIGURE 5.6   An example of a modern radar display.

Switch on and Set Up Functions

On and Off and Stand-by

There will be a main switch which provides power to the radar. This will be somewhere on the panel. In most modern sets the 'switch on' will start by carrying out a self diagnostic process or test to ensure all components are working correctly. Meanwhile the 'warming up' process is in progress. This can take three minutes before transmission can start, any attempt to transmit before this will be prevented. New technology (NT) radars do not need such a 'warm-up' and once the diagnostic test has been completed, are ready for transmission. Some sets show a 'count-down' to the 'ready to transmit' stage.

Switching on at this stage may, or may not set the scanner rotating. This will depend on the make of radar.

Once the warm-up has been completed the set is now on 'stand-by'. The display screen is active and many of the functions can now be set. The radar can now be set to transmit(and receive) by activating the 'transmit' (Tx) control, whatever form this takes. Very often it is a 'soft' control, activated by the roller ball and buttons.

In some sets the display will come on in a particular range scale, orientation and with range rings on.

Range selection
Several manufacturers have retained range scale selection as a 'hard' control, on the basis that is one of the functions that the OOW needs to frequently change quickly and often. This often takes the form of two buttons, one 'up' to increase the range scale and one 'down' to do the opposite. Otherwise it may be a form of 'soft' control operated by the roller ball and buttons. It may also be both.

Range rings
This is an 'On' or 'off' function, usually a 'soft' control.

VRM and EBL
Usually activated by 'soft' controls, but then manipulated by rotating knobs on the panel or keyboard. Sometimes they may also be manipulated by a 'drag and drop' process, using the roller ball.

Orientation
Almost always a 'soft' control. It changes the orientation of the display between 'Head Up', 'North Up' and 'Course Up'. There may be other orientation options such as 'Head Up' but with 'True' bearing scale.

Gain and Tuning
Usually 'soft' controls and usually with a bar indicator for 'tuning'. Tuning is absent in New Technology radars. Tuning usually can be set to 'automatic' (Automatic frequency Control (AFC)).

Clutter Controls
Again, these may be 'hard' or 'soft' controls. IMO requirements are

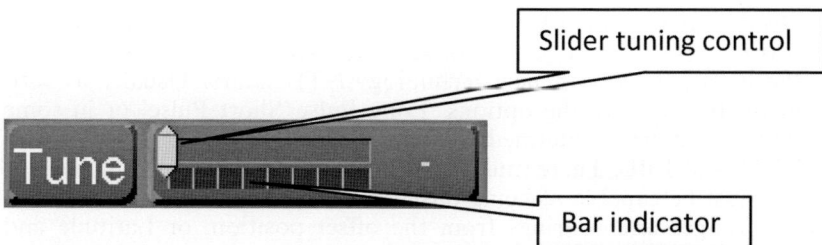

FIGURE 5.7   Example of a tuning control with bar indicator.

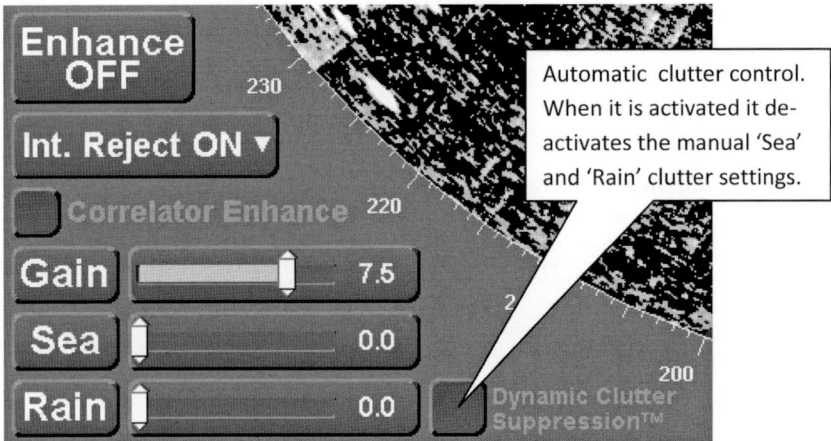

Enhance OFF

230

Int. Reject ON ▼

Automatic clutter control. When it is activated it de-activates the manual 'Sea' and 'Rain' clutter settings.

Correlator Enhance 220

Gain   7.5

Sea   0.0

2

200

Rain   0.0

Dynamic Clutter Suppression™

FIGURE 5.8   Example of gain and clutter controls.

such that the operators must be able to adjust and set the levels of these. There is usually an automatic clutter control which over-rides the manual settings.

Interference rejection
    This is usually a 'soft' control, either 'On' or 'off'.

Echo enhance
    This is usually a 'soft' control, either 'On' or 'off'.

Immediate Controls

Range scale selection switch or 'Up' or 'Down' buttons or 'soft' control.
    Display mode (Head Up, North Up Course Up and possibly others, relative or true motion, ground or sea stabilised).
    Pulse length. Not in new technology(NT) radars. Usually a 'soft' control. It may give the options 'Long Pulse/Short Pulse' or in some sets there may be an intermediate pulse option.
    VRM and EBL, There must be at least two of each. One or both of these must be capable of being offset from the origin, with the readout giving ranges and bearings from the offset position, or Latitude and Longitude of the indicated position.
    Range rings 'On' or 'Off'

Target acquisition function, roller ball and screen cursor with an 'acquire' button ('hard' or 'soft' or both).

Vector mode ('True' or 'Relative').

AIS 'On' or 'Off'

Target Association 'On' or 'Off'.

Less Immediate Controls

Vector time. May be a 'soft' 'slider', a virtual keyboard entry or a 'hard' control.

Trails (time/length) 'True' or 'relative'. The mode is often grouped with the display mode. The length usually has a 'soft' control.

History dots, 'On' or 'Off'.

Trial Manoeuvre. May be entered by virtual keyboard, or manipulated graphically on the PPI (drag and drop point/time of alteration and alteration of course).

Parallel index lines/navigation lines. At least two, usually four. These can often be differentiated by different dotted lines. Range from the origin and direction (bearing) may be manipulated by 'hard' controls, by 'dragging and dropping' or by keyboard entry (often a virtual keyboard).

Clutter adjustments. Sliders, 'hard' controls or automatic.

Tuning adjustment. Slider control or automatic.

Pre-drawn maps 'On' or 'Off'.

Chart underlay 'On' or 'Off'.

Chart adjustments, such as the degree of detail being shown are usually by 'drop down' menu.

Guard zones 'On' or 'Off'.

Screen brightness.

Daylight/dusk/night illumination setting.

Settings and Adjustments

CPA and TCPA limits. Usually a menu item. May be set by virtual keyboard or slider.

Guard zone limits. Often set graphically on the PPI.

Drawing maps. Drawn graphically on the PPI.

Past position (history) dots. Usually a 'soft' adjustment, by virtual keyboard to adjust the timing/spacing.

Trails length. Usually a 'soft' adjustment, by virtual keyboard to adjust the timing.

Display routes. This should be an 'On' or 'Off' function.

Fundamental Adjustments

These functions need to be accessed only occasionally. Such adjustments or functions should only be carried out by the officer whose principle duty is the navigation of the ship. It is usual to require a password to access these functions. Certain other functions and adjustments should only be carried out by a qualified technician for the particular radar equipment.

Ship parameters. These are the details of the ship which affect its navigations; its length, beam, draft, turning circle, CCRP and other static data of the ship.

Alarm settings. Some radars have a comprehensive list of alarm setting options.

Input connection/interface. Log, gyro, GPS,AIS, ECDIS, autopilot. The settings and connections for these various inputs can be checked for status or switched on or off.

Chart management. For chart radars which contain a library of navigational charts.

Interswitching between radar displays, scanners and transceivers.

Installation and maintenance functions. These are adjusted and set up during installation of the equipment and normally need no further action by ship's personnel. These functions should only be accessed by qualified technical personnel authorised by the makers of the equipment. Interference in these functions can have serious consequences for the effective working of the equipment.

Conclusions

The display is the 'human/machine interface'. A modern radar display is basically a computer screen presenting data and providing a means of controlling and adjusting the equipment. The incoming basic radar data is adjusted by various controls and settings decided by the operator or automatic functions by the computer. The resulting radar 'picture' is then presented on the screen. The basic radar picture is supported by a range of graphics and other functions including inputs from other systems in the ship's navigational equipment. The adjustment and setting of these inputs are also controlled at the display. Controls can be categorised into four levels; those which are almost constantly being manipulated in the normal course of watchkeeping; those which are used frequently; settings which are only occasionally adjusted and settings and adjustments which are only made on rare occasions.

Chapter 6

The Radar Horizon

This chapter looks at how far the radar can 'see'. This is examined under the following topics:

- The radar horizon in 'standard' atmospheric conditions,
- Factors affecting the range at which targets might be detected,
- The effect of different atmospheric conditions on the radar beam, and the consequent effect on detection ranges.

Marine radar transmission frequencies (approx. 10,000 and 3000 MHz), can for all practical purposes be considered to travel 'line of sight'. This means that taking all other factors into account such as the power of transmission, the target response characteristics and atmospheric attenuation the target will not be detected if it is below the *radar horizon*. There is some defraction, a characteristic of radio waves at the frequency of radar to follow the surface of the earth, but otherwise the close relationship between radar and visual horizon exists.

The effect of the atmosphere on the path of the radar transmissions is a further factor which must be taken into account when assessing the likelihood of detecting targets particularly when considering the appearance of coastlines.

Under standard atmospheric conditions, the radar beam tends to follow the surface of the earth slightly, more so than light waves.

In considering the distance to the radar horizon, we can consider three 'horizons', the geometric, the optical and the radar. The geometric is the straight line, from the point of observation (scanner or observer's

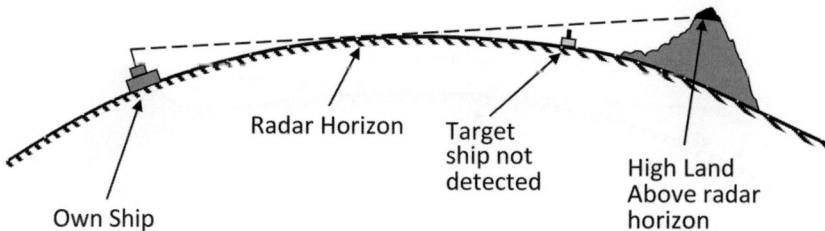

FIGURE 6.1   The theoretical radar horizon.

81

eye) tangential to the surface of the sea. The optical horizon is to the point where a theoretical object of no height on the surface could be seen by eye, and the radar horizon is that point where the radar could detect such a theoretical object.

The formulae are given, for comparison as follows:

Geometrical horizon $1.92\sqrt{h}$

Optical horizon $2.08\sqrt{h}$

Radar horizon $2.21\sqrt{h}$ ....(for 3 cm radar. 10 cm is slightly longer),

where h is the height of the point of observation (e.g. the scanner).

The possibility of detecting targets beyond the radar horizon will, in addition to all the other factors such as response characteristic, depend upon the height of the target (i.e. whether or not part of it extends above the horizon). Thus the theoretical detection range is the range from the scanner to the radar horizon added to the range from the target to the radar horizon of the target.

Detection range of a target $= 2.21\sqrt{h} + 2.21\sqrt{H}$

where h and H are heights of the scanner and target respectively in metres. The detection range is in nautical miles.

These formulae are theoretical on the assumption that:

- Standard atmospheric conditions prevail.
- The radar pulses are sufficiently powerful.
- The target response characteristics are such to return detectable echoes.
- The weather conditions, such as precipitation etc., through which the pulses have to travel, do not unduly attenuate the signals.

It is worth remarking that while scanner height is obviously a major factor in the detection range of targets, there comes a point where increasing the scanner height gains very little. To double the range of

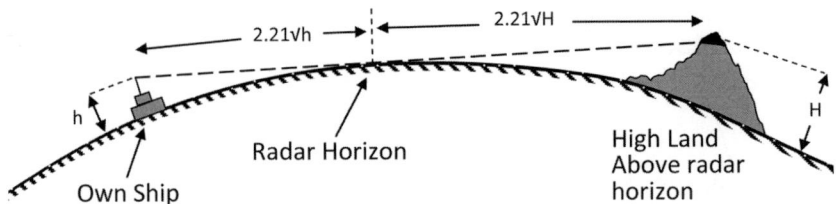

FIGURE 6.2   The theoretical detection range of targets.

the radar horizon, the scanner height would have to be quadrupled. For example a scanner height of 25m gives a theoretical radar horizon of 11.05 miles. To increase this to 22 miles would require:

$$22 = 2.21\sqrt{h}$$

$$\sqrt{h} = \frac{22}{2.21}$$

$$= 9.95$$

$$h = 99\,\text{m}.$$

Refraction in General

Refraction affects detection ranges or radar. The phenomenon of refraction should be well-known to every navigating officer. On occasions, in astro-navigation for example, he/she employs corrections in order to allow for the refraction of light rays passing from space through the atmosphere to the observer.

Refraction takes place when the velocity of a wave system, such as light or radio waves is changed. This happen when the wave front passes the boundary of two mediums (e.g. air and water) of different densities. One medium offers more resistance to the waves than the other and hence the velocity of the waves will change. The medium where the resistance is greatest is called the denser of the two. When the wave system meets the boundary between the mediums at right angles, no change in direction will place, but the wave will be accelerated or slowed down bodily. If, however the boundary is hit at an oblique angle, a change in the direction of wave system will follow. In the case that the oblique angle is very large, striking the interface at a small angle, and the wave system travels from a dense to a less dense medium, the change in direction of the wave front may be so large that the wave will return to the dense medium instead of going forward. This is called *total reflection* (see Figures 6.3 and 6.4). This phenomenon can be seen underwater where from below, only something above the surface immediately above the observer can be seen. Also between air and water a familiar phenomenon is when looking at a straight object such as long pole going down into clear water at an angle, there seems to be a bend in the pole as it passes the surface. This refraction also takes place where the medium is the same, but of different density in different areas. Air is subject to different density depending on different conditions (usually temperature) and the day to day effect of this is mirage and other optical illusions such as the appearance of 'water' on the road in

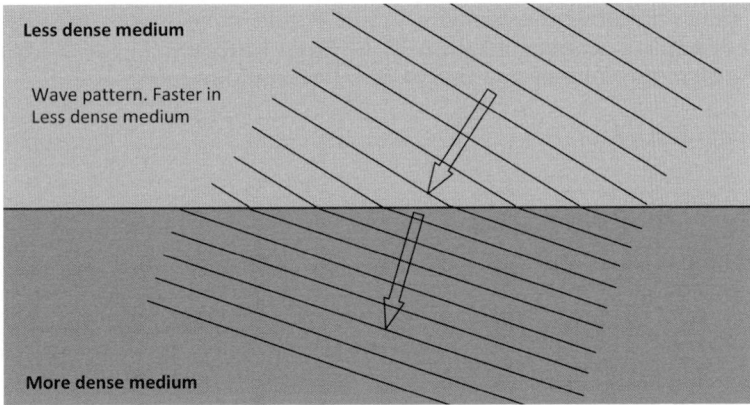

FIGURE 6.3   Refraction diagram 1.

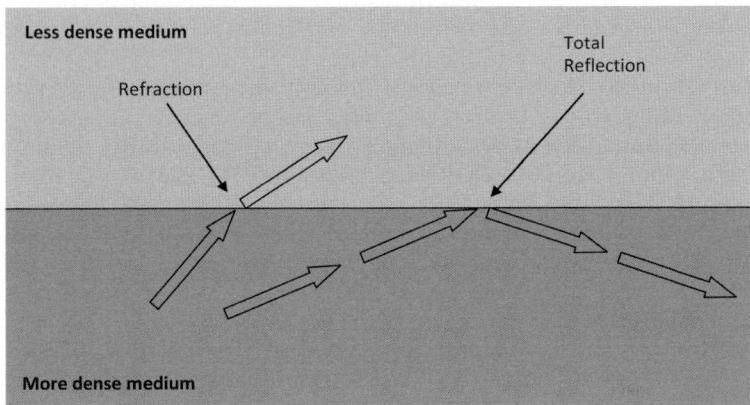

FIGURE 6.4   Refraction diagram 2.

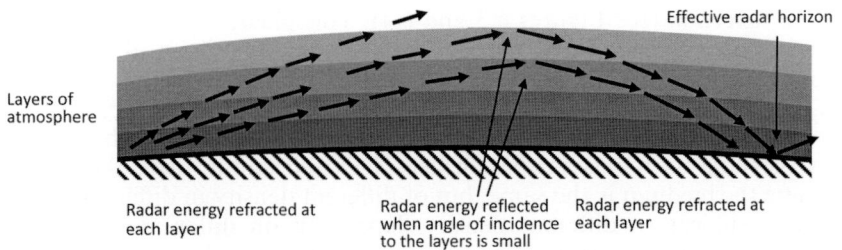

FIGURE 6.5   Refraction of the radar beam.

hot weather. Where the change in density is increasing or decreasing gradually, so the direction of the wave front gradually changes and it will follow a curved path. Note that the hollow part of the curve is always towards the denser medium.

In considering radar energy, being transmitted from a scanner comparatively close to the surface of the sea, and transmitted over a wide vertical beam, most of this will be subjected to some degree of refraction. That part of the beam which is sent out horizontally, under 'normal' conditions will not be refracted as it remains in the same density layer. However, the upper part of the beam, projected at between 10 and 20 degrees to the horizontal will be refracted. Most of this will be lost in space but at some point in the upper atmosphere that part of the beam which is at a shallow angle to the atmosphere layers will totally reflect. This will be duly refracted again back towards the surface of the earth, but unless there are conditions of super-refraction, they do not increase the range of detection (see Fig. 6.5).

Diffraction
This is the ability of certain wave-lengths to bend around objects in their path. This is one component of the range of the radar horizon, but should not be confused with the principles of refraction. Even light has a small degree of diffractive properties.

Standard atmospheric conditions, which never really exist in practice, are levels agreed by meteorologists so that standards can be agreed upon and data recorded, compared and tabulated. They are defined as follows:

Pressure = 1013 mb decreasing at 36 mb/1000 ft of height.

Temperature = 15°C decreasing at 2°C/1000 ft of height.

Relative humidity = 60% and constant with height.

These conditions give a refractive index of 1.00325 which decreases at 0.00013 units/1000 ft of height. While temperature decreases with height, and all other things being equal, cold air is denser than warm, but so does pressure decrease with height and the density of the air is less the higher one goes.

The term 'standard' should not be confused with the term 'normal', which is extremely subjective and imprecise. Standard conditions are rarely likely to be achieved in any particular part of the world.

These 'standard' conditions relate to the vertical composition of the atmosphere. The mariner cannot obtain a precise knowledge of this and so must rely on a more general appreciation of the weather conditions, the part of the world, the time of the year, meteorological data and forecasts and experience for guidance on the effects that they can have on the detection ranges of targets.

Under 'normal' conditions in most oceans frequented by shipping, super- rather than sub-refraction is more likely.

Sub Refraction

Sub-refraction occurs when the refractive index of the atmosphere decreases at a lesser rate with height than normal. The atmospheric conditions which can bring this about are:

1. increase in relative humidity with height, and/or
2. abnormal decrease in temperature with height.

Often these two conditions occur together. At sea, the surface layers of the atmosphere cannot be dry, but in certain areas of the world, such as polar regions, very cold air may be found over warm seas. In this case the lapse rate, the degree at which temperature decreases with height, is less than normal. Referring to Figure 6.5 showing the effect of refraction in the different layers of the atmosphere, there will be less refraction with the effect of bringing the effective horizon closer to the source of the radar energy. This will have the effect of lifting the radar beam up from the surface and reducing the distance to the effective radar horizon. Situations where sub-refraction may be encountered are, as stated, polar regions, where very cold air is present over comparatively warmers seas; in the Grand Banks area of the North Atlantic where cold polar air blows over the Gulf Stream; in areas where cold katabatic winds blow from mountains over warm seas, such as in the Gulf of Lions in winter-time. The onset of mid-latitude depressions may also cause sub-refraction mainly due to the high humidity. This is of concern as these conditions are often associated with poor visibility and the need for good radar visibility.

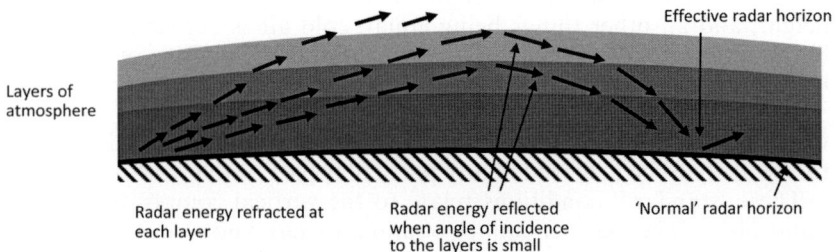

Effective radar horizon

Layers of atmosphere

Radar energy refracted at each layer

Radar energy reflected when angle of incidence to the layers is small

'Normal' radar horizon

FIGURE 6.6   Sub refraction.

Super-refraction

This of course is the opposite to sub-refraction, and of far less of concern to the mariner. If the rate of decrease in the refractive index of the atmosphere is greater than normal, the radar beam tends to be bent towards the surface, thus increasing the range of the radar horizon.

The atmospheric conditions which might bring about super-refraction are:

a decrease in relative humidity with height,
temperature falling more slowly with height than normal, or even increasing.

Such conditions are more often met with in tropical areas, but can be met in higher latitudes, often in high pressure, settled weather, even in winter-time. Under extreme conditions, land and shipping can be detected at extreme ranges such as ships and objects on the sea surface at 48 miles.

Ducting

This is a form of intense super-refraction. When there is a layer of warm air above a layer of cold air, usually called an 'inversion', the radar beam is effectively trapped between it and the sea surface; the

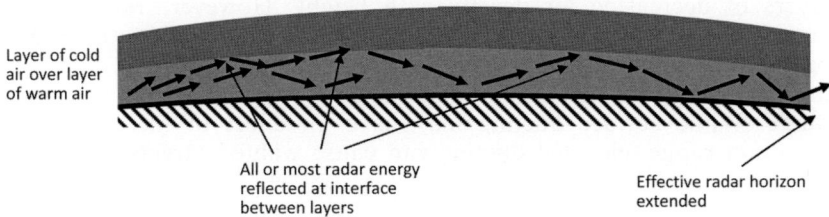

Layer of cold air over layer of warm air

All or most radar energy reflected at interface between layers

Effective radar horizon extended

FIGURE 6.7   Ducting.

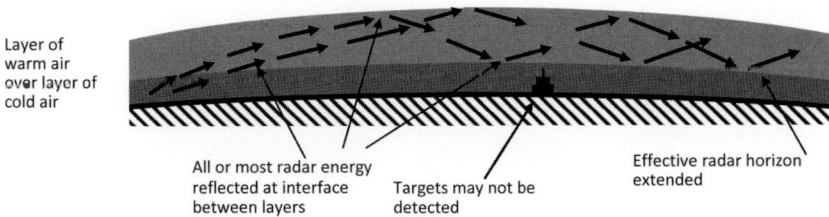

Layer of warm air over layer of cold air

All or most radar energy reflected at interface between layers

Targets may not be detected

Effective radar horizon extended

FIGURE 6.8   Elevated ducting.

refractive index between the layers is such that the radar beam is totally reflected. This can concentrate the beam and give it the ability to detect targets at extreme range. While this may seem to be a desirable condition, in fact is can be quite a disadvantage, as it usually produces second trace echoes (see Chapter 7 on false echoes) from land and other objects very far away and causing clutter and unwanted echoes which confuse the screen.

Such atmospheric conditions which can bring about ducting are cold settled conditions, often associated with high pressure.

Another form of ducting is elevated ducting. In this, there is a lower layer of less dense air above the surface of the sea, and a higher layer of less dense air, thus forming a 'sandwich' of denser air between two layers of less dense air. The radar beam is trapped between these two layers and exhibits much the same effects as surface ducting, but with the possibility that objects on the sea surface will not be detected.

Conclusions

Under standard atmospheric conditions the radar horizon is slightly further than the optical. The long range detection of targets, assuming good echo response, will depend on the height of the scanner combined with the height of the object. The radar beam is affected by the atmosphere, mainly by refraction /reflection as it passes through the layers of decreasing air density with height. However, varying and anomalous atmospheric conditions can have an effect on the radar beam in various ways, with super refraction making it possible to detect targets at very long ranges, and sub-refraction reducing the detection range. While surface ducting may have an effect of increasing the detection range, elevated ducting can cause wanted targets not being detected.

Chapter 7

False Echoes

This chapter examines false echoes produced by various means. These can be by

- Re-reflection off part of the ship's structure.
- Re-reflection off other ships or objects.
- Multiple echoes.
- Second trace echoes.
- Interference from other radars.
- Natural phenomena.

False echoes are produced by several different means. At best they are a nuisance and a distraction, at worst they may be mistaken for real echoes and cause some action to be taken which places the ship in danger. Between these two extremes they can clutter the screen and obscure genuine echoes.

In modern radars much effort and technology is applied to reducing false echoes, but some are of such a nature that, so far, has placed them beyond the capabilities of present technology to eliminate them.

Recognising false echoes is mainly a matter of experience. Many such echoes are less firm or more nebulous than real echoes and thus give an indication of their false character, but others can have all the appearance of real echoes. A knowledge of the causes of false echoes will help the observer to recognise them as such, but a situation can arise where even though an echo is suspected as being false it must be treated as being a real one until it is absolutely certain that it is not. This may mean stopping the ship, or turning away from the direction of the echo.

The siting of the scanner in relation to the ship's structure has much to do with certain false echoes, which is discussed in the section on scanner siting in Chapter 3. Sometimes scanners are sited in less than favourable positions for various reasons.

The proliferation of offshore wind farms, particularly in estuaries and near shipping lanes has caused concern, not so much because of the echoes that the wind generators produce themselves, but being strong radar targets they tend to re-echo off parts of the ship's structure and other objects such as passing ships and thus produce false echoes, sometimes in critical parts of the radar screen.

From ship's structure

This is also discussed in chapter 3 on the Scanner.

Where a part of the ship's structure protrudes into the radar beam, it may form a reflecting surface. Much will depend on shape, size, aspect and material of the part of the ship's structure, but samson posts, cranes, masts and funnels would be typical of such items. Much smaller items, close to the scanner can also provide a re-reflecting surface, such as VHF or other antennae or posts supporting them. The other factors are the echo strength of the real target and the disposition of it in relation to the reflecting surface of the part of ship's structure.

False echoes produced by re-reflection from the ship's structure must always lie in the direction of that part of the ship's structure causing the re-reflection in relation to the position of the scanner. For this reason all watchkeeping officers must be aware of the layout of the ship's structure in relation to the position of the scanner, and this includes the small items close to the scanner. However, it is remotely possible that under circumstances where a strong reflecting target is close by, that there will be a re-reflection from the side-lobes. This will of course give false echoes in directions other than that of the reflecting part of the ship's structure, but this usually appears as a smearing or curved elongation of the primary echo as is normal with side lobe effect (see comments on side lobe effect in Chapter 3 on the scanner).

Re-reflected echoes from parts of the ship's structure are generally at about the same range as the real target, assuming that the distance from the scanner to the re-reflecting surface is comparatively negligible. However, in large ships there may be a noticeable difference in range between the false and the real target where the re-reflecting surface is far removed from the scanner (see Figures 7.1, 7.2 and 7.3).

From other objects

Re-reflected echoes can be caused by good reflecting surfaces which are not part of the ship's structure. An example of such a situation might be where own ship is steaming near a large high sided vessel. The side of the vessel forms an excellent re-reflecting surface and other echoes of other ships, navigation marks and land can appear in the direction of the high sided vessel. Of course other objects can cause similar re-reflections and may not be as obvious as a nearby high-sided vessel.

Often the movement of such objects gives an indication of their false character. As own ship passes the object the false echo will move at a much greater rate than own ship's movement past the re-reflecting

Re-reflection from ship's structure   Target ship

Radar energy ———→

Scanner

Apparent
direction of
'other' target

Mast re-reflecting
target ship

FIGURE 7.1   Re-reflection from ship's structure.

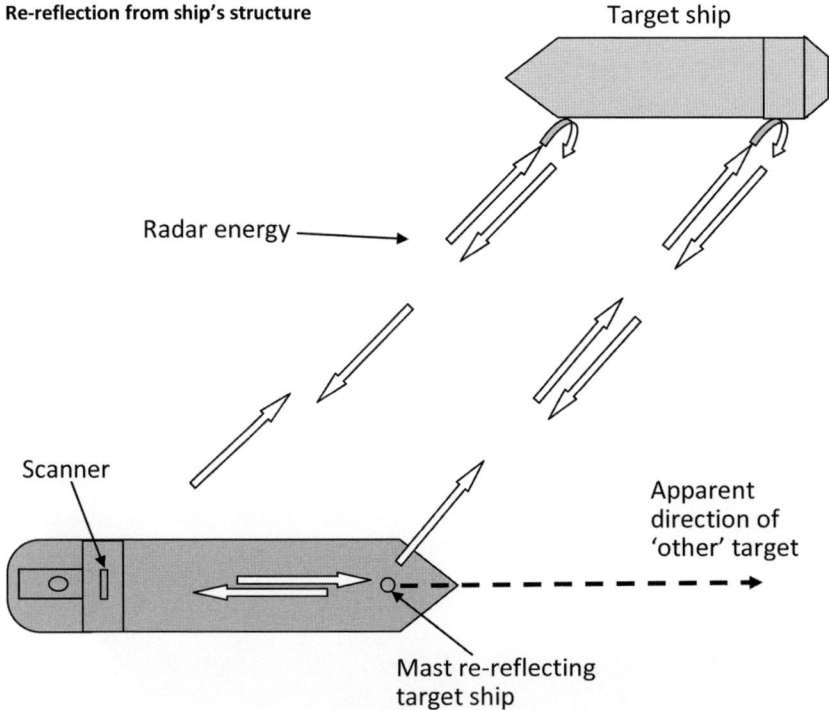

object, and may disappear altogether with the changing aspect of the
re-reflecting surface (see Figures 7.4 and 7.5).

Overhead cables

A situation which caused several groundings in the early days of
radar was where overhead power cables crossed the shipping channel.
These can cause a false echo, almost a mirror image of own ship. What
appears on the radar screen is an echo approaching own ship from the
opposite direction. When own ship alters course, to starboard, the other
echo also appears to alter course, but to port, putting it on a collision
course with own ship. There is no other indication of the presence of the
overhead power cables other than this one 'spot' echo keeping a
determined course to collide with own ship. This phenomenon should
not be confused with a similar one where a bridge or some other such

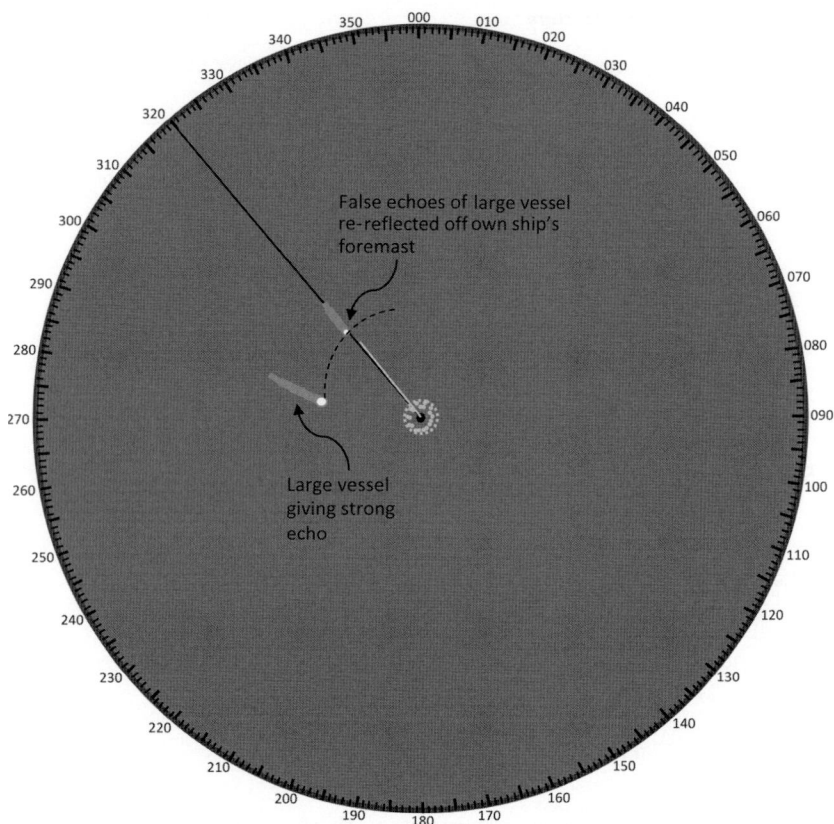

FIGURE 7.2   Re-reflection from ship's structure shown on radar screen.

good reflecting surface crosses the channel or shipping lane. In this case a re-reflection of own ship may occur, showing an echo approaching from the opposite side of the bridge.

At this stage mariners are generally aware of this phenomenon where overhead power cables are concerned and often there is a note on the chart or in the sailing directions warning of it where it occurs. There may be a similar warning regarding bridges (see Figures 7.6 and 7.7).

Second or multiple Echoes

This is a situation where own ship is close to a good reflecting surface, such as a high sided vessel. The radar energy from own ship is reflected

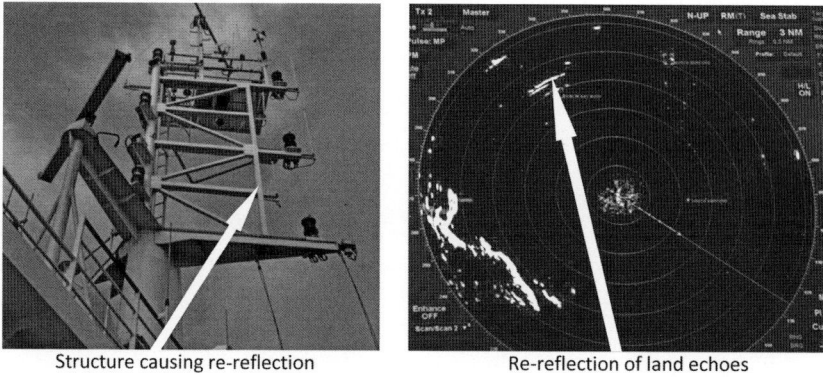

Structure causing re-reflection   Re-reflection of land echoes

FIGURE 7.3   Photos showing structure causing re-reflection and false echoes on the radar screen.

from the other ship, producing the normal target response, but then some of the energy is reflected off own ship's structure and again reflects off the other vessel, producing another echo, on the same bearing and at twice the range of the original echo. This process can repeat more than once, causing a series of echoes of diminishing strength, all on the same bearing and equidistant from each other (see Figures 7.8 and 7.9).

Second trace Echoes

These are produced under circumstances where echoes at a great range are received. The effect is usually associated with meteorological conditions where extra super-refraction occurs (see Chapter 6 on Radar Horizon). In effect it is the echo return in response to a previous pulse, not the one which has just left the scanner, and for which the amplifier is now receiving echoes. Say pulse A leaves the scanner. It produces echoes from the targets near the ship for which the amplifier is open, and then continues off into the distance. Then pulse B leaves the scanner and the amplifier opens to receive its echoes. However, echoes from far distant objects from pulse A arrive and appear on the screen in addition to the echoes from pulse B.

If a PRF of say 2000 pps (pulses per second) is in use on the 12 mile scale, then the PRP (pulse repetition period, (the time between one pulse and the next) is 500 m secs. This corresponds to a range of 40 miles, theoretically a target could be detected within this range (if it is above the radar horizon, strong enough to return a useable echo etc.). However, the set is on the 12 mile range so only echoes within 12 miles

Re-reflection from other structure

Apparent
'other' target

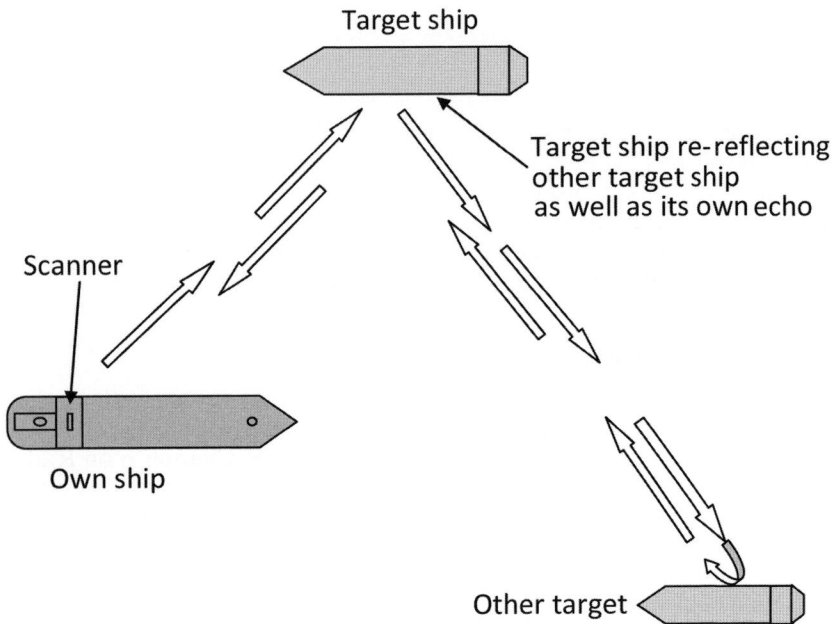

Target ship

Target ship re-reflecting
other target ship
as well as its own echo

Scanner

Own ship

Other target

FIGURE 7.4   Re-reflection from other structure.

will be displayed. Now take an object at over 40 miles range, say 46. The first pulse is striking this shortly after the next pulse has left the scanner, 500 m secs. later. The amplifier is now open again and the echo from the object at 46 miles arrives 75 m secs. after the second pulse has left the scanner. The amplifier has no way of distinguishing this from an

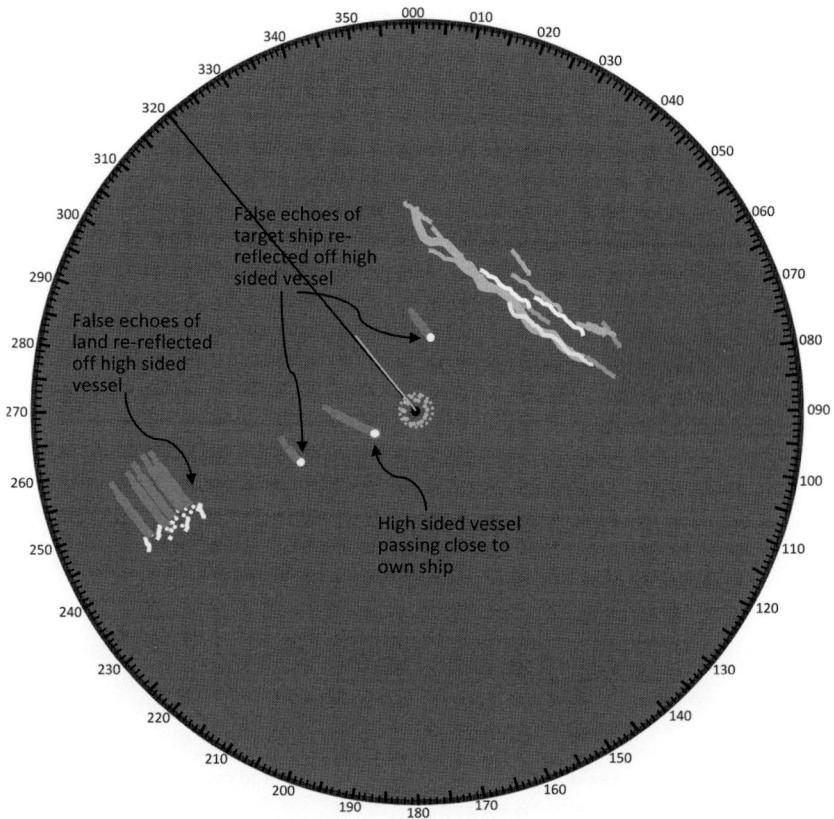

FIGURE 7.5   Re-reflection from other structure shown on radar screen.

echo from this second pulse and thus will display a target at 6 miles from the ship. This is displayed graphically below (see Figure 7.10).

Generally second trace return echoes can be recognised as such. They tend to be more nebulous or 'cloudy' than real ones. Many modern radar sets have eliminated the phenomenon to a large degree by varying the PRP for each pulse slightly and then allowing the correlation process to eliminate these echoes which appear in different positions with each pulse. However, if the far distant echo is large, such as land, it is possible that this process will not eliminate it altogether and observers should be aware of it. Adjusting the gain might eliminate such false echoes, but changing the PRF and thus the PRP should show up these echoes for what they are. The only way the observer has of

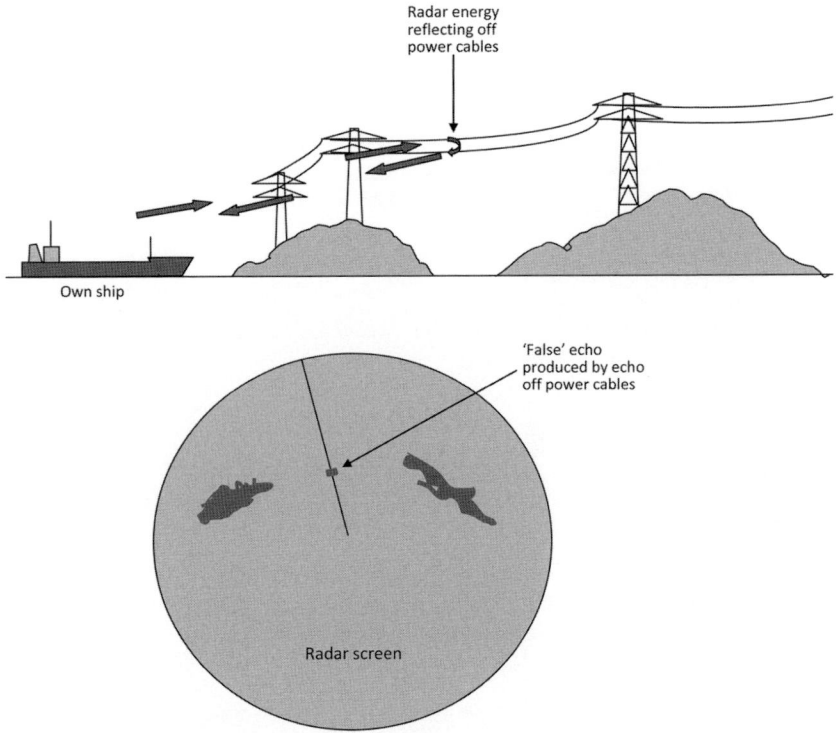

FIGURE 7.6   False echo produced by overhead cables.

changing the PRF is by changing range scale. If a suspected false echo on the 12 mile range scale disappears when the set is changed to 24 miles, or 6 miles (provided, of course it is within 6 miles), then it can be assumed that it is false.

Interference from other radars

Because of the nature of a magnetron, each radar set transmits on a slightly different radio frequency. This has the beneficial side effect reducing the interference of one set with another. Nonetheless interference from other radar sets, transmitting on a similar frequency as own ship, can be seen on the screen. In modern sets this interference can largely be eliminated by an interference rejection process. This is

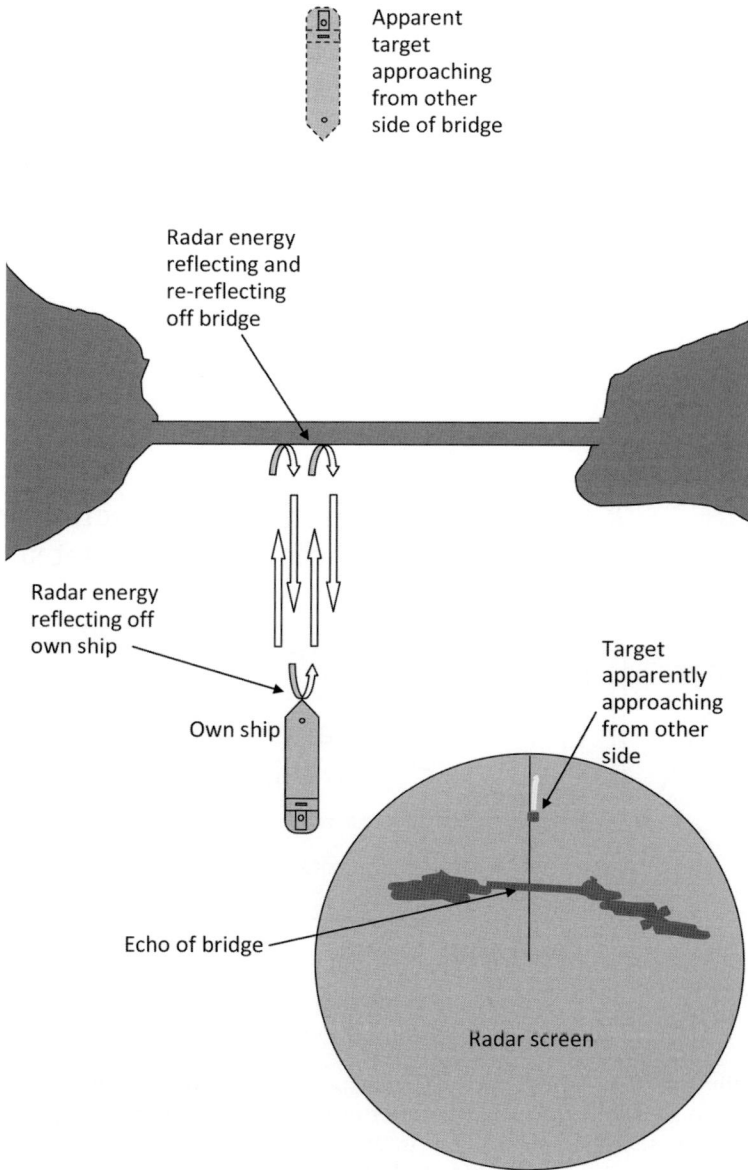

Apparent target approaching from other side of bridge

Radar energy reflecting and re-reflecting off bridge

Radar energy reflecting off own ship

Own ship

Target apparently approaching from other side

Echo of bridge

Radar screen

FIGURE 7.7   Re-reflection from overhead bridge.

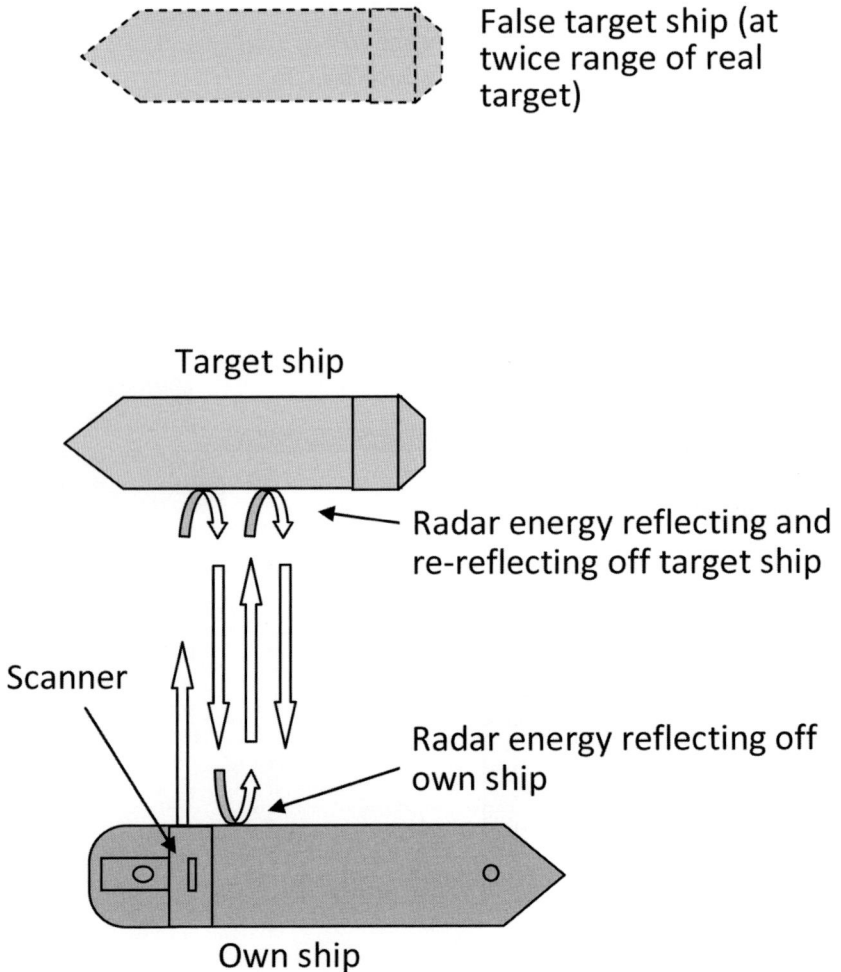

False target ship (at twice range of real target)

Target ship

Radar energy reflecting and re-reflecting off target ship

Scanner

Radar energy reflecting off own ship

Own ship

FIGURE 7.8   Multiple echoes.

usually left 'ON' under most circumstances, but the operator has the ability to switch it off, usually under circumstances where small targets are being searched for. The interference from other radars is a characteristic curved dotted radial line, intermittent in character, radiating from the origin.

When the interference reject function is 'Off', other random and

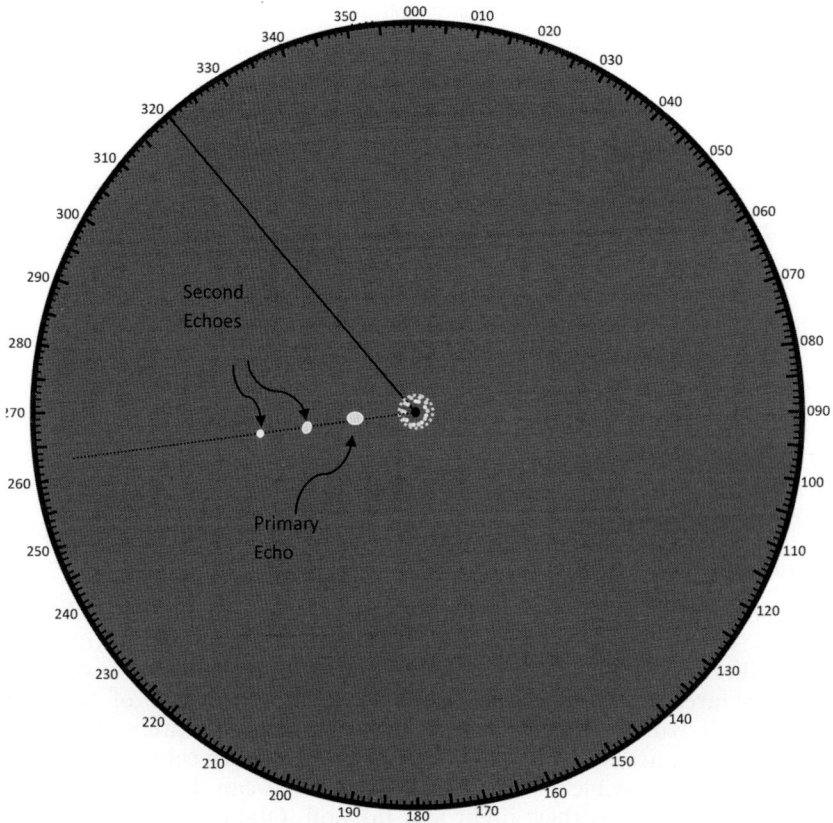

FIGURE 7.9   Multiple echoes shown on radar screen.

haphazard false echoes can appear, from a variety of sources. They may be a nuisance but rarely can be mistaken for real echoes due to their random nature.

New technology (NT) radars have no magnetron and therefore the remarks above do not apply. These are 'coherent' sets meaning that they transmit and receive on very precisely tuned frequencies. Therefore a different approach is needed to ensure that one radar set does not interfere with another. This is achieved by the nature of the pulse coding of each pulse, and a slightly different assigned frequency for each radar set.

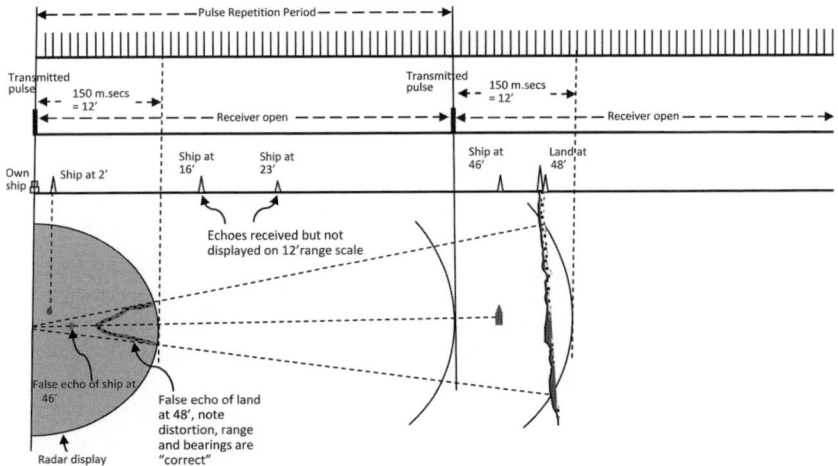

FIGURE 7.10   Diagram showing second trace return.

Natural Phenomena

It should also be recognised that apart from sea and rain clutter, which is discussed elsewhere, that such phenomena as tide rips and overfalls can appear as fairly consistent echoes. Breaking seas, or even different areas of clutter can be caused by the presence of reefs or underwater banks. In tidal waters different areas of clutter can result from different areas of tidal flow. Sea-birds can return echoes which are particularly noticeable in very calm conditions as can their disturbances of the sea surface as they alight and take-off. Often if they congregate in a definite spot, such as when they are fishing, they can produce a large strong echo. These, of course are not really false echoes, they indicate the presence of something real. Some of them can be useful as those indicating the presence and location of reefs and banks, but more often than not they clutter the screen and at times can be mistaken for dangerous echoes.

Conclusions

This chapter has examined false echoes. The possibility exists that such false echoes can be mistaken for something real, such as another ship or a natural feature leading to an action which could endanger the ship. Mostly false echoes are a nuisance and a distraction. They may obscure 'wanted' targets, or they may otherwise clutter the screen to the extent that it interferes with proper detection and tracking of targets. Operating the interference rejection, clutter controls and gain control can reduce or eliminate many false echoes, but OOWs should be aware of the effect such actions have on the detection of weak wanted targets. Some false echoes may not be eliminated by adjustment of the controls, such as re-reflected echoes and second echoes, but ships' officers should be able to recognise them by being aware of how they are caused.

Chapter 8

Target Response Characteristics

The ability of the radar system to detect and display a given target depends on a large number of factors. This chapter examines those physical attributes of radar targets or objects which may return an echo and which affect their ability to reflect radar energy. In examining the factors which affect the response an object may give, these factors have been broken down into simplified descriptions. It should be realised that most actual targets will contain of a number of these factors.

Many familiar characteristics of the behaviour of a beam of light and reflecting surfaces can be attributed to radar energy and the reflection of it. In this chapter the following factors are examined:

- the aspect or angle the reflecting surface presents to the radar beam,
- the effect the texture of the reflecting surface has on detection,
- how the material of construction affects detection of a target,
- how size and shape effect detection, and
- RACONs, target enhancers (RTEs) and SARTs.

Aspect

This is probably the most important factor contributing to echo return. This is the angle a reflecting surface presents to the radar beam. A good reflecting surface, at right angles to the beam, both in the horizontal and vertical sense, will return a good echo. The same reflecting surface, at an angle to the radar beam will tend to reflect the radar energy away from the direction of the radar beam and will thus present a poor echo or none at all. This can be compared to a beam of light being shone on a good mirror. If the mirror is at right angles in every sense to the beam of light, then it will be reflected back to the light source; if the mirror is at an angle the beam will be reflected in some other direction, depending on the angle of the mirror to the beam (see Figure 8.1).

Surface Texture

As discussed above smooth surfaces reflect radar energy in much the same way as a mirror reflects light. A rough surfaces tend to scatter

Good Aspect

Radar energy
reflected directly
back

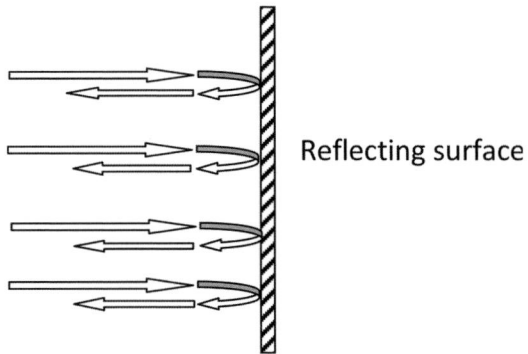

Reflecting surface

Poor Aspect

Radar energy
reflected away

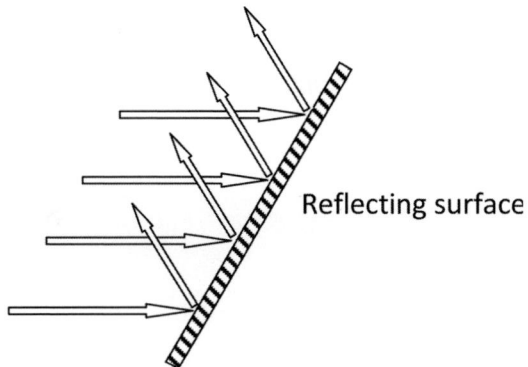

Reflecting surface

FIGURE 8.1 Aspect.

energy. The degree of 'roughness' is proportional to the wave-length of
the energy striking the reflecting surface. Light, being of a much higher
frequency will be scattered in all directions by a surface that will appear
'smooth' to radar energy. Using the light analogy, a light beam striking
a white painted masonry surface will show well when seen from any
direction, in other words the light is scattered in all directions. Similarly
a surface that is 'rough' to radar energy will scatter the reflected energy
in all directions.

**Rough Texture,
Good Aspect**

Radar energy
reflected in various
directions

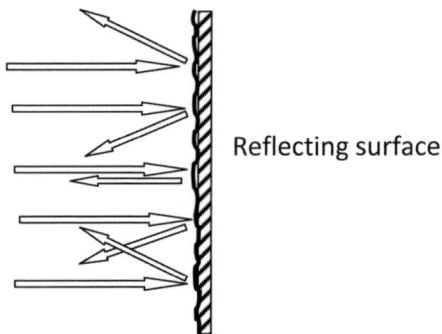

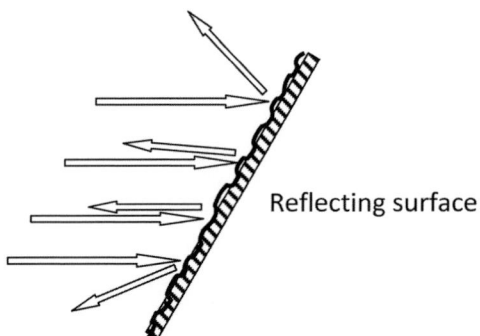

Reflecting surface

**Rough Texture,
Poor Aspect**

Radar energy
reflected in various
directions

Reflecting surface

FIGURE 8.2.   Surface Texture.

Material

Generally materials which are good conductors of electricity are also those which are good reflectors of radar energy. This is because a good conductor absorbs the radar energy and re-radiates it at the same wavelength. Other materials absorb the radar energy and do not re-radiate it but convert it to heat (which is what happens in a micro-wave oven). Some of the reflection coefficient of various substances are as follows:

Steel Reflection coefficient 1.0
Sea water 0.8
Ice, with no salt 0.32
Land with short grass 0.7 – 0.8
Land with long grass 0.1 – 0.4
Land with scrub or trees < 0.1.

Approximate relative densities of some common ship/boat materials		
Material	relative density	reflective property
Wood (soft woods to hard woods)	0.3 -0.9	poor
GRP	1.5 – 2.0	poor
Aluminium	2.7	poor
Steel	7.7	good
Copper, Brass phosphor bronze	8.0	good

FIGURE 8.3   Table of approximate relative densities of some common ship/boat building materials.

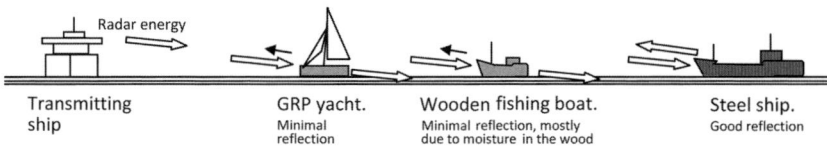

Transmitting ship

GRP yacht.
Minimal reflection

Wooden fishing boat.
Minimal reflection, mostly due to moisture in the wood

Steel ship.
Good reflection

FIGURE 8.4   Materials.

The density of the material also plays an important role in re-radiation of the radar energy and can give some indication of its likely response characteristics. Some relative densities of popular boat/ship building materials are given in Figure 8.3 above. It should be remembered of course that even a boat built with a poor reflecting material will be constructed with other components such as engines, steel fittings and rigging and such like features which will give some improved response. Nonetheless the wise owner of such a vessel will have some form of echo enhancer fitted.

Size

Obviously the size of a target or a reflecting surface will have major influence on the amount of energy that is reflected. The more energy reflected from each pulse, the better the echo. This point has to be considered in the case where two reflecting surfaces are presenting the same area to the radar beam, but one reflecting surface is higher than the other. As the radar beam is much wider vertically (taller) than it is horizontally, at any instant the return will be much stronger from the tall reflecting surface than the low wide one. However, in considering this point, the echo of a particular target, such as a ship, is made up of many echoes from multiple pulses, and the wide target is likely to appear bigger than the tall narrow one. Because of the stronger echo return from the tall narrow target, it is likely to be detected at longer range than the low wide one of the same reflecting area.

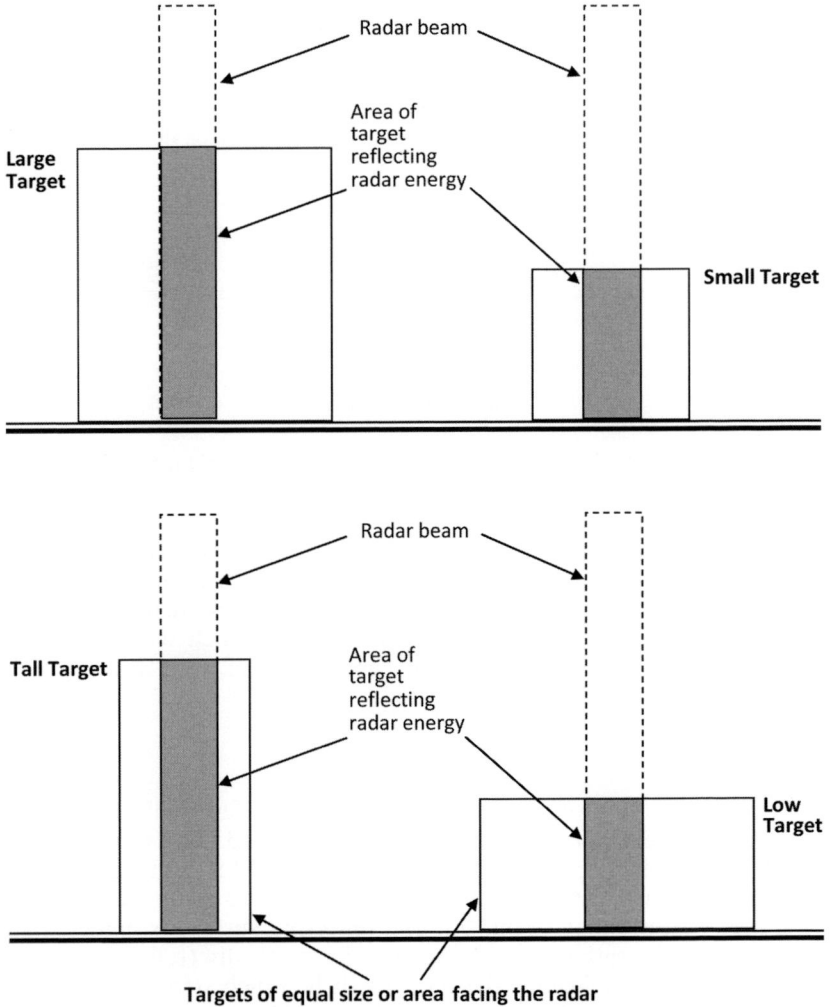

FIGURE 8.5   Size.

Shape

Shape has an influence in so far as an object presents a reflecting surface to the radar beam. Most targets are complex series of reflecting surfaces with varying degrees of the factors discussed above. To analyse this topic we consider a few basic shapes and how the shape influences the response to the radar beam. For the purpose of this discussion it is assumed that the material and texture in all cases present good response when perpendicular to the radar beam.

A cube shaped object will provide a good response, provided one of its faces is perpendicular to the radar beam, otherwise the faces present poor aspect and the cube has no surface at right angles to the radar beam.

A cylindrical object will present a comparatively small reflecting surface. This will be a vertical strip facing the direction of the radar beam.

A spherical object will present a small reflecting surface as a disc facing the direction of the radar beam.

A conical object will not have any surface at right angles to the radar beam.

In considering these shapes it should be realised that movement, as in a sea will affect the response; the aspect of the reflecting surface, if any, will change as the object moves. In the case of the conical object it may occasionally present a small reflecting surface perpendicular to the radar beam (see Figures 8.6a, 8.6b and 8.6c).

Radar Cross Section

This is often quoted to give some indication of the echoing strength of a target or object. It can be considered as the equivalent echoing area. It is not, as may be assumed the area of a flat specular (mirror-like) reflecting surface placed perpendicular to the radar beam. It is a surface which theoretically scatters the radar energy evenly in all directions (isotropically). A sphere can be considered to meet this requirement, so the figure quoted is the cross-sectional area of a sphere that would give the same radar return as the object in question.

Figures quoted for the radar cross section of some objects are as follows:

Metal sphere of radius r metres $= \pi r^2$

Metal plate, are A sq. metres $= 4\pi A^2 / \lambda^2$ ($\lambda =$ radar wave-length in cm).

Sphere or circular target

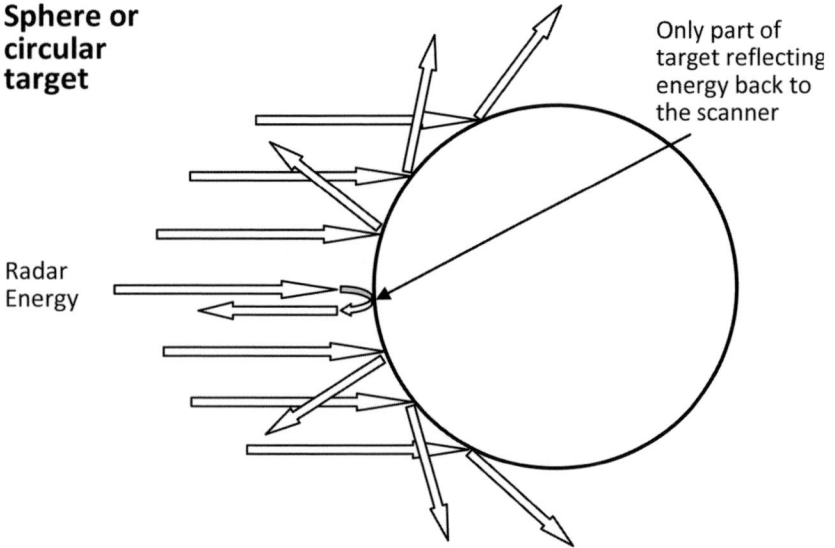

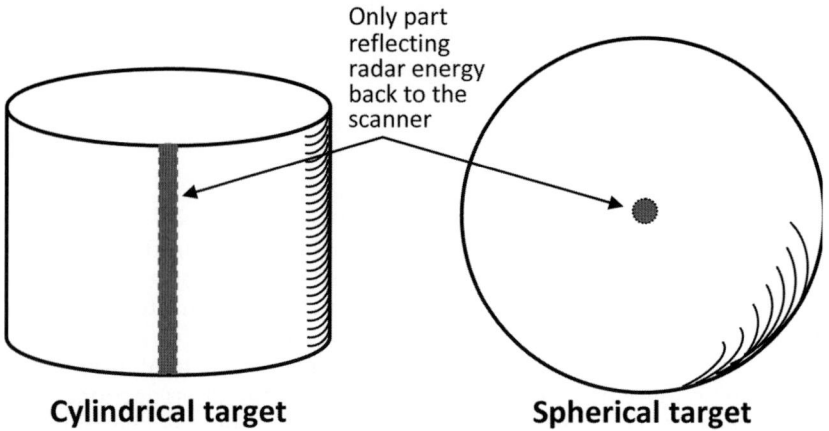

Only part of target reflecting energy back to the scanner

Radar Energy

Only part reflecting radar energy back to the scanner

Cylindrical target

Spherical target

FIGURE 8.6a   Round shapes.

**Conical
Shaped
Target**

No part
reflects
radar energy
back to the
scanner

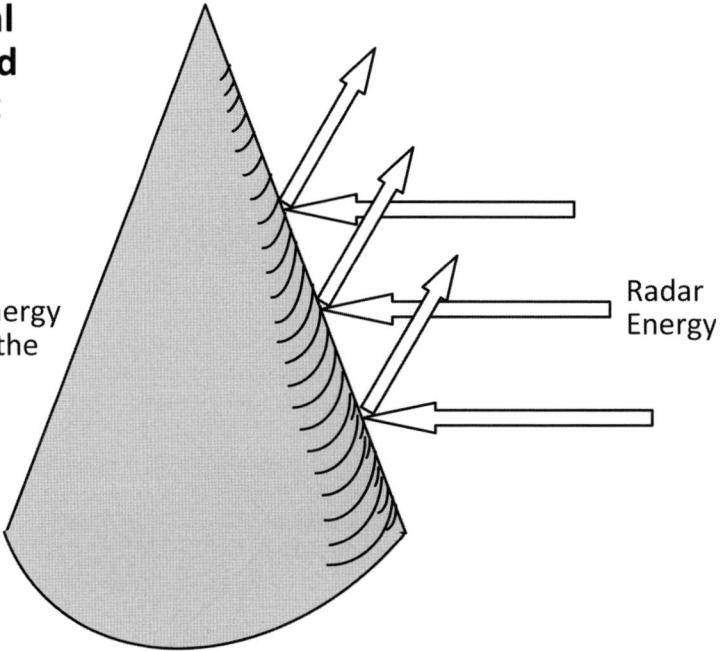

Radar
Energy

FIGURE 8.6b   Conical shape.

Using this standard, it is interesting to compare the radar cross section of the sphere with that of the flat plate, of the same area, placed perpendicular to the radar beam. The area in each case is 1m^2. The radar wave-length is $3.2\,\text{cm}$.

$$\frac{Radar\ cross\ section\ of\ the\ plate}{Radar\ cross\ section\ of\ the\ sphere}$$

$$= \frac{4\pi \dfrac{A^2}{\lambda^2}}{\pi r^2}.$$

But $\pi r^2 = 1 = A$.
   Then

$$= \frac{4\pi A}{\lambda^2}.$$

Cube or rectangular target

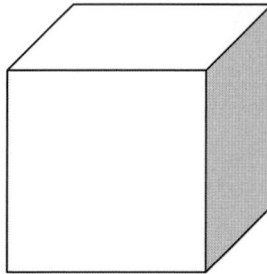

Echo response depends on the aspect of the target.

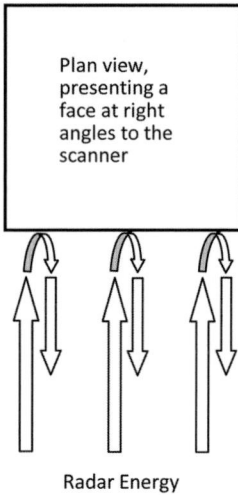

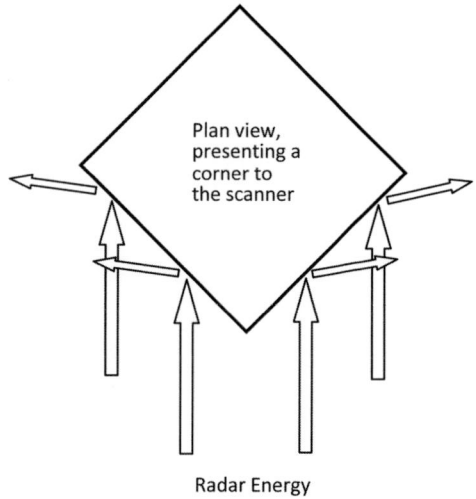

Plan view, presenting a face at right angles to the scanner

Plan view, presenting a corner to the scanner

Radar Energy

Radar Energy

FIGURE 8.6c   Square and cube shape.

$$A = 1\,m = 100 \times 100\,cm = 10,000$$
$$= \frac{3 \times 3.1417 \times 10,000}{3.2 \times 3.2}$$
$$= 12,273.$$

The flat plate has over 12,000 times the reflective properties than the sphere.

In the IMO Performance Standards the range performance is quoted in terms of a navigational buoy having an effective echoing area of

$10 \, m^2$. This, in effect is a sphere with that cross section of $10 \, m^2$, or a diameter of 3.6 m.

Some maximum ranges at which various maritime objects have been detected, giving an idea of their radar echo response assuming normal propagation and a scanner height of 15 m:

Object	Approximate range
Hills and mountains	15 – 40 miles
Coastline 60 m high	approx. 20 miles
Coastline 6 m high	approx. 7 miles
Piers and breakwaters	5 – 10 miles
Low sandy coastline	1 – 5 miles
Ship of 5000 grt	approx. 10 miles
Fishing vessels	3 – 5 miles
Small wooden boats	1 – 4 miles
Large conical buoy	2 miles
Large conical buoy with Radar reflector	5 miles
Icebergs	3 – 15 miles
Growlers < 1 m height	0 – 3 miles.

Radar Reflectors

Using some of the principles discussed above, particularly material, texture and aspect, devices have been designed from the earliest days of radar to enhance the response of certain targets which in themselves do not return good echo responses. Take, for example, a conical buoy constructed of GRP (glass reinforced plastic or fibreglass). Its shape and material are not conducive to returning a good echo. Therefore it needs something to enhance its response characteristics, and something which is not too large in relation to its size.

The 'corner reflector' in various configurations has been the main form of passive reflector for may years. In this three plates of good radar reflecting material are placed mutually perpendicular to each other, forming a rectangular corner. Any radar energy entering this corner will be reflected directly back, no matter what direction it comes from.

Using this principle, the two most common forms of deployment of corner reflectors is in arrays called 'pentagonal cluster' where five such corners are arranged in a circle, and 'octahedral cluster' where four plates form eight corners.

In deploying such corner reflectors, the 'corner' is most effective when the axis of symmetry is in the direction of the incoming radar pulse,

Radar energy entering the
corner from different
directions

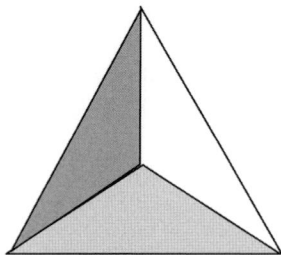

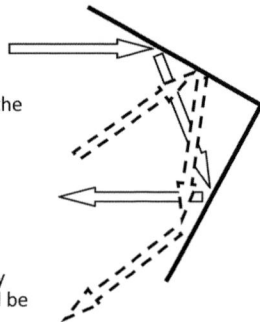

Corner Reflector. Three
reflective plates mutually
perpendicular to each other

Principle. Radar energy
entering the corner will be
reflected directly back

FIGURE 8.7   Corner reflector.

horizontal, for all practical purposes. The axis of symmetry is the imaginary line emanating from the apex of the corner and equidistant from all three sides. This means that the pentagonal cluster has each of the five corners with the bottom plate tilted from the horizontal by 35°. In the case of the octahedral cluster, there are two corners apparently wasted, one facing up and the other down, but the remaining six are then correctly deployed.

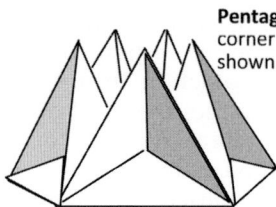

Pentagonal Cluster. Five
corner reflectors arranged as
shown.

Octahedral cluster . Three
reflective plates forming
eight corners; one facing up,
one facing down and six
facing horizontally

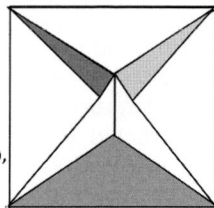

FIGURE 8.8   Pantagonal cluster, octahedral cluster.

Further uses of the principle of the corner reflectors is where a series of pentagonal clusters are stacked, each cluster being rotated a few degrees in relation to the other so that the reflective properties are increased and the sectors of reduced effect are eliminated. Such a form of stack reflectors is usually encased in a plastic casing and is a favourite for pleasure craft and other smaller vessels.

Another form of passive reflector is the Lunenburg Lens. This is a sphere constructed of a series of concentric shells of increasing refractive index. As the radar energy enters the sphere it is refracted to focus on a point on the opposite surface of the sphere. A metallic band

Lunenburg Lens

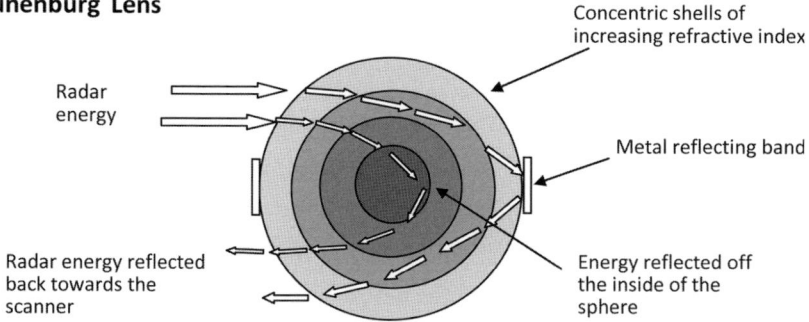

FIGURE 8.9 Lunenburg Lens.

around the equator of the sphere reflects the energy which is then refracted again so that it returns directly in the direction from which it came.

Active Echo Enhancers

These are devices which transmit a signal at the radar frequency and are therefore visible on the radar screen. They normally are triggered by a ships radar beam striking them. There are two main types in use to enhance and/or identify navigational aids. The first of these is the RACON. This gives a periodic 'flash' on the radar screen emanating some small distance on the far side of the echo of the aid containing the RACON. Some of the applications of RACONs are:

- to identify aids to navigation, both seaborne (e.g. buoys) and land-based (e.g. lighthouses),
- to identify landfall or positions on inconspicuous coastlines,
- to indicate navigable spans under bridges,
- to identify offshore oil platforms and similar structures,
- to identify and warn of environmentally-sensitive areas (such as coral reefs),
- to mark new and uncharted hazards (these should use the Morse identifier "D"), and
- there are other similar applications for which they can be used.

Frequently the RACON flash will show before the echo of the actual aid (buoy, light-float, light-ship). On closer range the echo of the aid will be seen, and when it shows, the flash will appear a little distance just beyond it. This gap should not exceed 100 m by IMO regulations; it is usually in the range of 75 m. Range should therefore be measured from the actual echo and not from the start of the RACON flash. In some cases the flash can be coded with a morse letter to identify the RACON flash of one aid to navigation from that of another. The RACON flash of buoys fitted with RACON can be often detected at ranges of up to 15 miles.

Because of the variation in frequency between one radar and another, within the radar bands (3 cm or 10 cm) the RACON will sweep the whole band-width. This will achieve two desirable effects, one, it will respond to a particular ship's radar and show a flash on it, and it will disappear for a couple of minutes before appearing again, thus ensuring that any targets which the flash might obscure are detected. Typically a RACON flash will show for one or two rotations of the scanner (normal scanner rotation rate) and then disappear for about 2 minutes before appearing again. Because each radar has a unique transmission frequency, it is unlikely that a ship will pick up the RACON flash triggered by another ship.

A 'frequency agile' RACON will detect the frequency of the incoming radar pulse and will tune itself to that frequency and transmit a coded response at that frequency. It again will be programmed to show a flash only every minute or so for the reasons stated above. The receiving, tuning and transmitting takes about 4 micro-seconds, which results in a range error of about 60 metres. This accounts for the small gap between the object and its RACON flash.

Mo(A) 10s
Dublin Bay
Racon(M) AIS

Chart symbol for buoy with RACON and AIS.

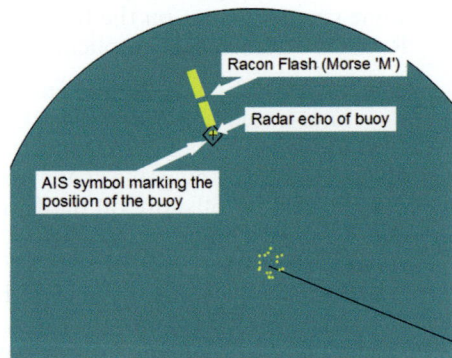

Racon Flash (Morse 'M')

Radar echo of buoy

AIS symbol marking the position of the buoy

Radar screen showing echo of buoy with RACON flash and AIS symbol.

FIGURE 8.10

In observing RACON signals the following points should be noted:

- RACONs which appear on each scan could mask other targets and so the RACON is made to be 'silent' for as short period in each minute.
- The application of FTC and some video processing techniques including 'interference suppression' can mutilate RACON signals and should be switched off.
- With frequency agile RACONs it is essential that the receiver (radar) is in tune, especially if the receiver bandwidth is narrow.

In seeking RACON signals the following can help their detection and correct identification:

- Switch anti-rain clutter control off before observing.
- If the RACON is attached to a buoy, the flash may obscure the buoy's echo in which case range can be obtained during off period.
- Anti-rain clutter control may be used to detect strong targets obscured by the flash provided that the target signal is stronger than the RACON signal.
- Beacon signal may be coded (usually a Morse code single letter) up to six digits.
- Details are listed in the Admiralty List of Radio Signals (ALRS).
- RACON signals should not be used for tuning a radar set.

RACON flashes can be sometimes reflected off nearby strong targets thus showing a flash on the wrong bearing. Also at close range side-lobes can trigger the RACON transponder.

Radar Target Enhancers

A radar target enhancer (RTE) is a device which as the name suggests effectively increases the echo response of a target which in itself presents a poor radar target. An RTE consists of a microwave frequency amplifier with separate transmit and receive antennas. The incoming radar pulses are, received, amplified and retransmitted at the same frequency (within the 3 cm band), without changing their length or introducing any significant delay. The ITU Recommendation specifies that delay and stretching of the output should not exceed 10 percent of the length of the received pulse, or 0.1 micro seconds, whichever is greater. This ensures that the return appears at the exact location of the target on the display and secondly to fulfil the requirement of IMO Resolution A.615(15) by avoiding any enlargement of the return, which could give a misleading impression of the nature of the target and could also result in cluttering of the display. Therefore the effect of the device is to improve consistency of response, rather than increase size. Some radar processors increase the number of pixels used on the display to mark the target, depending on the persistence and strength of return signals. This can result in a slightly larger apparent return, but does not give the impression of a large vessel.

Trials with RTEs have shown that they are very effective on small craft which have little intrinsic echo return. Tests carried out with a rigid-inflatable inshore-lifeboat showed consistent returns up to 7 miles in quite large seas. Tests with navigation buoys have shown that an RTE will increase the range at which the buoy can be consistently detected from 2-4 miles when equipped with a passive reflector, to up to 6-10 miles, depending on sea state. RTEs have been shown to improve target detection in both sea and rain clutter and are unaffected by anticlutter processing, except at extreme levels, when all passive returns are suppressed.

With powerful radar pulses, the amplifier of an RTE will saturate. This means that it will reach a point where it will not amplify the incoming pulse to any greater extent. The effect of this is that the enhancement over the normal response of the target will diminish as the range closes. This will depend upon the actual range and the power of the transmitted pulses, but a buoy with a passive reflector will have no appreciable enhancement from the RTE in under 2 miles in low powered radar and below 4 miles for a more powerful one. For a small boat with little intrinsic echo return there could be enhancement down to 1 mile. This can be a useful feature as it means that as the ship closes such a target it will tend to take on something more like the size and strength of echo that characterises the object itself.

Search and Rescue Transponder

This is a device that is also triggered by a received radar pulse. Like other devices it sweeps the 3 cm band. It does this very rapidly, taking 7.5 micro-seconds for each sweep, with a 0.4 micro-second return sweep. This cycle is carried out for 12 sweeps. This has the effect of producing a series of 12 dots on a ship's X-band radar screen. The line of dots indicates the direction of the transponder. However, because of the delay (0.5 micro-seconds) in responding to the interrogating radar pulse, the range of the transponder cannot be accurately measured from the series of dots, the nearest dot does not indicate the position of the transponder. What does occur is that as the transponder is approached, the dots will start to widen into arcs, which in turn will widen as the range closes until they form complete circles around the origin on the radar screen. At this point, the transponder is very close, probably

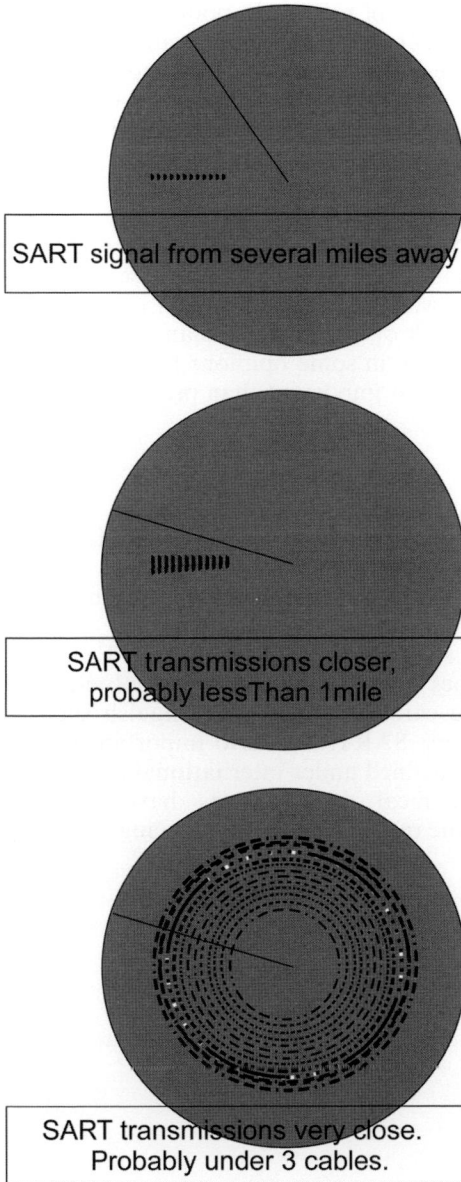

SART signal from several miles away

SART transmissions closer,
probably lessThan 1mile

SART transmissions very close.
Probably under 3 cables.

FIGURE 8.11 SART transmissions.

within a matter of tens of metres. In poor visibility care should be taken not to run down the craft deploying the transponder.

Such transponders are carried in all seagoing ships, under SOLAS/ GMDSS regulations. A typical SART has a battery life of 96 hours and can transmit for 8 hours. It should activate at 5 miles when interrogated by a ship's radar whose scanner height is 15 m and at 30 miles when activated by an aircraft's radar at 3000 feet. Ships have often picked up SART transmissions at 8 miles or more.

The effect of New Technology (NT) radar on RACONs RTEs and SARTs is still under discussion. As discussed in Chapter 18 on NT radar, the transmission of the radar signal is very precisely controlled. This increase the risk of a ship other than the observing one triggering the transponder. Also, in some opinions the advent of ECDIS and AIS has made such aids less important than previously.

Conclusions

The size, shape, material, texture and aspect of targets play a major part in the detection of such targets by radar. Using the factors which contribute to such echo response, devices can be constructed which enhance the echo response of targets which would otherwise not show strong or detectable echoes.

Transponders such as RACONs and RTEs are used to further enhance the echoes of important navigational aids such as buoys and light-ships. These can also help to distinguish such aids. Search and rescue transponders, SARTs, are also important for search and rescue purposes, and are defined under international regulations.

Continuing technical developments have altered the degree of importance of some of the detection enhancing aids.

Chapter 9

Setting up a Marine Radar

This chapter outlines a systematic way of switching on and adjusting the radar controls for optimum performance. In this it outlines:

- The checks before switch-on.
- The setting of controls prior to adjustment.
- The check of inputs to the radar.
- The adjustment of controls for optimum performance.
- The periodic performance checks during use.

Additionally this chapter looks at other controls and adjustments normally found on modern sets. A modern radar set is a complex piece of electronic and computer equipment. While the IMO require some degree of standardisation, there is a wide variety of equipment and controls to:

a. Comply with the basic IMO requirements, and
b. provide additional useful features extra to those requirements.

Some makes of radar will have quite an extensive keyboard of 'hard' control keys and adjustment knobs, others will have a minimum of such with most of the switches and adjustments in the form of 'soft' controls to be set by the roller ball and three buttons.

The use and adjustment of several of these controls are discussed in more detail in other chapters.

Setting Up

The term 'setting up' refers to switching on the radar for the first time, for example in preparation for leaving port. However, many of the steps or stages mentioned here should also be carried out periodically at sea, for example on taking over the watch. The object of these procedures is to ensure that all targets which are capable of being detected by the radar are detected and that all the 'raw data' is being displayed on the radar screen. To achieve this, the setting up procedure should be done in a systematic way. These procedures must be done before using any of the interference controls or auto clutter controls. (The use of these controls are outlined in the relevant sections).

It must be emphasised that should the radar not be set up correctly and there is maladjustment of the controls then targets may not be detected, with the subsequent dangerous consequences.

Ensure that all health and safety requirements are in place. All panelling to the radar is in place and that there are no notices advising that the set is non operational or not available. The following sequence is a guide; it is important that the radar operator's manual is consulted as there are variations in makes of radar sets, but the adoption of a procedure for setting up, incorporating the instructions in the manual should be formalised for the ship or radar set and subsequently followed every time.

1. Check that the **scanner is free** to rotate and that there in nobody in the near vicinity. Also check that there are for example no halyards loose, near, or entangled in the scanner. This is to ensure that nobody is injured by the powerful radiation produced by the scanner or that the rotation motor or gear train is not damaged by being obstructed in any way (see Figure 9.1).

2. Ensure all the **controls are turned to minimum**. This includes gain, tuning, all clutter controls etc. This, in the old CRT sets, ensured that no damage occurred to the tube or components of the radar set by the sudden full brightness coming on to the screen. However it is still good practice today as it makes sure that the operator has all the control at the same starting point and no control is inadvertently affecting the radar display which could affect the adjustment. Depending on the make of the radar, some of these adjustments may not become available until the set is switched on, but once they become available they should be turned to the minimum.

3. **Power on** the radar. The switch for this is normally on the display, and power is normally available. The main switch providing power to the radar may not necessarily be in the wheelhouse. Some radar sets have an independent switch for the scanner or most likely the scanner may be activated by the radar being set to 'transmit'. However, in some makes the scanner rotates even when the set is on 'stand-by'. In this regard, ships' officers should be fully familiar with the power supply, fuses or trip switches to the radar (and all navigational instruments in their particular ship).

4. The radar is now in a 'warming up' mode. Under IMO specifications this 'warm up' mode must not be more than 4 minutes from cold and no more than 15 seconds once warmed up, i.e. from 'stand-by to 'transmit'. New Technology (NT) radars may not have such a warm-up period. Most radars will

FIGURE 9.1 Radar mast.

not allow you to activate any transmissions until this period of warming up is completed. During this time most radars will also complete a self test to ensure that all the systems and computer elements are working correctly. If any element should fail then a code or error message will be displayed. Successful completion will usually be followed by a confirmation that all self tests have been completed.

5. Now **adjust the brilliance** to the ambient light. Too high a setting will destroy any contrast on the screen; too low will not allow enough contrast to distinguish potential targets and will make it difficult to read any of the data on the screen. The brilliance adjustment may be a hardware control or knob, probably near the screen, or a 'soft' control on the screen.

6. Ensure the following inputs are **matching the source** data, if fitted:
   a. Gyro compass
   b. GPS or DGPS
   c. GLONASS
   d. Loran C or E Loran
   e. Kalman filter
   f. Doppler log
   g. Electro-magnetic Log/s
   h. Chart system
   i. AIS
   j. Any other instruments interfaced with the radar.

7. Ensure **required input data from the various instruments, particularly gyro and log is being correctly displayed**.

8. **Set the orientation** of the display to the preferred mode. The choice is usually 'Head Up', 'North Up' or 'Course Up' and mode 'Relative Motion' or 'True Motion'. In modern sets often the mode of vector is selected with the motion mode such as 'True Motion Relative Vectors' (TM(R)). For example select relative motion, relative vectors, and north up.

9. **Set trails to 'Off'**.

10. **Set the range** preferably to 6 or 12 mile range whether in port or at sea. This provided the best scope for adjusting the setting of the various control discussed below.

11. **Set the pulse length** to 'medium' or 'long' depending on what is available on the particular set. However if a smaller range scale is in used then a short pulse length is required or a long pulse if a higher range scale is used. Note that some radars will not allow changes to the pulse length until the set is transmitting. Usually there is a choice of 'long' or 'short' pulse length for each range

scale (see Chapter 2 on The Transmitter for more on pulse length). Some radars will require you to access a menu system to allow manual changing of pulse length as this is usually completed automatically, dependent on the range in use. There is no adjustment for pulse length in NT radars.

12. Select radar to **transmit**
13. Re adjust radar **brilliance** if required.
14. Look at the text in the radar screen to determine if the radar is in **focus**. Switch on range rings ensure focus of the radar display.
15. Using the range rings check the **centring** of the radar screen. Observe the outer most range ring and ensure that it is equidistant with the outer rim of the radar display. Also check the number of range rings. For 6 or 12 mile range scales there should be 6 rings. Centring is important, even in modern sets where the bearings taken by the EBL are usually read of in the EBL

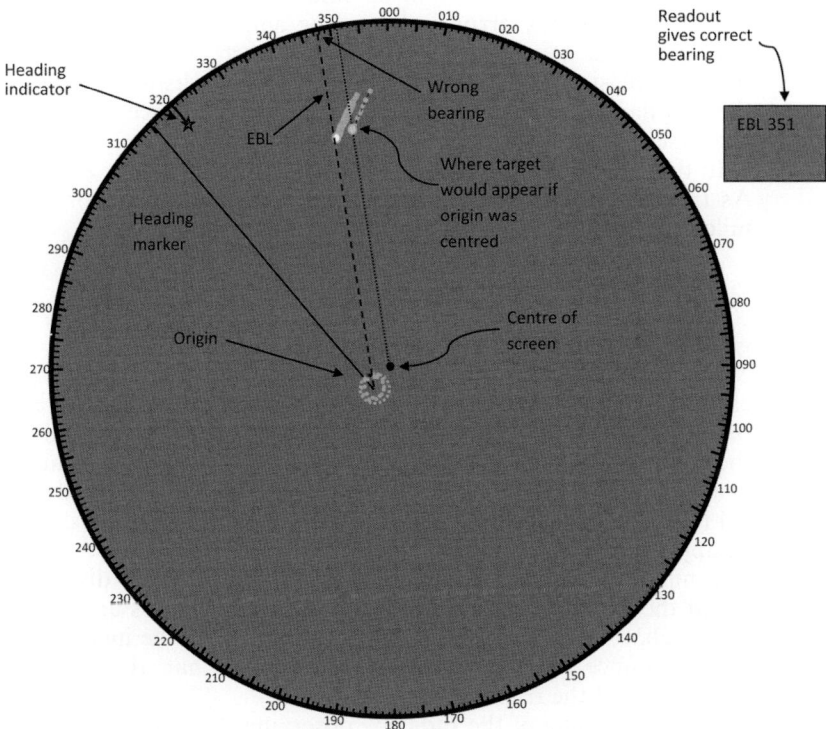

FIGURE 9.2 Origin not centred.

read-out, but may be read off the bearing scale around the display screen. In such a case, if the origin is slightly off centre then an erroneous bearing could be read off. In some makes of radar there is a square or rectangular radar display area. Centring can be checked by using the VRM to ensure the origin is centred. However, as the bearing scale on these radars is aligned with the origin, centring is not critical on such sets (see Figure 9.2).

16. With the range rings still on, activate the variable range marker (**VRM**) and check the digital read out with the inner and outer-most range ring. A further check should be carried out where possible, such as when the ship is secured alongside and there are suitable targets which are also shown on the chart. The range should be measured off the chart and compared with that on the radar using the VRM (this of course will probably mean changing range scale). Any error found should be noted and displayed near the radar display, and applied to all ranges. Restore the range scale to 6 or 12 miles. Another check for range measurement index error is when the ship is passing between two headlands to measure the range of each. Any discrepancy between the sum of the two ranges and the total distance between the headlands, measured off the chart will be twice the index error (see Figures 9.3a and 9.3b).

In the diagrams the distance between the headlands is Z miles. As the ship passes through the entrance the range to port is x miles and that to starboard is y miles. Therefore

$$(x + \text{index error}) + (y + \text{index error}) = Z$$
$$2 \times \text{index error} = Z - (x + y)$$
$$\text{Index error} = (Z - (x + y))/2.$$

Suppose the distance between the headlands, measured off the chart is 4 miles. The range to port from the radar is 1.8 miles and that to starboard is 2.6 miles. The total distance measured by the radar is then 4.4 miles. The difference between the charted distance between headlands and that measured by the radar is 0.4 miles. The index error is therefore – 0.2 miles and this should be applied to any range taken from this radar.

Similarly if three ranges taken from objects well dispersed about the ship are taken, and the arcs of these ranges are drawn on the chart and do not intersect at a point, then the index error is the distance from the centre of the 'cocked hat' thus formed to the sides of the triangle (see Figure 9.4).

17. Check the **heading** of the radar by observing the heading marker. This should coincide with the zero if 'Head Up' or the ship's

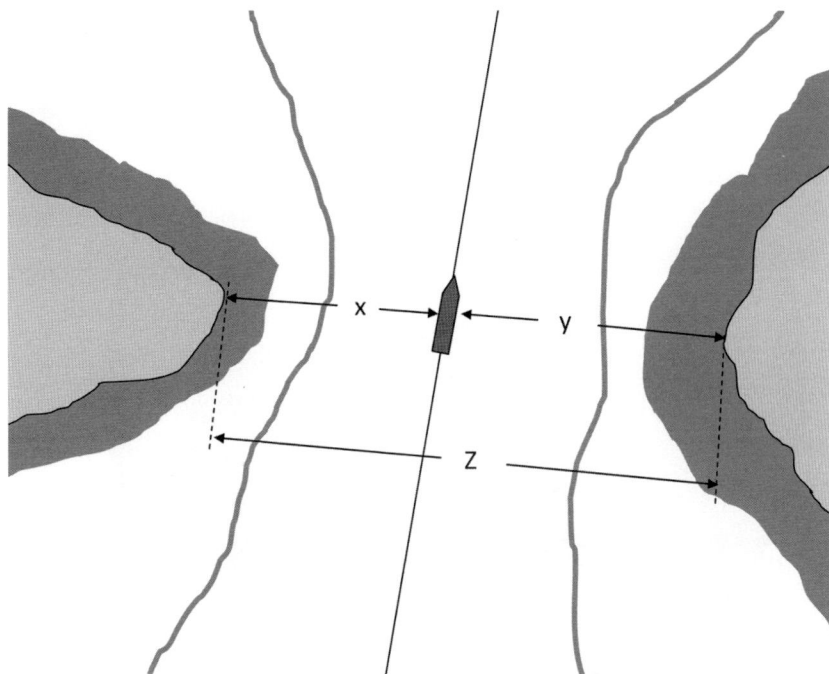

FIGURE 9.3a   Radar index error chart.

current gyro heading if 'North Up' on the outer scale on the radar screen. Note that some radars' bearing scales do not compensate for own ship being off-centre (see the note on centring above).

18. Using the Electronic bearing line (**EBL**) check the digital readout at 000°, 090°, 180° and 270° by revolving the EBL around the radar display.

19. Correctly adjust the **gain control**. Starting with the gain at zero or minimum (fully left position) it is slowly increased. This increases the sensitivity of the radar receiver. If this control is not set correctly then the radar may not be sensitive enough to show weaker targets. If set too high then weaker target may be obscured by the 'noise', the speckled background that increases as the 'gain' is increased.

Continue increasing the gain until the screen is showing a fine speckling throughout the radar display. With modern radars the

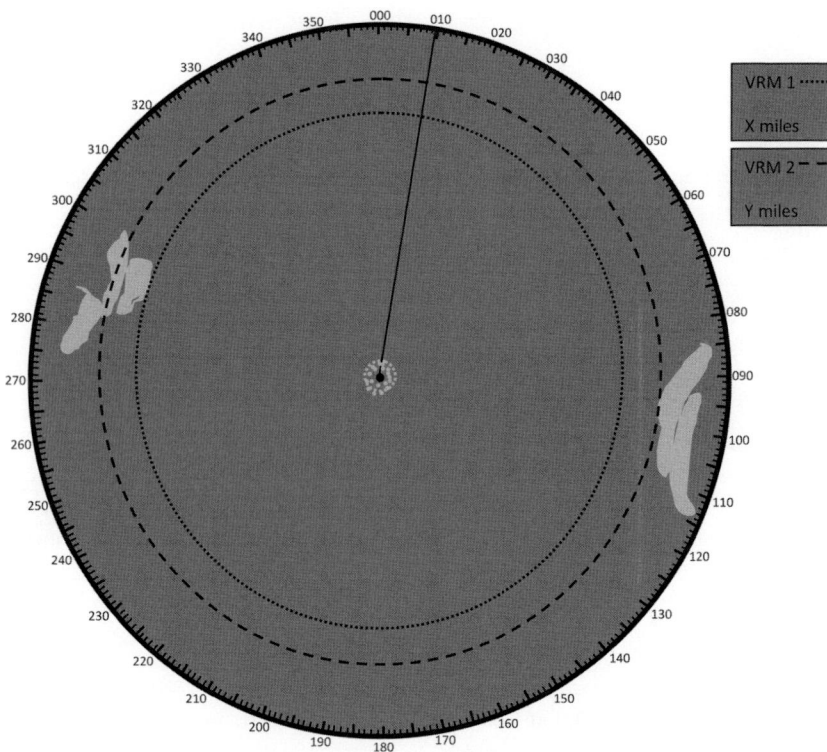

FIGURE 9.3b   Radar screen checking index error.

gain control can be reduced slightly from this point until only occasional speckling is seen. The gain control is now correctly set. However it may need to be adjusted after this setup procedure and at require intervals thereafter (see Figures 9.5a, 9.5b and 9.5c).

20. The **tuning** is the final setting to adjust in this procedure and probably the most important. This ensures that the radar is correctly tuned to the transmitted signal. If not set correctly then the radar may not show some targets at all, only stronger ones. In situations with land and other strong targets in the vicinity the set can be off tune unknown to the observer as such strong echoes are being displayed giving the appearance that the set is working correctly and it is only when the tuning is adjusted that the full range of smaller targets becomes visible.

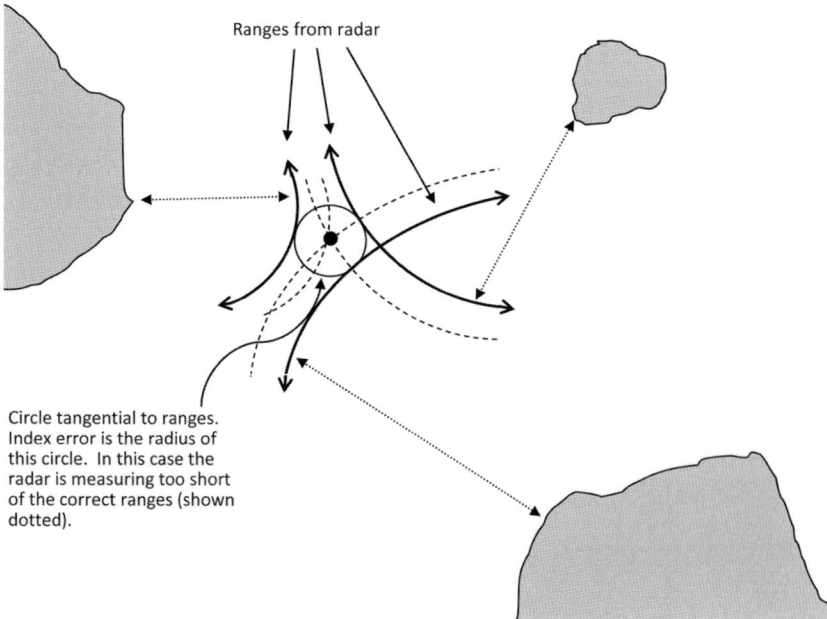

Ranges from radar

Circle tangential to ranges.
Index error is the radius of
this circle. In this case the
radar is measuring too short
of the correct ranges (shown
dotted).

FIGURE 9.4   Radar index error by three ranges.

Set the tuning control to manual. Start with the control fully
to the left and increase the tuning control slowly and watch for
maximum response from one of the potential targets listed
below. In using weak land targets, usually the strong land targets
close to the origin are visible even when the tuning is off-tune.
Select a weak target on the outer edge of the group of targets and
watch it as the tuning is increased. As this happens, other weak
targets, usually further out from the origin will appear, move to
one of these to see how it strengthens, but once again more weak
targets will appear as the tuning comes close to the 'In-tune'
position. Continue rotating the tuning control (or 'dragging' it if
a 'soft' slider control) to the right, bringing up new weak targets
until the last ones to appear just disappear. You have just gone
beyond the optimum position for tuning and the control should
be moved to the left to regain those last weak targets (see Figures
9.6a, 9.6b, 9.6c and 9.6d).

During this operation the tuning indicator should be observed.
Usually this is a bar indicator that shows the greater extent when
the set is in tune; compare its extent when the last weak targets

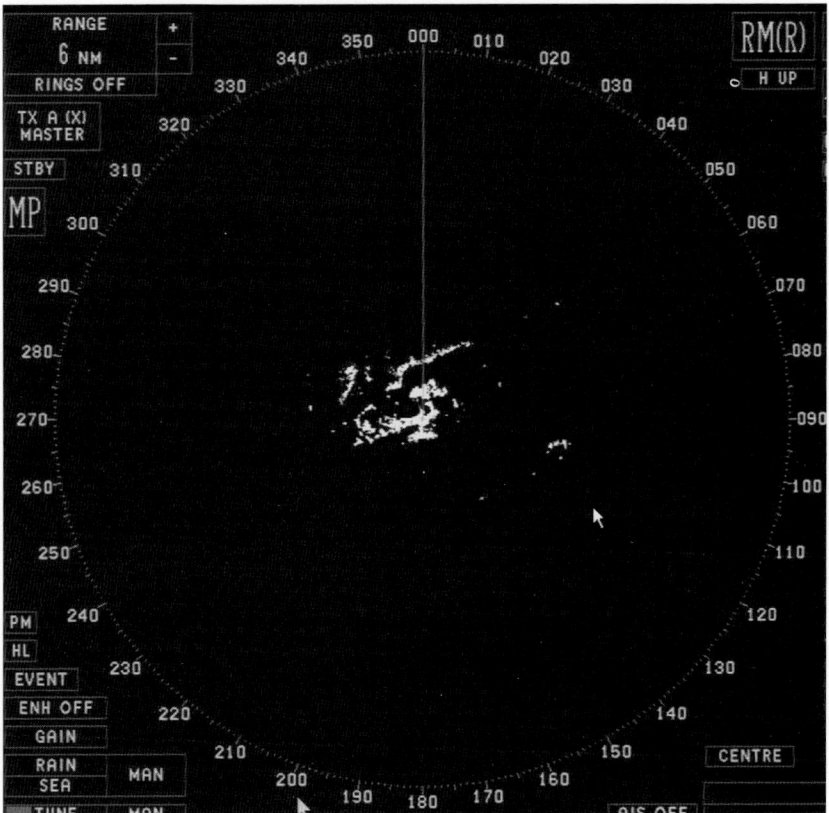

FIGURE 9.5a   Radar screen with gain right down.

are showing. It is not unusual to find a small discrepancy, that
is that the last weak targets are not showing when the tuning
indicator is at its greatest extent.

Most modern sets have automatic tuning. In spite of this, the
above procedure should be followed, and when the optimum
position by manual tuning is achieved, switch to 'Auto' and
observe any difference in the appearance or strength of weak
targets. There may be some slight discrepancy, but usually the
'Auto' is quite good in ensuring small or weak targets are
detected, but look at the comparison of the images in Figures
9.7a, 9.7b and 9.7c below.

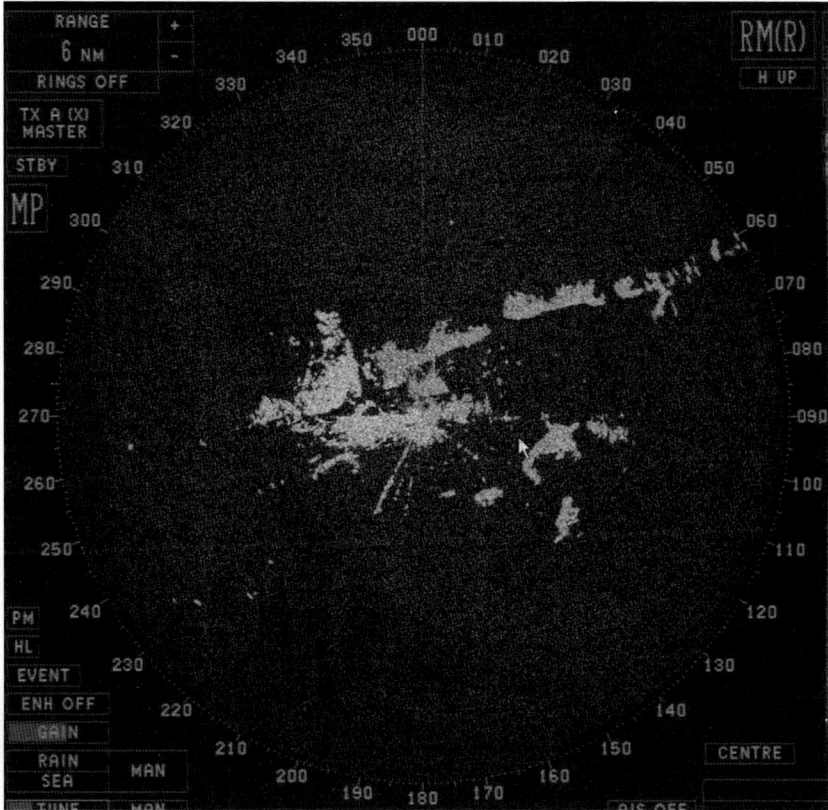

FIGURE 9.5b   Radar screen with too much gain.

In some sets the scope of the tuning control may need to be re-set. This occurs if it is found that the optimum tuning position is near one end of the scope or the other. In other words, it is found that the weak targets are apparent near the start of the tuning procedure and then disappear and don't reappear at all, or only appear just as the control is near the fully right position. The optimum position should be somewhere in the middle. This adjustment can sometimes be found in several menu steps, but should be adjusted only after the operator's manual has been carefully studied.

The following are methods of manually tuning a radar (in order of importance)

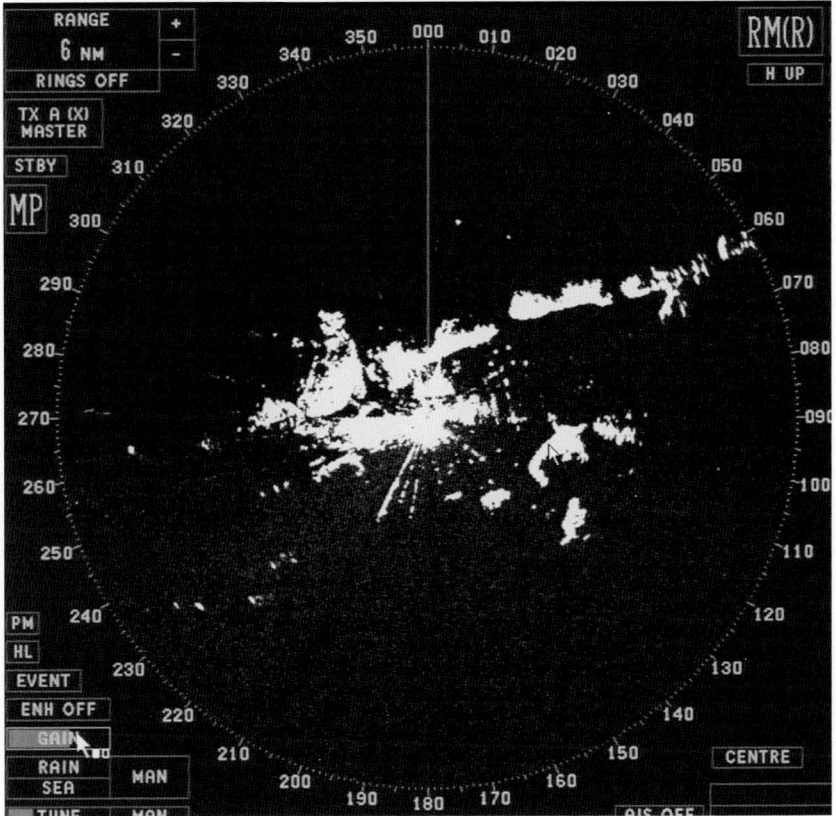

FIGURE 9.5c   Radar screen with correct gain setting.

      a. Weak land targets.
      b. Maximum sea clutter.
      c. Maximum rain clutter.
      d. Performance monitor.
      e. Constellation of buoys.
      f. Tuning indicator.
    There is no tuning control in NT radar sets.

    This sequence of setting up a radar is designed to produce a procedure so that the operation of the set will be checked in a systematic way and that things will not be overlooked. It will ensure that the radar is showing as much and as many targets as is possible, i.e.

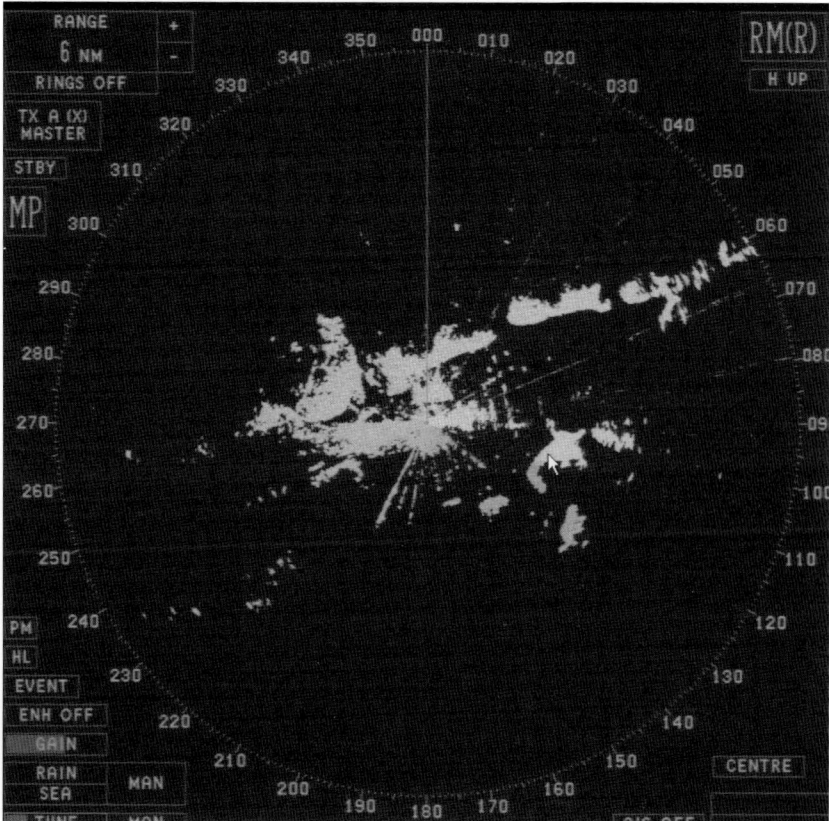

FIGURE 9.6a   Stage 1 of tuning adjustment.

maximum 'raw data'. It is only after this procedure has been followed that anti clutter controls and interference rejection circuits may be activated and used. If manual tuning is used after this procedure then it should be checked approx. every 15 minutes for the first hour then every hour thereafter.

Other Controls

Many of the functions discussed here are also dealt with in the chapter on The Display. They are included here to be checked as part of the set up procedure.

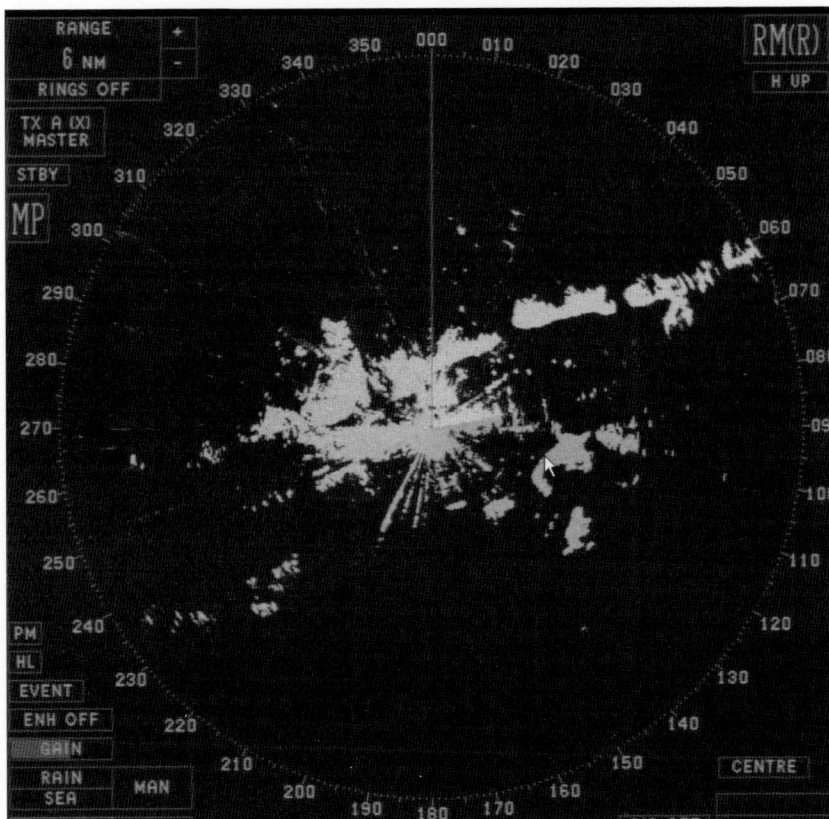

FIGURE 9.6b   Stage 2 of tuning adjustment.

Display Colours. There is a function to change the display lighting and colour range for night-time or daylight. Some makes may have intermediate selections such as for dusk.

Interference rejection. This eliminates interference from other radar sets. Such interference can sometimes be seen as random curved dotted spokes emanating from the origin.

Echo enhance. This function lengthens received echoes in range. It makes it possible to distinguish weak echoes. Its appearance is similar to increasing the pulse length, but without doing so and the position of the target is the leading edge of the echo.

Trails. Target trails, showing the immediate past positions of the

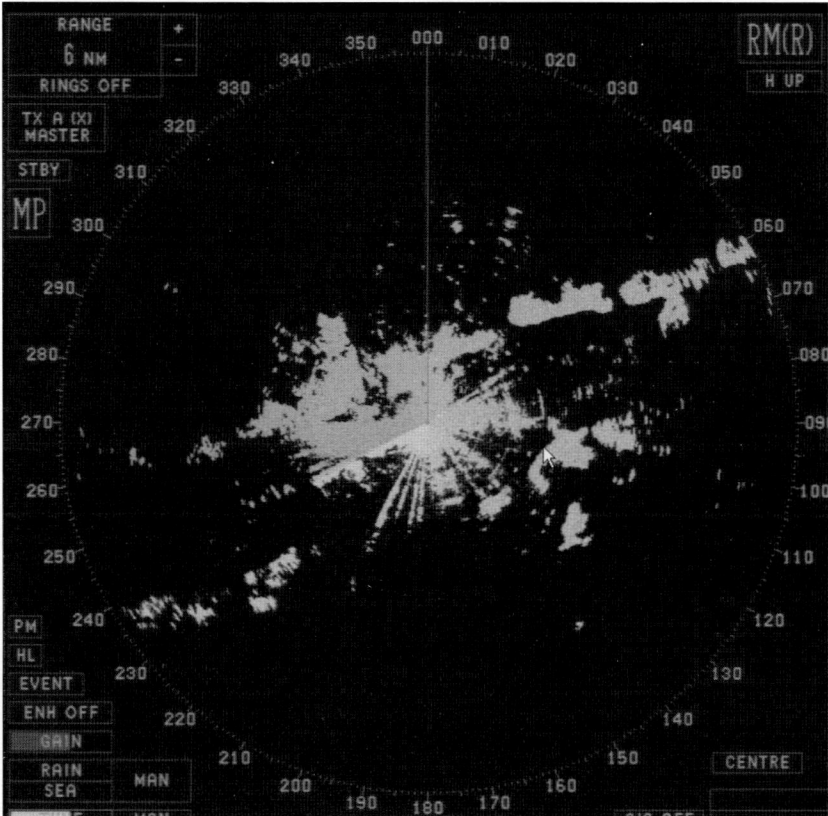

FIGURE 9.6c   Stage 3 of tuning adjustment.

target echoes can be set to 'true' or 'relative'. This in most modern sets is independent of the display mode; one can have relative trails when in true motion, or visa-versa. The length of the trails is usually adjustable, usually as a time such as 1 minute, 3 minutes, 6 minutes and 'continuous' where the trail does not fade out at all, or 'Off' where no trails are shown.

Heading Marker. The heading marker or heading 'flash' can be switched off. However this can only be done by holding the control or 'right button'. In other words it cannot be switched off permanently.

Sea/Ground stabilisation. See the chapters on plotting and collision avoidance. Ground stabilised mode derives the own ship's movement

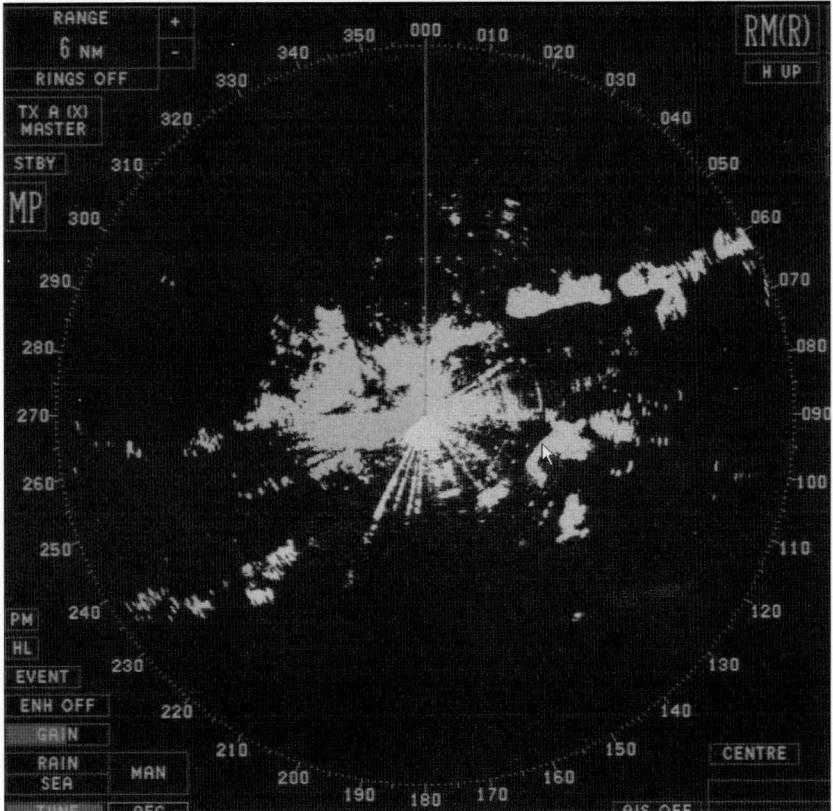

FIGURE 9.6d   Stage 4 of tuning adjustment.

from GPS (it can also be derived from a designated tracked radar target). Sea stabilised derives the ship's movement from the ship's water track log and gyro compass.

Cursor. This is often a 'mouse' pointer arrow when not on the radar screen, that is, in the areas around the screen to access different functions and menus, and changes to a cross when on the radar screen for various functions such as mark targets to be 'acquired' or drag PiLs (Parallel Index Lines), or off centre the display. The position of the cursor is displayed in a readout in the area around the screen, giving either range and bearing from the origin or Latitude and Longitude (presuming GPS is input), or both.

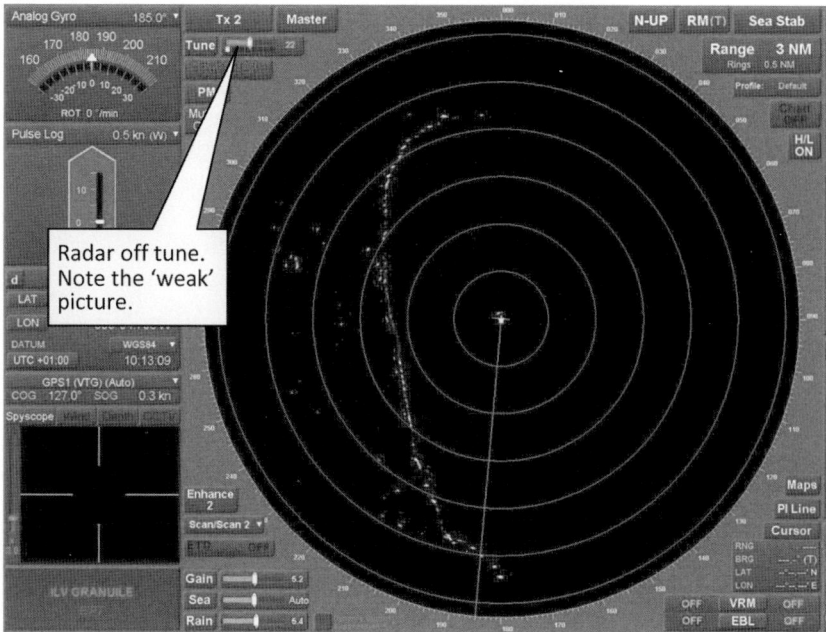

Radar off tune. Note the 'weak' picture.

FIGURE 9.7a Comparison of manual and automatic tuning.

EBL. There are usually two, distinguished from each other by different dotted lines, although some makes of radar may have more. These can usually be rotated by 'hard' controls and sometimes also by 'dragging' with the cursor. They can often be offset from the origin to a designated position by the cursor. This location of the offset EBL may remain relative to the origin, for example if set to a relative bearing of 045° at 1 mile, then if own ship alters course it should still be at a relative bearing of 045° at 1 mile. It can also be 'dropped' in which case it should stay in the geographical position in which it is dropped. However, makes of radar differ in this and care must be taken until one is familiar with the function on a particular set. The bearing is read out at the side of the screen in the EBL 'box' and may give the bearing as 'true' or 'relative' or both. Sometimes there is a little range measuring circle on the EBL which appears if the VRM circle is not activated, but the range of this is controlled by the VRM range adjustment.

VRM. There may also be two (or more), distinguished by different

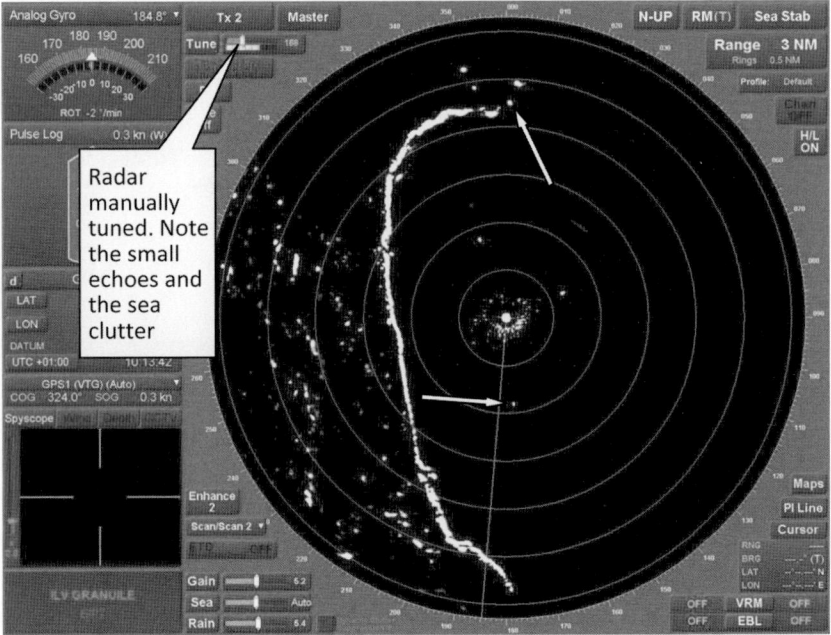

FIGURE 9.7b   Comparison of manual and automatic tuning.

dotted circles. These also are usually manipulated by a 'hard' control but often also by 'dragging'. These often can also be offset from the origin to a position designated by the cursor. The same proviso in relation to the origin as that for the EBL applies. The range is read off in the VRM 'box' at the side of the screen. Some makes of radar allow range to be given in units other than nautical miles. This is usually set by a function in a 'menu'. Great care needs to be taken if this is changed from nautical miles as a serious error could ensue if some operator was unaware of this.

Acquire. This designates a target to be acquired by the plotting computer. It functions slightly differently in different makes of radar, such as activating the function and then every target 'clicked' using the cursor will be acquired, or pressing the 'acquire' button each time the cursor is placed over a target.

Vectors. These normally appear once a target has been acquired. The adjustments to be made are:

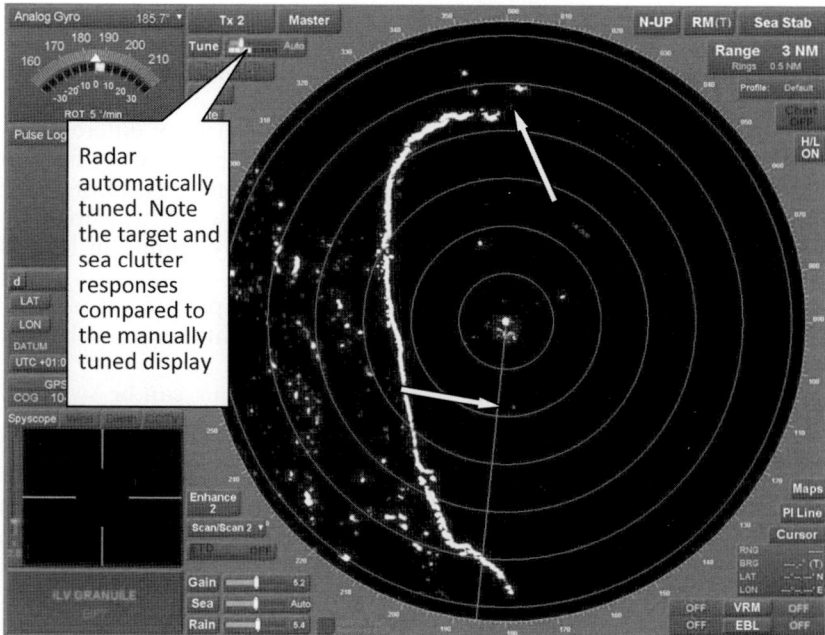

FIGURE 9.7c   Comparison of manual and automatic tuning.

- Vector length. This is set in time. It may be variable up to, often, one hour, or in 3 minute steps. The vector length/time setting is applied to all vectors.
- Vector mode. This is either 'true' or 'relative'. In some makes of radar, the default vector mode is that of the display mode; if the display is in relative motion, the vectors will default to 'relative'. 'True' vectors may be selected while the display mode is relative motion but after a few minutes will automatically revert to 'relative'. The opposite holds true with such radars when the display mode is true motion.

Past Positions/Track history. This function shows, by a series of dots, the past motion of the acquired target. Such history dots should be 'true' although it has been found that some makes of radar will show them as 'relative' if the display mode is relative motion. The number of dots is usually six, and the sometimes the frequency, or spacing of the dots can be set by the operator (e.g. 3 minutes, 6 minutes, 12 minutes).

In some makes, the dots will only start to generate once the function has been initiated; in others the full range of dots will appear immediately.

Parallel Index or Navigation lines. A least two and maybe more can be switched on. These can be rotated in azimuth and moved to any distance from the origin. They are often distinguished by different dot patterns. In positioning these care should be taken that they are fixed in azimuth which is normally what is required, as some makes of radar have a default setting of 'relative' which means that when the ship alters course, so do the navigation lines. They are usually infinite in length; the operator cannot adjust their length.

Maps. The drawing of maps is discussed on Chapter 16, 'Radar for Navigation'. These can be accessed in a storage facility and switched on or off as required.

Chart. In a 'chart radar' the vector electronic chart can be switched on or off by a single operator function. The adjustment of the chart display for such things as degree of detail, safety contour etc. is usually in a menu function.

AIS. This can be 'On' or 'Off'. If on, all stations transmitting AIS signals, within range will appear on the screen. Ships will appear as small green triangles and aids to navigation (AtoNs) as faint diamonds with a cross. Once AIS is 'on' a ship target can be activated (or acquired) in which case a vector, and usually a name will appear for that target. The AIS vector will have the attributes assigned by the operator for vectors ('true'/'relative', vector length). AIS and radar acquired targets can be 'associated' or 'fused' or some such term to describe combining the vectors and data.

Conclusions

This chapter has detailed the sequential steps to be taken in setting up a modern radar display. Such a series of steps are important to ensure that all wanted targets are detected by the radar. The other steps are to ensure that the navigational data being presented by the radar is correct such as bearings and ranges of objects. Other features discussed are to ensure the correct anti-collision data is presented such as vectors and trails.

These are the main functions which are immediately available in a modern radar sets. There are several other functions and adjustment which are seldom required and are accessed through a few menu steps (such as changing range measurement from miles to metres or yards). Most of these should only be changed by someone who really knows what he/she is doing and fully understands the implication of such adjustments.

Chapter 10

Radar Plotting

In this chapter the following topics are discussed:

- the need for manual plotting,
- the techniques used in relative motion plotting to determine the threat of a target,
- the determination of manoeuvres by targets,
- the effect on the plot of own ship's manoeuvres,
- allowing for time delays in such manoeuvres, and
- collision points and 'Potential Areas of Danger' (PADs).

In the modern context, it is often asked why is there a need for ships' watchkeeping officers to develop the knowledge and skills to carry out manual anti-collision plotting? The computerised plotting systems can carry out the function more effectively and efficiently than any human plotter. The reasons are as follows:

1. In the event of the breakdown or malfunction of the automatic plotting system, a ship's watchkeeping officer must be able to handle the situation professionally and effectively.
2. Manual plotting develops the appreciation of relative motion; it helps the ship's officer to recognise what is happening on the radar display with regard to collision avoidance and navigation generally. It helps him/her to recognise the veracity or otherwise of the data that an automatic plotting aid may be presenting.

In the practical sense, plotting is prone to error and developing the necessary skill and confidence requires practice. This should only be done under tutorial conditions, either in a school or college with the necessary simulation facilities, or aboard ship as a separate function from the watchkeeping role, such as a cadet or class of cadets under the instruction of an officer who is not the OOW. In very light traffic situations, in clear weather an OOW might consider maintaining his/her skills by doing a manual plot, and comparing it with the automatic plotting aid, but this should not distract from the keeping a proper lookout and all the requirements of good watchkeeping.

In class-rooms, plotting exercises are initially done in a theoretical sense, usually where no time limit is imposed on the student. As stated above, although such exercises may seem far removed from the real and fast moving world of a modern ship's bridge, they are necessary to understand the basic principles of the plotting processes. But an essential part of such training is then to carry out plotting exercises in real time, on a simulator. This gives the students the experience of dealing with fast moving situations, realising the limitations of manual plotting and developing their skill in doing it quickly and accurately.

The relative motion plot

This is the most basic and fundamental part of the plotting process. One's own ship (OS) is fixed at the centre of the screen, or plotting diagram. A line, from the centre to the graduated edge of the plotting diagram represents the 'Ship's Head', the direction of motion of the ship. A stationary target plotted on such a diagram will appear to move in the opposite direction, and at a rate equal to own ship's speed. Moving targets will have an apparent motion which is a combination of their own motion and that of own ship.

Let us look first of all at a stationary target, say a ship lying stopped in the water, or a free floating buoy. If we plot a series of plots on the plotting diagram, over a period of, say, six minutes, we will get an apparent motion of the target opposite to own ship's course and a speed equal to own ship. In a plot the convention is that the first plot is designated 'O' and the last plot 'A'. However, to resolve the relative motion plot, when the target is not stationary, so that we can get the true movement of the target, we have to resolve a vector triangle. In the case of the stationary target mentioned above, the apparent movement is in fact the motion of own ship. The track followed by the stationary target is called the '**zero speed line**' and is designated 'OW' (an aid to memory is that this is 'Own (ship's) Way'). In any plot we can consider that at the start of the plot the target puts an imaginary free floating buoy over the side. 'OW' then is the track traced out on our relative motion plot by this imaginary buoy (see Fig. 10.1).

Time of course is an important factor in resolving the relative motion plot. The accuracy of time measurement therefore in actually carrying out a plot is important. The time over which the plot is carried out, that is the time between the first plot and the last is the '**Plotting Interval**'. The more plots which are done the better as this reduces the error, but in practical terms, the plotting interval cannot be too long; decisions need to be made before the target gets too close. For convenience plots can be carried out at three minute intervals, plotting three times to produce a plotting interval of six minutes. Six minutes being a decimal of an hour, the vectors will be a decimal of speed. If own ship is making

Plot of a Stationery Target

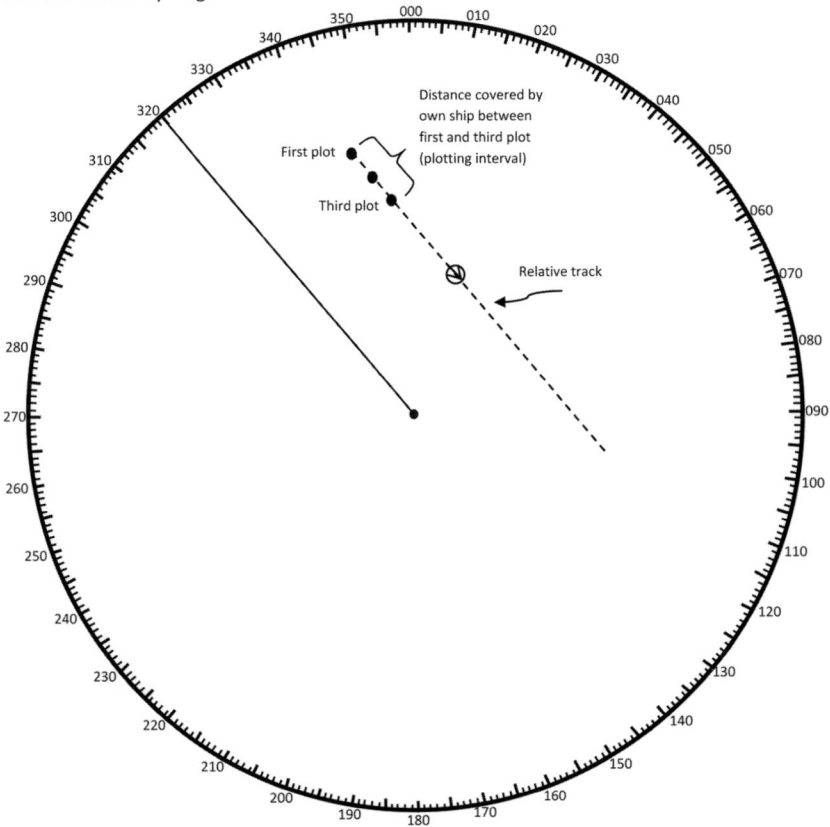

FIGURE 10.1   Track of stationary target.

15 knots, and if the plotting interval is six minutes, then OW will be 1.5 miles.

In considering the plot where the target is not stationary a series of plots traces out the relative motion of the target. This is a combination of own ship's motion, and that of the target. The first plot is 'O' and the last is 'A'. These two plots are joined by a line which is extended past the centre of the diagram. This is the relative track and is further marked by an encircled arrow pointing in the direction of motion. From 'O' a line is marked off parallel to the heading line, but in the opposite direction. This is the '**Zero Speed Line**' which is designated 'OW' as

Plot of a Moving Target

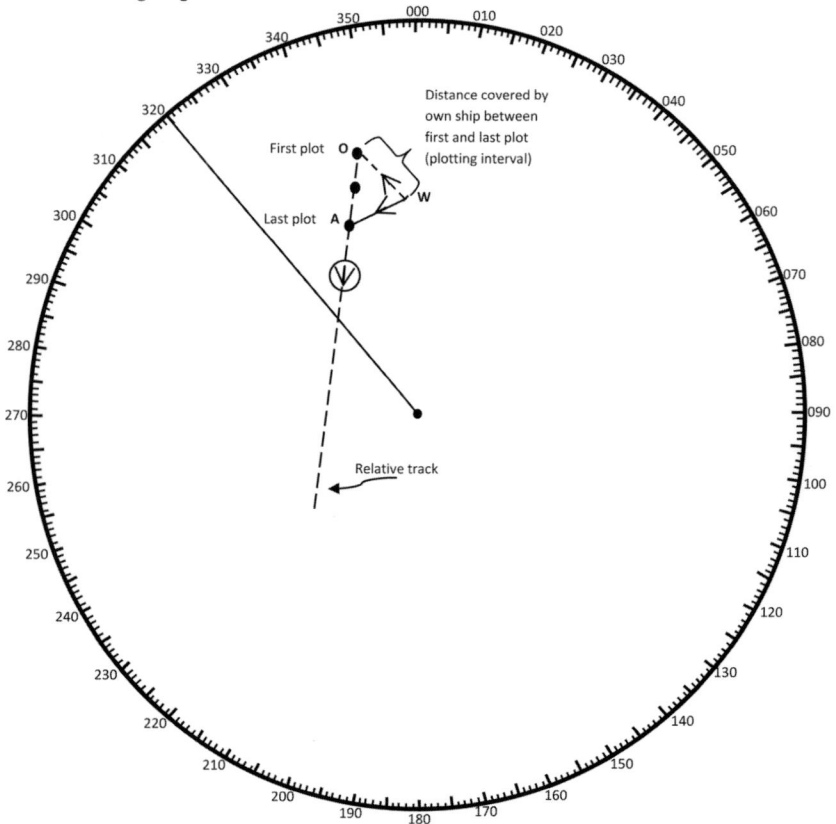

FIGURE 10.2   Track of a moving target.

explained above. This is the track of the imaginary buoy. Its length is the distance covered by own ship during the plotting interval. If the plotting interval is 6 minutes, and own ship is making 12 knots, it will be 1.2 miles in length, if own ship is making 15 knots, then 'OW' will be 1.5 miles. 'OW' is marked by an arrow pointing in the direction in which own ship is heading, i.e. from 'W' to 'O'.

Meanwhile the target is steaming away from the point at which she dropped the imaginary buoy, which at the end of the plotting interval is 'W'. She is now at 'A', so 'WA' represents the true motion of the target; 'W' to 'A' shows the direction she is moving in, and the length of 'WA' is the distance she covered in the plotting interval, so if the length is 1.6 miles, and the plotting interval is 6 minutes, she is making 16 knots (As an aid to memory 'WA' is 'Way of Another').

Obviously, if the plotting interval is other than 6 minutes, then a small calculation will have to be done for 'OW' and 'WA'; if for example the plotting interval is 10 minutes, this is one sixth of an hour, and if own ship is making 12 knots, then 'OW' will be a sixth of the speed (the distance covered in one hour), that is 2 miles. If the resultant 'WA' is found to be 1.4 miles, and the plotting interval is 10 minutes, then the speed of the target is 1.4 multiplied by six, which is 8.4 knots. Two things of importance can now be immediately deduced; one is **Closest Point of Approach** (CPA) and the other is **Time of Closest Point**

Plot of a Moving Target

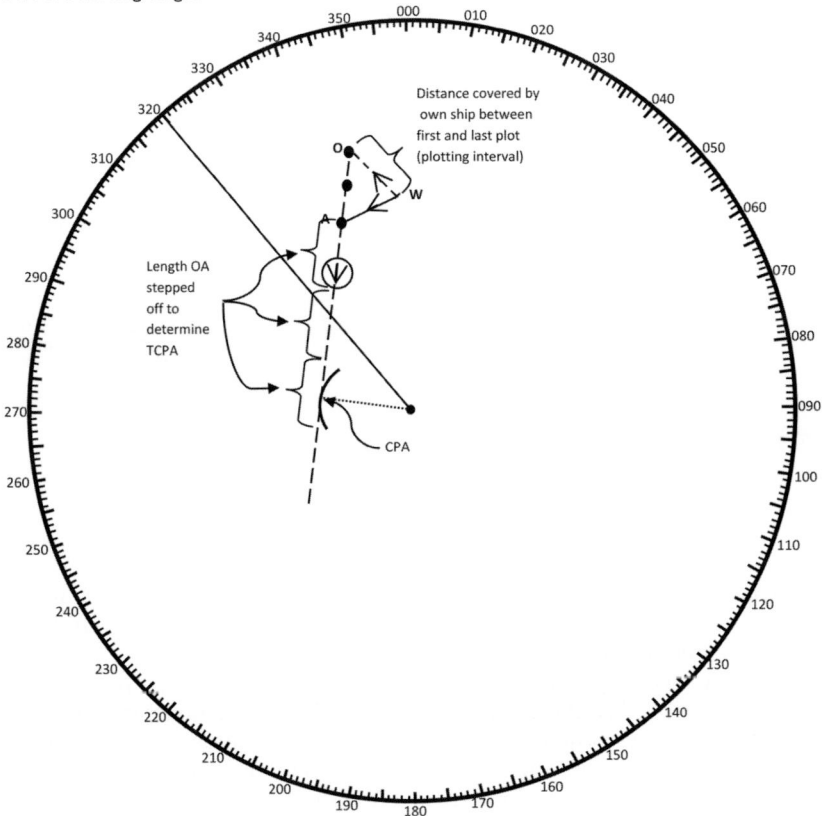

FIGURE 10.3 Relative Plot showing OA, CPA and TCPA.

Aspect

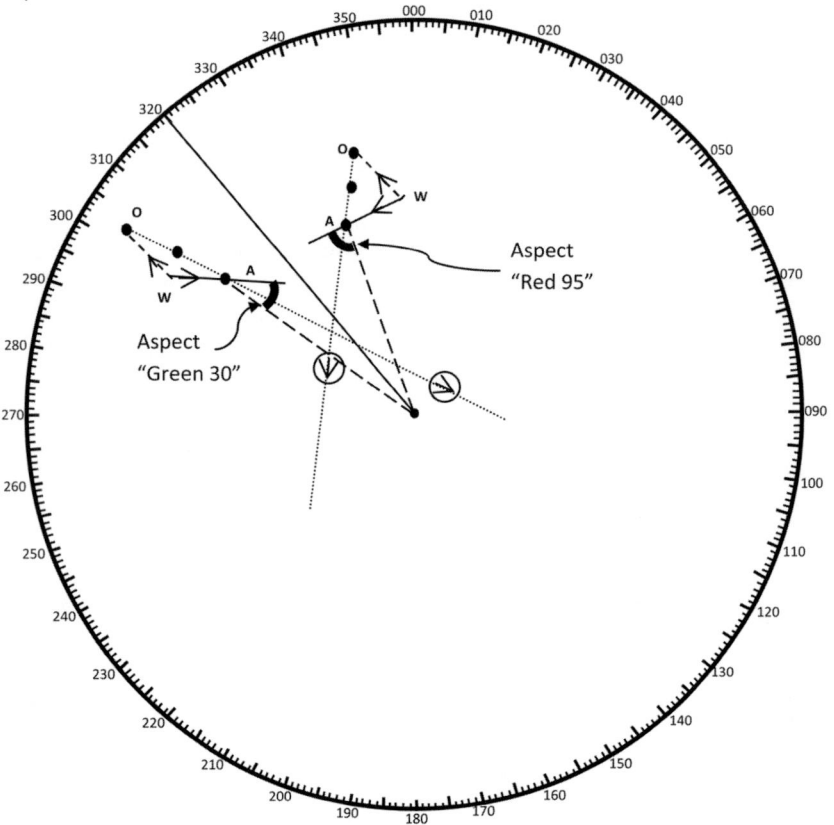

FIGURE 10.4   Aspect.

of Approach (TCPA) see Figure 10.3. If the plotting interval was 6 minutes in Figure 10.3, then the TCPA is approximately 15 minutes after the time of the last plot.

We now have a good deal of information to assess the situation and risk of collision. We have the CPA, the TCPA, the course and speed of the target. There is one last piece of information which we can get from the plot which will help in the assessment. This is '**Aspect**'. This is the relative bearing of Own Ship from the target; where does he 'see' us. Oddly enough, this is one piece of information most automatic plotting aids do not provide digitally. It can be determined by inspection of the true vectors, but it is rarely (never?) presented in the readout. 'Aspect' is the angle between the targets true heading, the 'WA' line, and the last bearing from own ship (see Figure No. 10.4). This has been traditionally designated 'Red' or 'Green' from '0' to '180'. In naval context the term **'angle on the bow'** is used for the same thing and using 'Port Bow' or 'Starboard Bow' instead of 'Red' and 'Green'.

Report

Once the plot has been completed and the various data established the report can be made.

The information is delivered in a standard reporting format. This is as follows:

1. Designation of target (e.g. Target No. X, or 'Target Alpha').
2. Last range and bearing.
3. Trend in bearing and range (e.g. bearing drawing left or right or steady; range closing, opening or steady).
4. CPA.
5. TCPA (either in actual time, or in number of minutes since last plot).
6. Target course.
7. Target speed.
8. Aspect.

This should be in a spoken format, a report from the radar plotter to the con. There may be slight variations in this format, depending on the particular preferences in ships, companies or services, but the basic information should be presented in this format and in the order of that format.

Of course all this data is historical, the information only tells us what the situation was up to the time of the last plot and what we can deduce from this. The decisions to be made must be on the basis that if the target alters course or speed, then the information we have is redundant. This is an important point to remember, even with automatic plotting aids; none of them can predict the future. For this reason it is essential to continue plotting until the target is past and clear. This continuation should not only establish that the target is continuing on its original track, but that its relative speed also remains the same. Any change in direction, or speed of the relative motion means that the target has altered course and/or speed. Only a new plot will establish precisely what change has occurred.

Figure 10.5 shows the relative motion plots of two targets. In the case of the target on the starboard bow, the target has obviously changed something of either its course, speed or both; there is an obvious change in direction in the relative track. If this target is being observed on the radar screen in relative motion, with trails on, a bend in the trail will be seen, and the observer alerted to the fact that the target vessel has made some alteration of course and/or speed (and in this case not what might be assumed from simply observing the trail). The target on the port bow, however shows no change of direction of its relative track. On the radar screen in relative motion, there will be no apparent change of direction in the target's trail. It is only by plotting that it can be determined what action the target has taken.

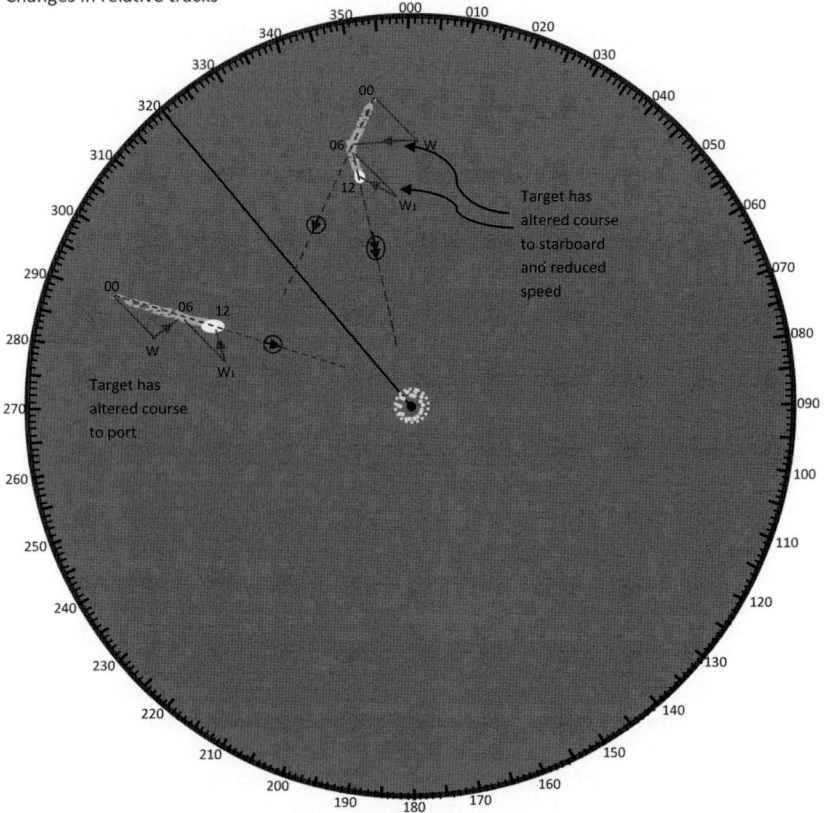

FIGURE 10.5   Showing changes of relative track indicate manoeuvres by targets.

The basic OWA plot is usually fairly obvious when the target is approaching at a broad angle on either bow. What tends to pose more of a problem to plotters are the situations where targets are approaching from astern or on the quarters or where there is a small rate of approach (see Figure 10.6). In Figure 10.7 there are four targets on either bow. One of these plots on the same position for each of three plots, two others show relative tracks approaching own ship and one shows a relative track moving away from own ship. One of these is a stationary target, one is being overtaken, one is on the same course and speed as own ship and one is faster than own ship.

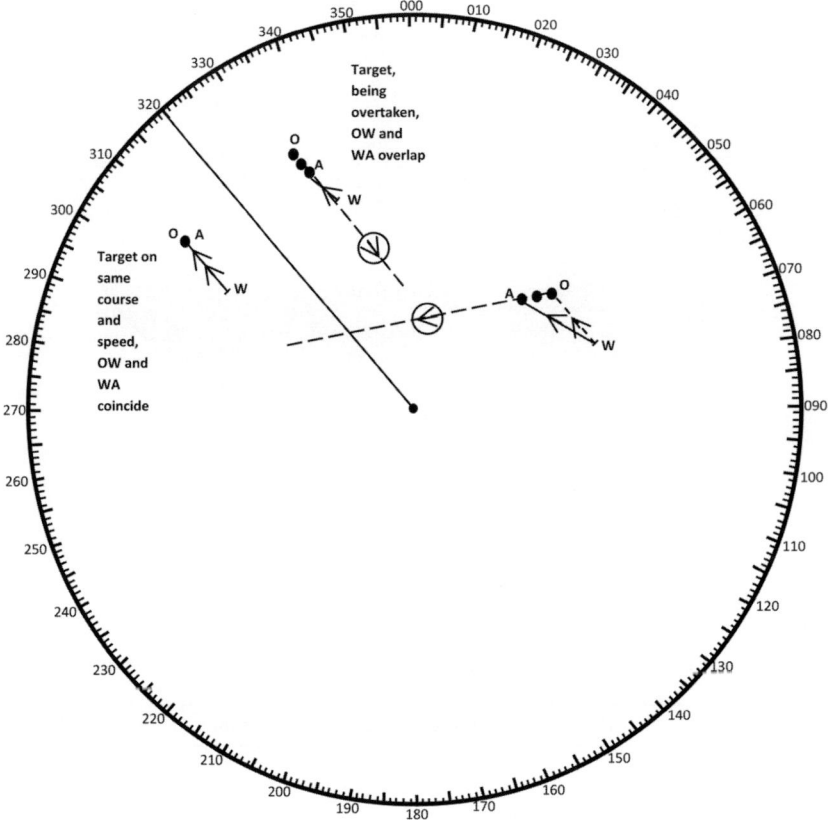

FIGURE 10.6   Various plots 1.

Plot of a stationary and other targets

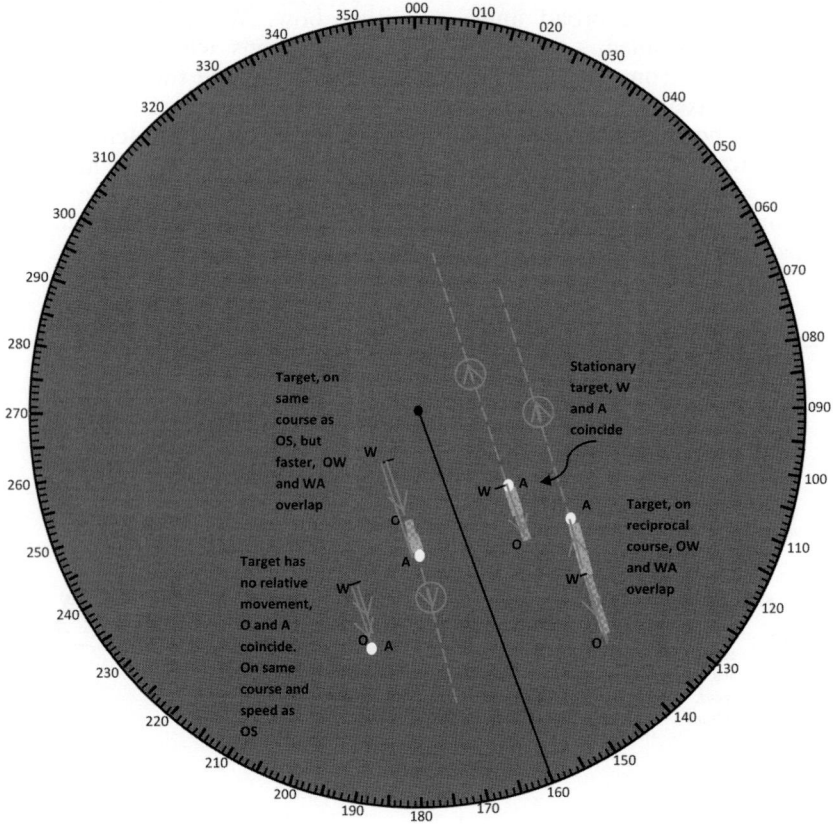

FIGURE 10.7   Various plots 2.

Let us now look at an actual plot. Own ship is steering 200° at a speed of 14 knots. The following observations are made:

Time	Target A		Target B	
	Bearing	Range (Miles)	Bearing	Range (Miles)
2210	251	9	156.5	10.4
2215	254	7.8	158	9.3
2220	259.5	6.2	161	8.0

Plotting Example

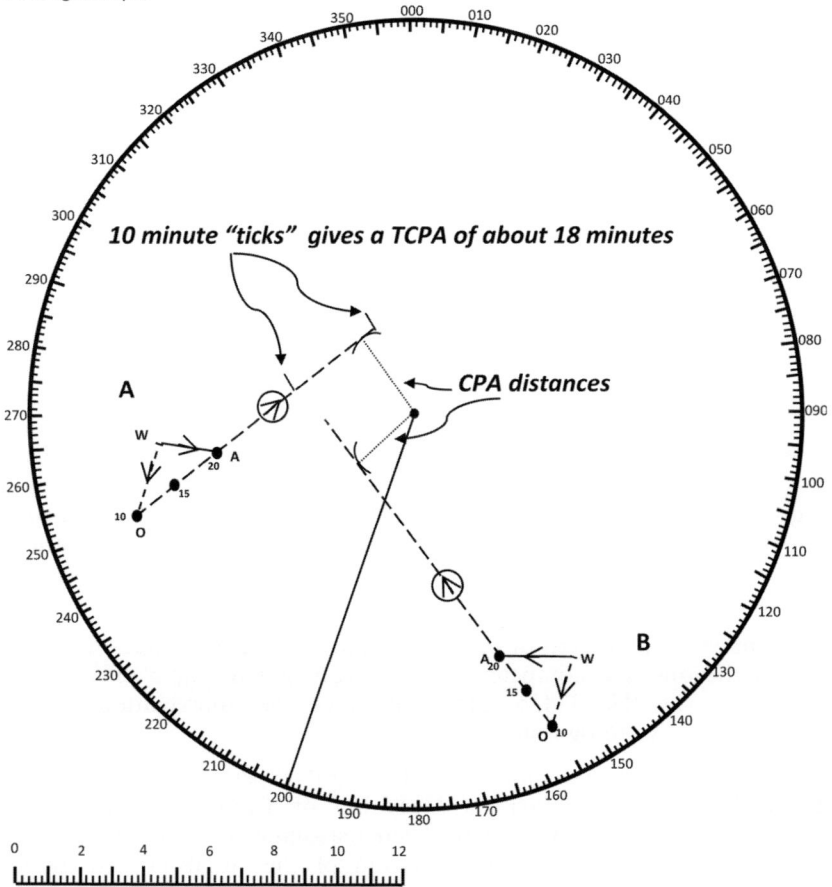

FIGURE 10.8 The example plotted.

Figure 10.8 shows the plot for these targets. The reports are as follows:

	Target A	Target B
Bearing	259° drawing right	161° drawing right
Range	6.2 miles, closing	8.0 miles, closing
CPA	2.9 miles	2.2 miles
TCPA	18 minutes	25 minutes
Target Course	098°	271°
Target Speed	12 knots	13.2 knots
Aspect	Red 30°	Green 45°

Head-up Plots

Before going any further it is worth mentioning the 'Ship's Head Up' plot. In practical terms in the modern context this may be rarely a requirement, and most of the exercises in this book have assumed a 'North Up' display, where there is gyro or compass stabilisation. However, several 'Head Up' exercises have been included for the following reasons:

In the event of a gyro or compass input failure, the situation should not be beyond the competence of a professional ship's officer, and Practising the 'Head Up' plot reinforces the appreciation of the situation relative to own ship.

It is also worth mentioning that several experts in modern navigational techniques advocate a 'Head Up' or 'Course Up' display for both radar and electronic charts. Their reasoning is that such a display orientation gives a better appreciation of the situation (situational awareness) relative to own ship, and with electronic chart displays one is no longer restricted to the 'North Up' orientation of traditional paper charts. Thus, on the displays, both chart and radar, what is ahead appears centrally in the upper part of the screen, what is on the left hand part of the screen is out to port, and what is in the bottom of the screen is astern. This debate, 'North Up' versus 'Head/Course Up' has gone on for many years and whichever is adopted is presently a matter of preference. Of course ship's watchkeeping officers should be fully conversant with all modes of display.

If gyro or compass stabilisation is lost, of course 'Course Up' doesn't work either. Therefore one is thrown on back on basic 'Head Up'. One of the major problems in practical plotting in 'Head Up', in real time situations, whether at sea or in a simulator, is 'Yaw' (see Figure 10.9). As own ship yaws from side to side in a seaway, either real or simulated, the target echoes on the screen shuffle back and forth correspondingly. For this reason taking the bearings for the plot has to be done with great care; either the ship's head has to be noted at the precise time of taking the bearing, or the bearing taken at the moment the ship is precisely on course which makes plotting at regular intervals impossible. On the face of it such detail may seem unnecessarily fussy, given all the other errors that are inherent in manual plotting, but these can bring about major errors in bearing and in timing and if not done accurately enough the ensuing plot can be seriously in error.

Effect of yaw in Head Up mode

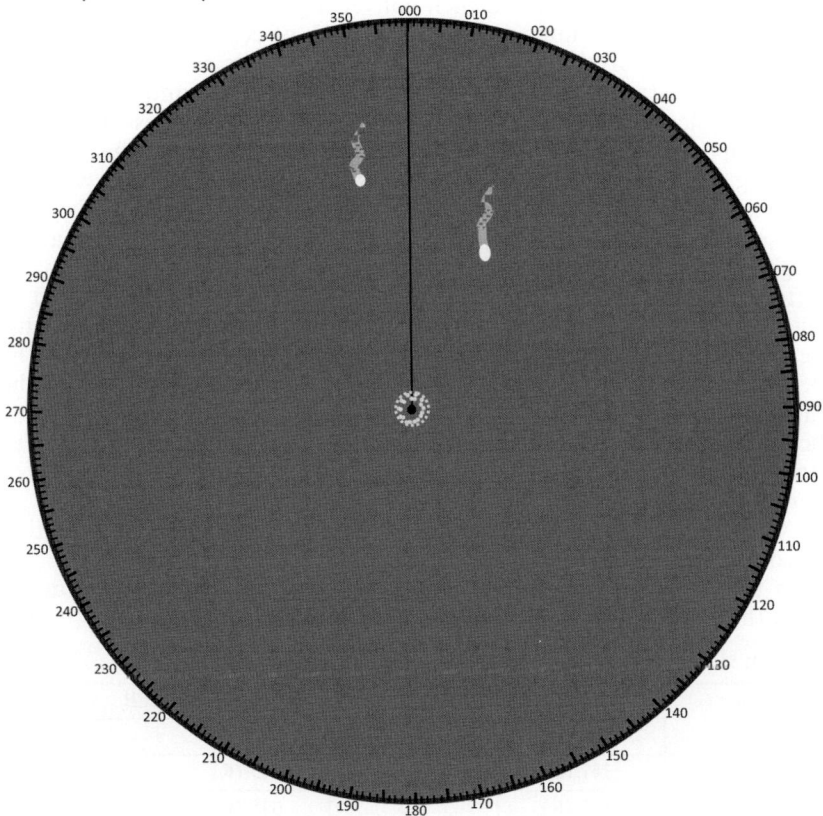

FIGURE 10.9   Showing effect of 'Yaw'.

In presenting the 'Head Up' plotting exercises, we have dispensed with the old format of 'Red' and 'Green', whereby bearings on the port side of the ship were designated from 'Red 0' (ahead) to 'Red 180' going anti-clockwise, and those on the starboard side were designated from 'Green 0' to 'Green 180', going clockwise. Instead we use the more logical 360° notation going from 000°(R) (ahead) to 359°(R) (one degree on the port bow), going clockwise. A target, for example, on the port beam will bear 270°(R).

In working such a 'head Up' plot the heading marker will of course be at 000° on the plotting diagram, no matter what the heading. The heading should be written at the top of the plot. Just above the 000°.

Sometimes people prefer to convert a 'Head Up' plot to a 'North Up' one, by inserting the heading marker on the heading on the bearing scale in the normal way, and then converting all the target bearings to 'True'. This really is unnecessary work (and, of course prone to error), as initially the only bit of information which needs to be converted to 'True' is the targets' course when the OWA triangle is completed.

A complication in doing 'Head Up' plots however does arise when own ship alters course. The plot remains orientated to the original heading, but a continuation of the plot giving relative bearings will give those bearings relative to the new heading. This often catches people out, as the target seems to jump improbably across the plot. Take a target that was bearing 350°(R) before a 30° alteration to starboard. After the alteration it will bear about 320°(R) and will be reported as such. However in plotting it, the alteration of course to starboard must be added to the relative bearing (or to port subtracted) to convert the bearing to the original orientation of the plot. In other words it will be plotted at 320° + 30° on the plotting diagram. Such a complication may be less of a problem for those who have converted their bearings to 'true', but remembering to convert the bearings after the alteration of course using the new heading.

An example of continuing the plot in the head-up mode after alteration of course.

In the diagram in figure 10.10, target A on the port bow and target B on the starboard bow are plotted as follows:

Own ship alters course 90° to starboard at 1021

Time	Target A		Target B	
1015	340°(R)	9.4	045°(R)	10.2
1018	341°(R)	8.7	045°(R)	9.4
1021	342°(R)	8.0	045°(R)	8.7
1024	248°(R)	8.0	309°(R)	8.0

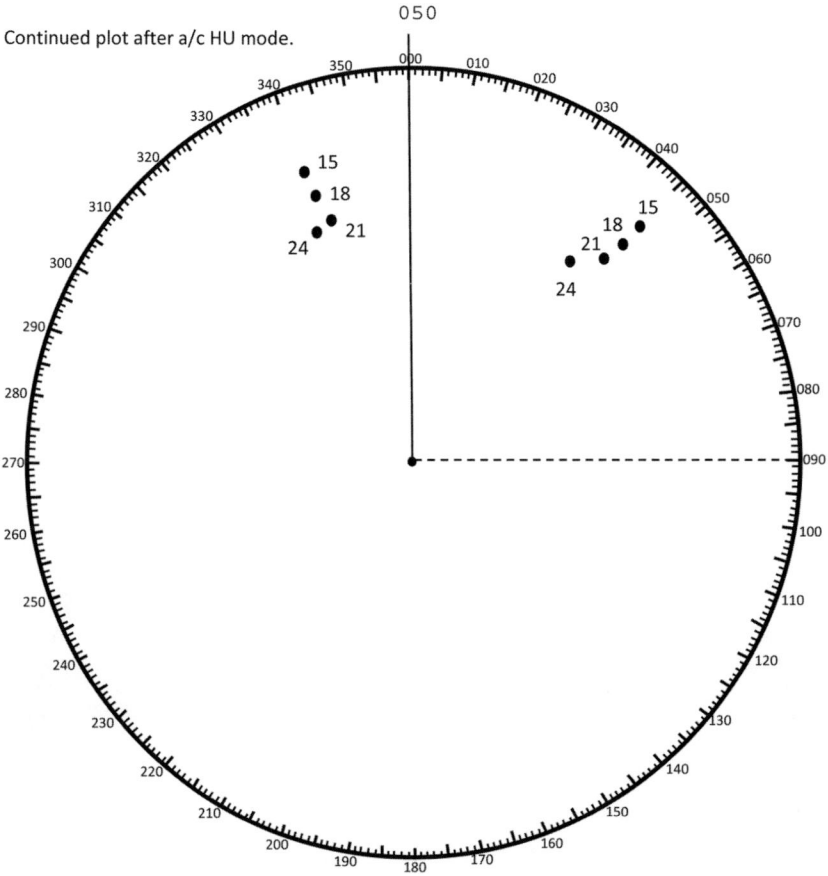

FIGURE 10.10   Alteration of course in Head Up mode.

Own ship Alters Course or Speed

Having established the basic OWA plot, this can be used to see the effect of own ship's manoeuvres on the situation, in other words, a trial manoeuvre. In the first instance, let us look at an alteration of heading.

Using W as centre, O is rotated to the proposed new heading to a position O_1. The O_1A is the new relative motion of the target. All this, of course is on the assumption that the target continues on its original course and speed. If own ship could instantaneously alter course, without loss of speed at the time of the last plot, then the target would move down O_1A extended. This, of course is unrealistic; in the first instance, time has passed between the last plot and the time of altering course, and secondly, the actual alteration of course in itself takes time and is usually associated with a drop in speed which takes time for own ship to regain her original speed. For the time being we will ignore the second of these considerations and assume that when own ship alters course it has immediate effect. Therefore, if the last plot was at say 1012, and the alteration of course was at 1015, then the target should have continued along the original OA line extended for another

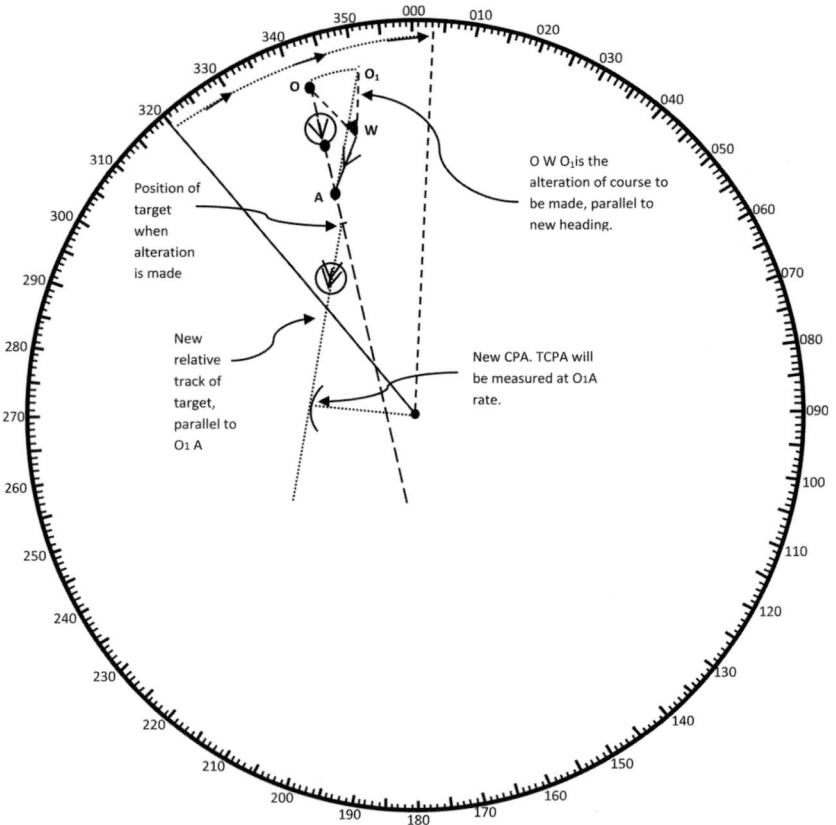

FIGURE 10.11   Own ship alters course.

Alteration of speed.

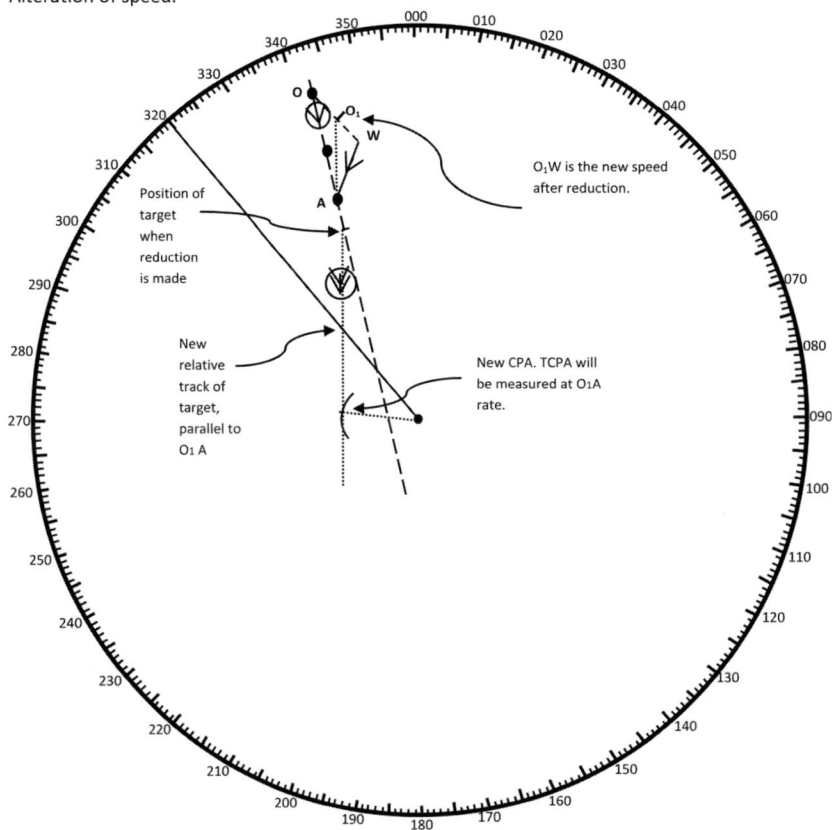

FIGURE 10.12   Own ship reduces speed.

three minutes and it is from this point that the O_1A line is drawn, or more correctly, a line parallel to O_1A (see Figure 10.11).

Alteration in owns ship's speed is handles in exactly the same way as an alteration in course. If speed is reduced, then the OW line will be shortened to reflect the new speed. Sometimes plotters are confused as to which way to shorten the OW line, but it should be remembered that W is fixed, the WA vector is the target's vector and it will not be affected by own ship's alterations. In our diagram, own ship is making a speed of 15 knots, and once again the plotting interval is 6 minutes; therefore OW is 1.5 miles. At six minutes after the last plot speed is reduced to 10 knots, assuming the reduction in speed to have immediate effect. O_1W, representing the new speed of 10 knots is now 1.0 miles. O_1A now represents the predicted relative motion of the target after the reduction is speed. The new relative track of the target, parallel to O_1A is marked off from the position of the target 6 minutes after the last plot (see Figure 10.12).

An increase in speed, which is probably the less likely occurrence, is handled in exactly the same way except that O_1 will move further away from W than O.

As an exercise in understanding relative motion plotting, it is sometimes asked what alteration of course (or speed) is needed to get a target to pass clear by so many miles. In this case, the a tangent to the clearance circle is laid off through the point on the OA line extended where the target will be at the time of the proposed alteration of course (or speed). This then is transferred back through A in the OWA triangle and extended back far enough beyond O. The OW line is then rotated until it cuts this transferred line at O_1. O_1W represents the new course to steer to get the target to pass clear by the required amount.

While this technique is useful for illustrating the effect an alteration in own ship's course can have on the target's relative track, and it has some practical use in navigation in tidal waters (see 'Current and tidal stream' below), it is not to be encouraged in actual anti-collision situations. This is because the alterations of course suggested by such an exercise can often be small, and small alterations of heading are to be avoided in developing close quarter situations.

A far more practical exercise where a required clearance is required is to make a bold alteration of course (or speed), and then resume the original course (or speed) so that the target passes no closer than the required clearance.

In the example here, own ship is steering $060°$ at 22 knots. A target on the starboard bow is plotted as follows (see Fig. 10.13):

Time	Bearing (Degrees True)	Range (Miles)
2100	082	11.0
2103	082½	9.8
2106	083	8.3

It is required to pass the target at not less than 2 miles. With an effective time of altering course at 2109, and making an alteration of $40°$ to starboard, at what time may the original course be resumed?

Presuming the target has continued on her course and speed, when own ship resumes her original course (and speed) the target should

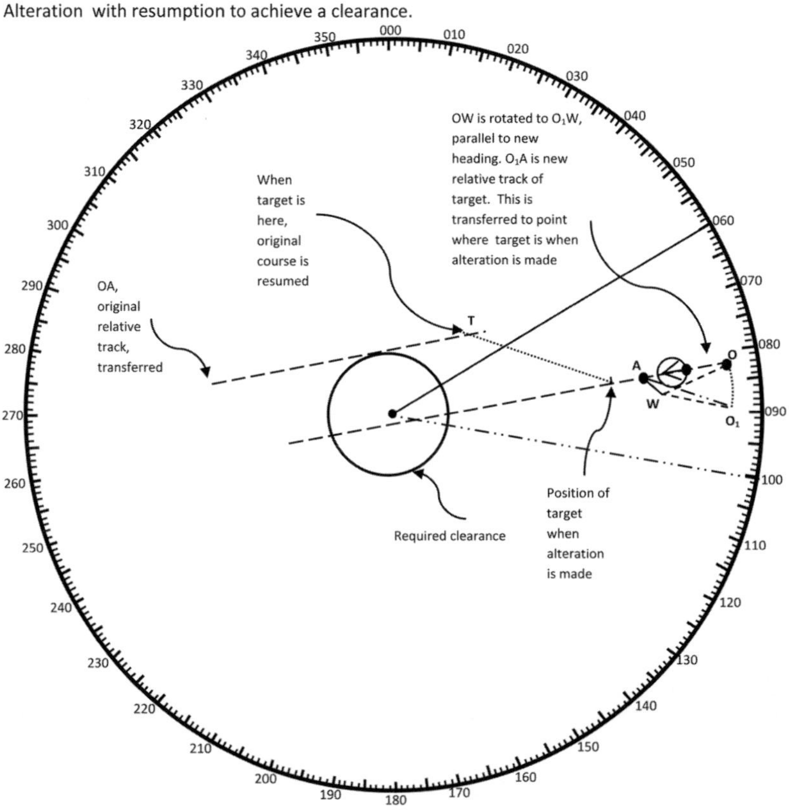

Alteration with resumption to achieve a clearance.

FIGURE 10.13   Alteration with resumption.

resume her original relative track. In the above example, the target is plotted in the normal way, and the report made. Then, the effect of a 40° alteration to starboard is plotted on the original OWA triangle, giving the new relative motion, O_1A. This is transferred to the position where the target should be at 2109, which is the effective time of the alteration and extended a reasonable distance.

Now a line parallel to the original OA is drawn as a tangent to the 2 mile clearance circle, and cutting the transferred O_1A line (at T in the diagram). This represents the time when course may be resumed, bearing in mind that the time measurement will be at the O_1A rate, that is, in this case the O_1A represents 6 minutes while own ship is steering 160°. From the diagram it will be found that course may be resumed in about 9 minutes (2118) after the alteration.

An alteration in speed (usually a reduction) is treated in exactly the same way.

If there is more than one target being plotted, then of course the effect of the alteration of course by own ship and the resumption of course will have to be applied to all.

In carrying out such plotting exercises in real time, aboard ship or in a simulator, it will be found that the target rarely moves precisely as predicted. This is due to the inherent errors such as the ship's manoeuvring characteristics not being allowed for, small errors in plotting, small errors in timing and so on. It is only with practice, and

Relative v True motion plots

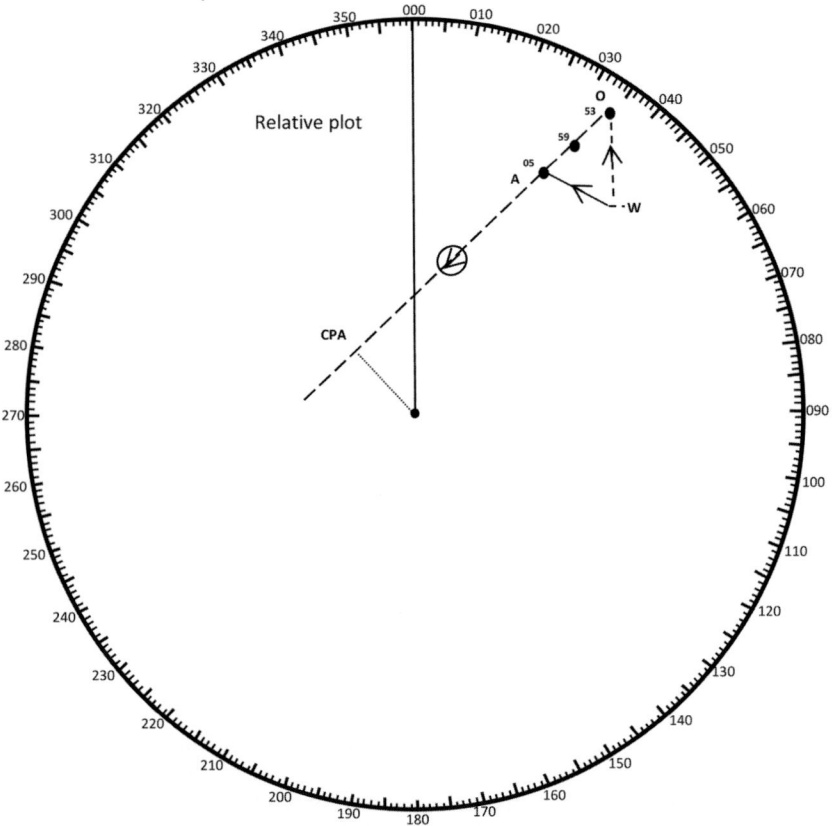

FIGURE 10.14a   Relative v True, relative plot.

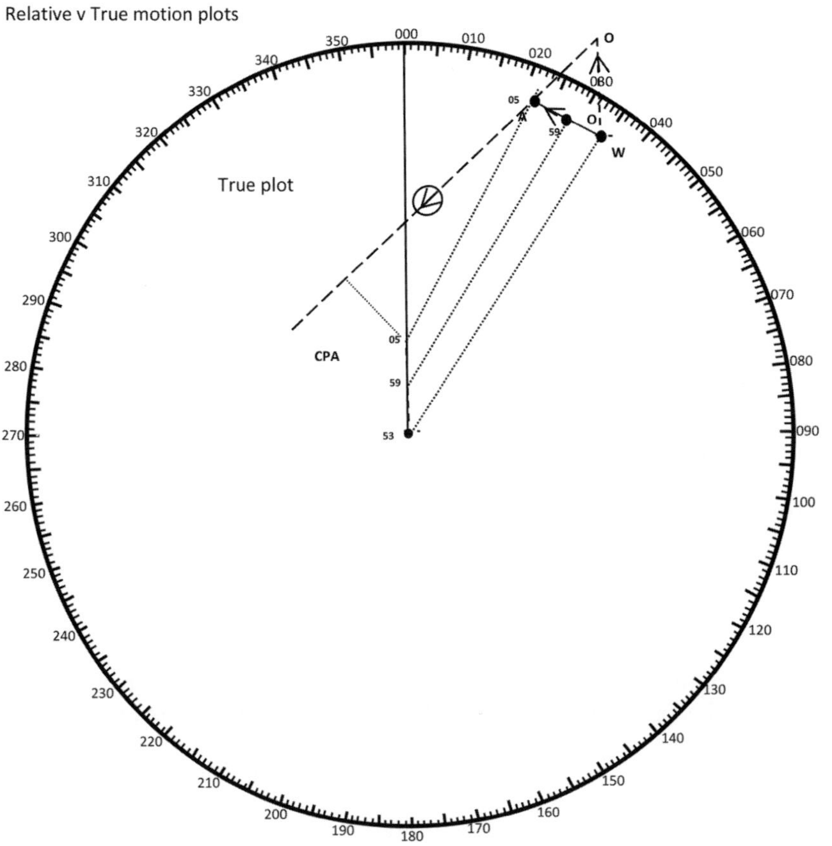

Relative v True motion plots

True plot

FIGURE 10.14b   Relative v True, true plot.

allowing for such errors that confidence will grow in the technique. Also, in practising such exercises the lessons learned will be transferred mentally to ARPA.

The True Plot

'True Motion' as a mode of radar PPI display has been around for many years, and is discussed in the chapters on practical operations. There are certain advantages which a true motion display has over a relative motion display, but plotting a true motion plot has very little to recommend it. However, we will look at the process and discuss what information can be gleaned from it.

When given a series of ranges and bearings for a given target, each plot has to be laid off from a different position for own ship. Such a series of plots will immediately give the target's true course, distance made during the plotting interval and thus its speed and aspect; in other words in provides WA of the OWA triangle. Construction is needed to find CPA and TCPA.

In the Figures 10.14a and 10.14b following, the same situation is plotted:

 a. As a relative motion plot, and
 b. As a true motion plot.

It can be seen that for each plot in the true motion plot, the plot is laid off from different positions, i.e. from the advanced positions of own ship as its position (the origin) tracks across the screen. The first plot however is not O, it is W. We now have the WA line, the target's vector. From W, the WO must now be laid off in the direction of own ship, not in the reverse direction, as in the relative motion plot. This, of course will be the same as the distance the origin (own ship) has advanced during the plotting interval. We can now complete the OWA triangle and establish the relative motion track and find CPA. The CPA is given where the relative track (OA extended) passes the last position of the origin, that is at the time of the last plot. In order to see if the target is keeping to this relative track, it will have to be advanced as both the target and the origin (own ship) move across the screen.

As can seen, as stated above, for the doubtful advantage of immediately establishing the true vector of the target, the true plot has very little to recommend it. It is more prone to error, and the difficulty of tracking the target in relation to the relative motion track is obvious. These comments should not be taken as referring to a true motion display and its uses which are discussed elsewhere.

Current and tidal stream

If a known geographically stationary object, such as a moored buoy or isolated rock is tracked and plotted and W and A do not coincide, then the ship is being set one way or the other. This is due to the current or tidal stream, although it can also be due to leeway. The amount or rate of such set and/or leeway can be established from the plot, the WA gives the amount of set in the plotting interval but the direction in which own ship is being set is the opposite, i.e. AW.

An example will make this clear.

With own ship steering course 193 (T) 7½ knots the following observations were taken of a target, which was known to be a light vessel:

Time	Bearing (True)	Range (Miles)
1553	263	9.0
1601	267½	8.3
1609	272	7.7

Set and Drift Example

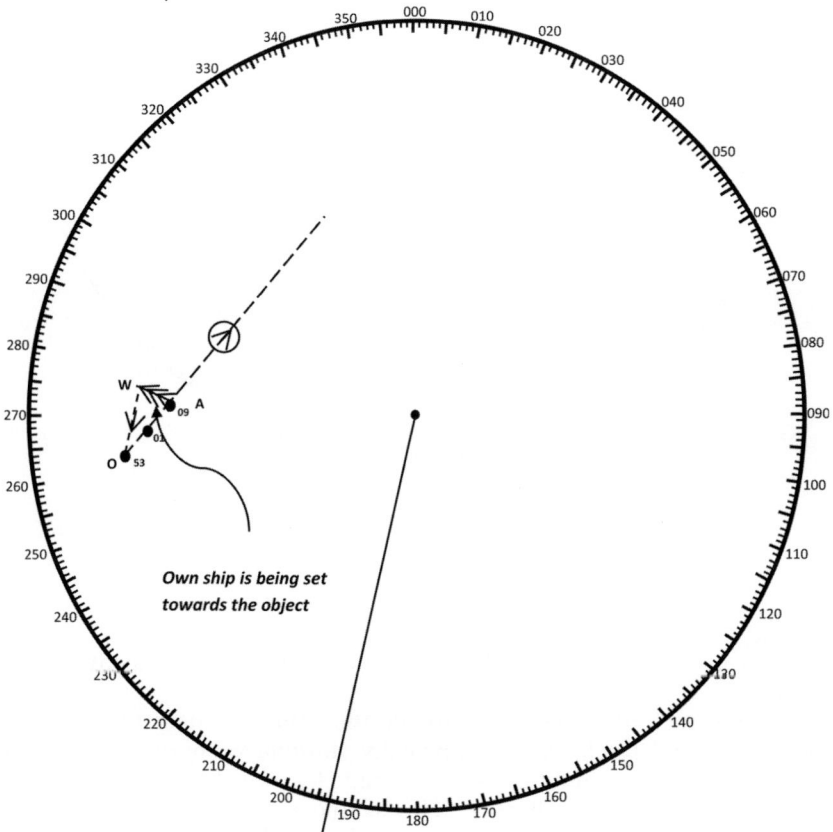

FIGURE 10.15   Set and drift.

Set and Drift Example

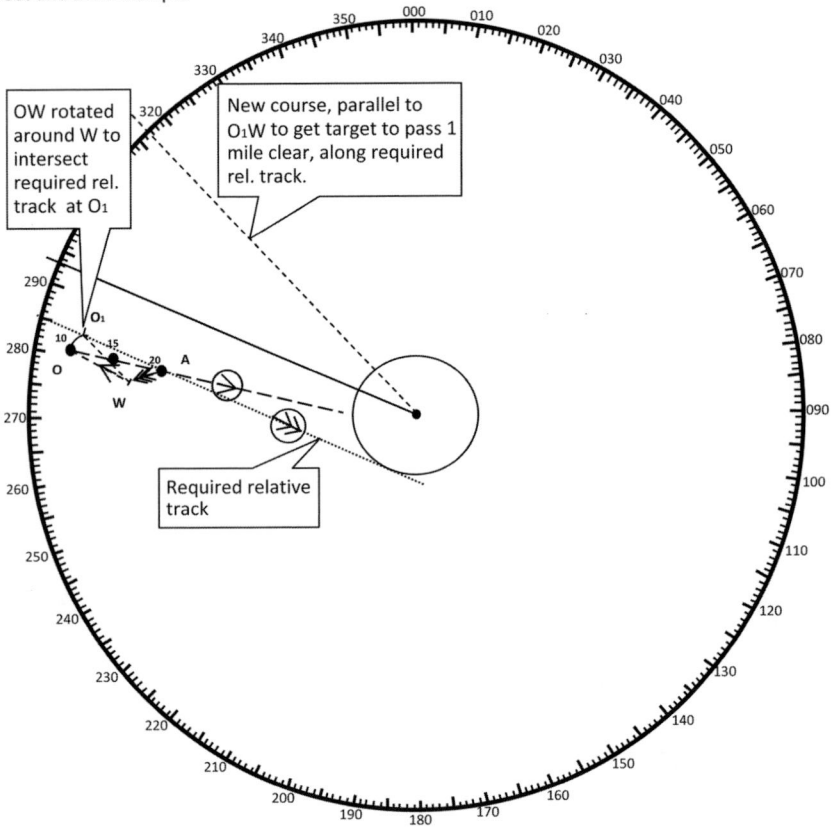

FIGURE 10.16   Course to steer with set and drift.

Find the set, drift and rate of the tide.

Answer: Set is 295° 1.0 miles in 16 minutes which gives a rate of 3.5 knots.

Plotting such an object can provide the course to steer to achieve a desired ground track. A geographically stationary object must follow the reverse of the ground track being made by own ship, when observed on a relative plot or display. This is the OA track of the object. Therefore if the OWA triangle is worked to find a particular OA, then the course to steer to achieve this is provided.

With own ship steering course 293 (T) 12 knots the following observations were taken of a target, which was known to be a light vessel:

Time	Bearing (True)	Range (Miles)
1810	281	10.8
1815	280½	9.4
1820	280	8.0

Find the set and drift, and the course to steer at 1820 so that the light vessel will pass 2 miles to port.

The target is plotted in the usual way and the set is found to be 251° at 6 knots.

The required relative track is then marked off from A (or a later position of the target along the OA line extended) and extended back through A. OW is then rotated around W to intersect the new relative track cutting it at O_1. O_1W now is the own ship vector to achieve the required relative track, and this gives a course of 317°.

Potential Collision Points (PCPs), Potential Areas of Danger (PADs) or safety Domains

While CPA and TCPA give a good indication of the possibility of a close quarter situation, they do not give much advice on the action to be taken on keeping away from one. Certain makes of ARPA have a process by which an area, a circle or oval, could be shown on the screen ahead of the target and once own ship did not enter this area, then the target would be avoided by whatever distance had been set by the operator (e.g. 2 miles). This function does not seem to be available in more modern sets, but the concept of a safety domain, whereby keeping outside it will keep the target at a given distance (CPA) is useful. In the practical sense it is unlikely that such an area could be constructed manually on the bridge of a ship in an actual collision avoidance situation, but it is useful to examine such constructions to develop an idea on how such areas appear.

Another ARPA function which exists on some makes, is an indication of Potential Collision Points (PCPs). These are positions that show an interception point that own ship could steer to intercept the target, presuming it maintained course and speed. If one of these PCPs appears on the heading marker then a collision will occur if no avoiding action is taken by either own ship or the target.

Let us examine the PCP function first. We have looked at the construction of the relative motion triangle OWA, and also how it is used to achieve a required CPA, as in the section on 'Current and tidal stream'. We have also seen that if the OA line extended passes through the centre of the plot, then the target is on a collision course with own ship if both maintain course and speed. Using these facts we can now look at the situation where a point, or in some cases, two points can be displayed ahead of the target's true vector which indicate the collision (or interception) point. It is also possible that there will be no collision point, that the speed of the target and its

Potential Collosion Points

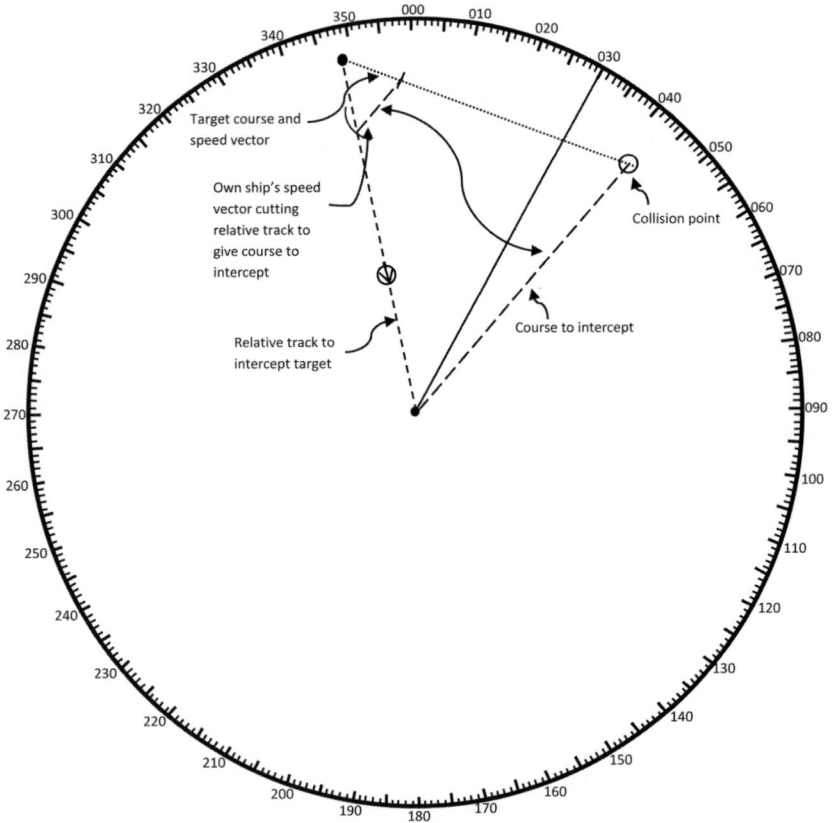

Target course and speed vector

Own ship's speed vector cutting relative track to give course to intercept

Collision point

Course to intercept

Relative track to intercept target

FIGURE 10.17a   Single collision point.

Potential Collosion Points

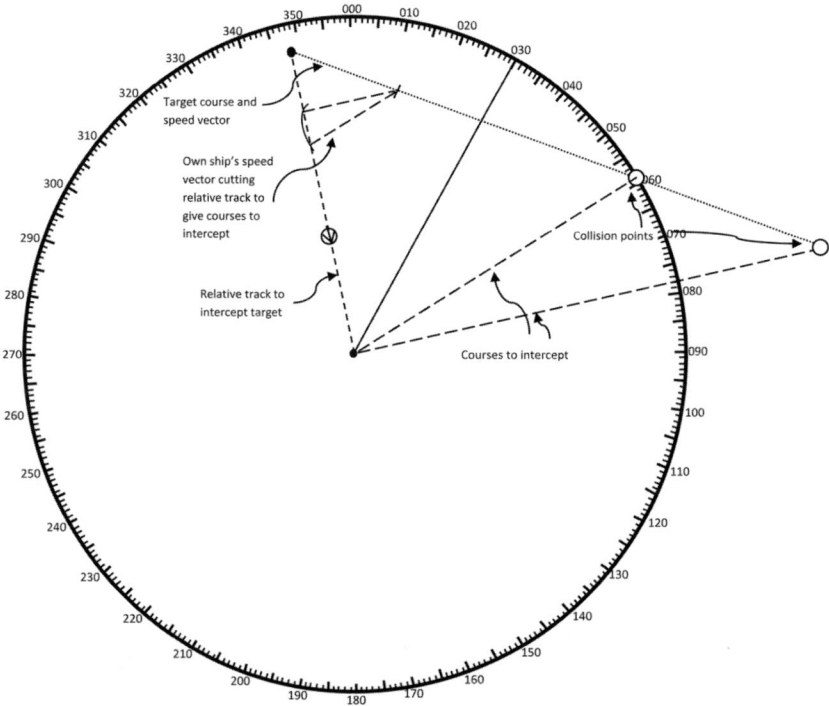

FIGURE 10.17b   Two collision points.

position relative to own ship will mean that own ship can never catch up with it (always presuming it maintains course and speed) (see Figures 10.17a and b).

Potential Areas of Danger (PADs)

If we use the construction in the PCP diagram to find a course to steer to get the target to pass by a required amount, say 2 miles, we can extend these courses to the target's true track. Therefore, if own ship steers outside either of these two courses it will avoid the target by at least two miles. If a zone two miles on either side of the target's true track, between the intersections of the two courses to steer is added, an area around the PCP can be shown that own ship must remain outside to keep the 2 mile clearance (see Figure 10.18).

Potential Area of Danger

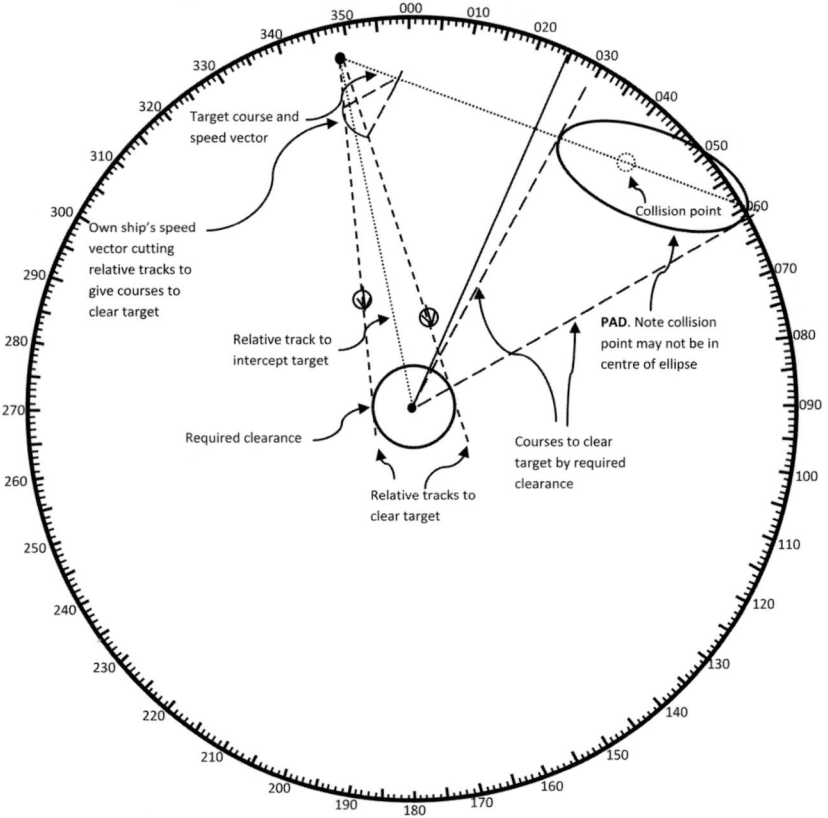

FIGURE 10.18   Construction of PADs.

The Headreach Problem

When own ship stops her engines, or even puts them astern in order
to take all way off, she will 'headreach' by a certain amount. This is the
distance she will cover during the process of stopping. To be able to
account for this in practice needs knowledge of the ship's manoeuvring
characteristics. This can be very variable as the only information
usually available is from the trials carried out when the ship was new
and this in itself was done under particular conditions. On any day the
conditions are unlikely to be the same as they were on the day of the
trials; at the very least, the loaded condition of the ship is unlikely to be

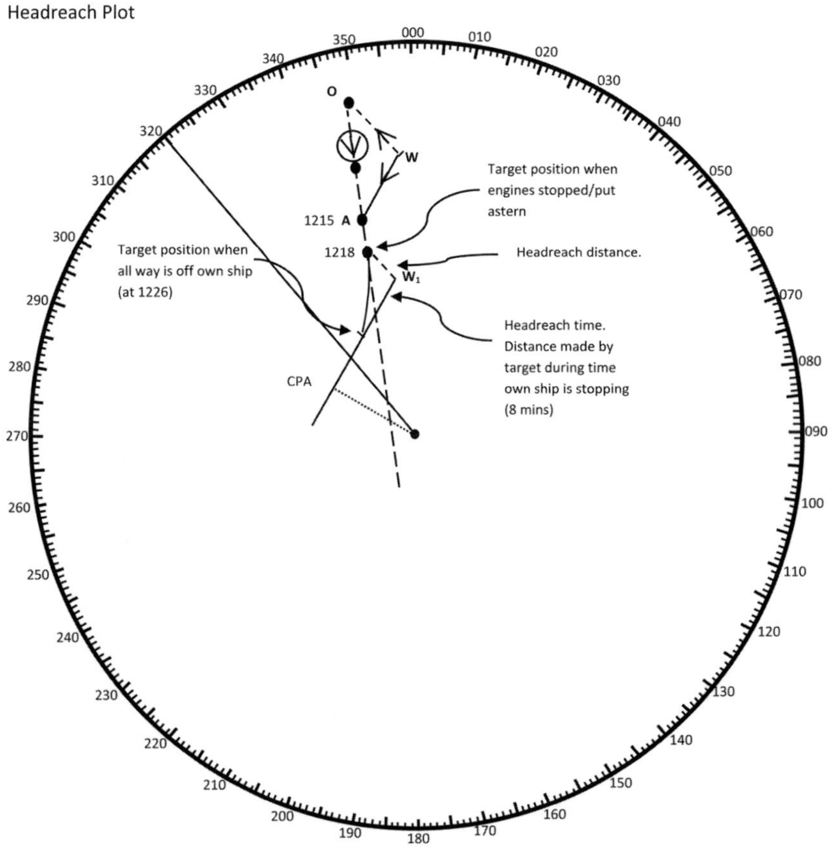

Headreach Plot

Target position when
engines stopped/put
astern

Headreach distance.

Target position when
all way is off own ship
(at 1226)

Headreach time.
Distance made by
target during time
own ship is stopping
(8 mins)

CPA

FIGURE 10.19 Plotting headreach.

the same. Weather and sea conditions and depth of water under the keel can have profound effects on the manoeuvring characteristics.

For this reason, in order to illustrate headreach and get a more accurate plot, assumptions and approximations have to be made. In our plotting exercises the time to come to a stop is given and the distance covered during this time. It is also usually assumed that the ship will remain on the same heading during the stopping process, something which rarely happens in practice, the ship's head usually falls off one way or the other.

Let us refer to Figure 10.19. At 1215 the last plot was made and at

Headreach, time to stop Plot

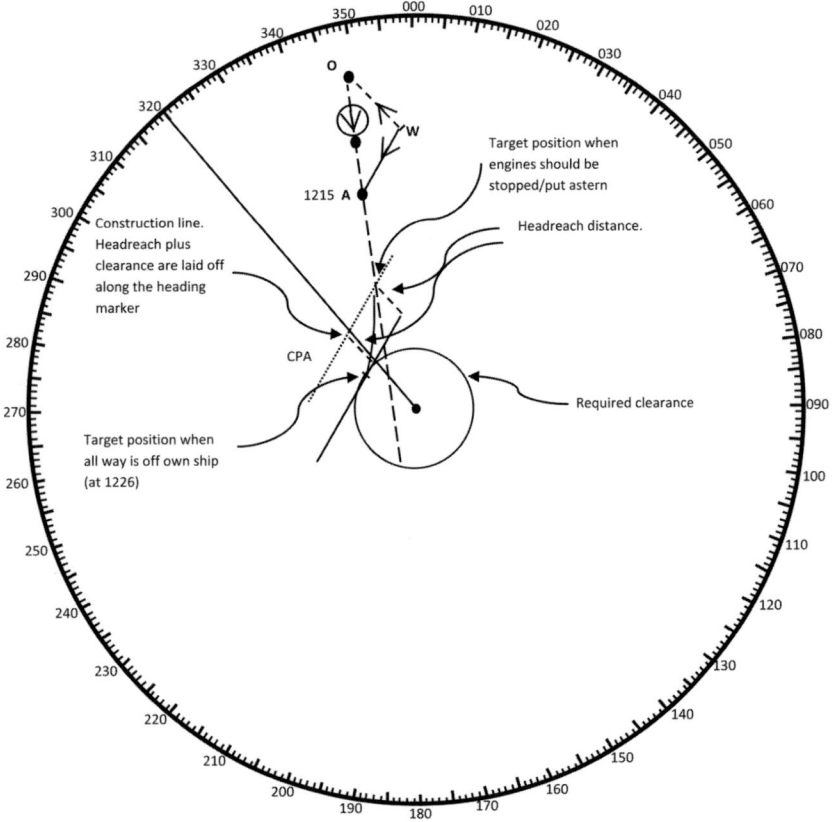

FIGURE 10.20   Headreach and time to stop.

1218 engines were stopped. The ship is expected to carry her way for 8 minutes over a distance of 3 miles.

The zero speed line, a length of 3 miles, is laid off from the 1218 position on the OA (relative track) line extended. If the target vessel was to launch a float with a radar reflector when she gets to the 1218 position, its echo would appear to track down this zero speed line, getting slower and slower, until is finally comes to a stop at W_1. This reflects own ship's decreasing motion as it comes to a stop.

Meanwhile the target vessel continues on its course and speed from the position where it supposedly launched the float. To illustrate this, a line parallel to WA in the OWA triangle is laid off from W_1. The time that own ship takes to stop is 8 minutes, so where the target vessel will be at the end of this time is laid off from W_1 along the line parallel to WA, at the target's speed. The transition from the relative track to the true (as own ship is now stopped in the water) track of the target can be illustrated by a curve, if necessary, as shown.

Thereafter, the target vessel's echo continues on her true course and speed on the plot while own ship lies stopped.

A variation of this is where the question asks at what time engines should be stopped so that the target vessel does not pass closer that so

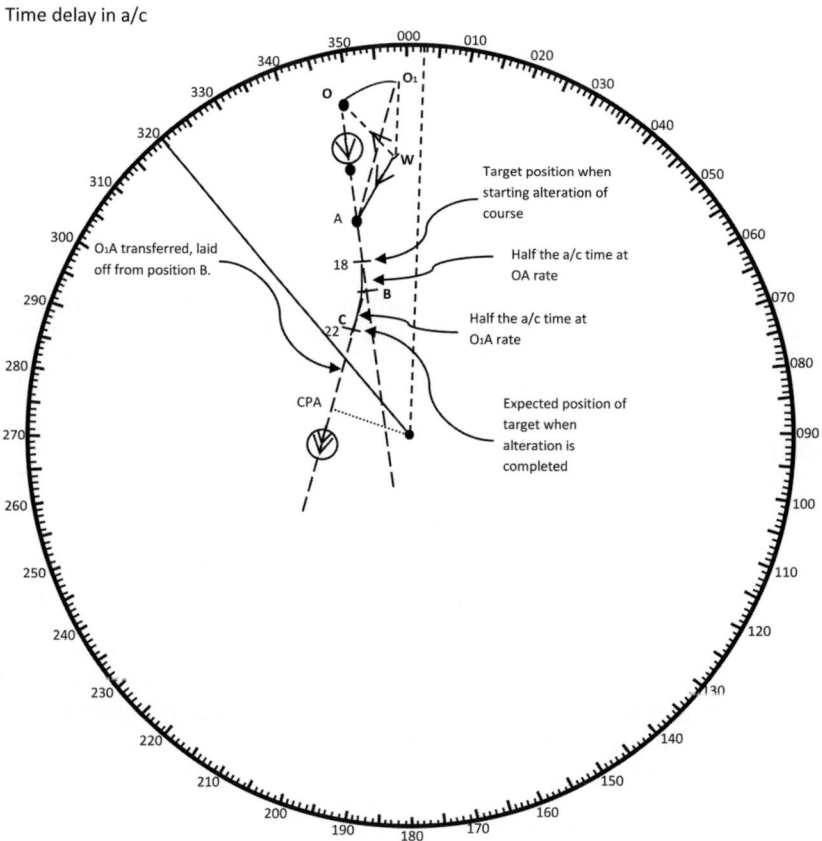

FIGURE 10.21   Allowing for alteration of course.

many miles, or so many miles ahead, given the headreach distance and time (although time is not necessary to answer the basic question unless it goes on to ask where the target will be when all way is off).

In this exercise, the clearance distance is laid off, either along the heading marker, or as a range circle, and the headreach distance is added to this. A line parallel to the WA line is laid off, tangential to the circle, or through the point on the heading marker and made to cut the OA line extended. Where this intersection occurs is the point that when the target reaches it, engines should be stopped.

Allowing for alterations in course and speed

As stated above when own ship alters course or speed this never takes place instantaneously, there is an appreciable time between the start of the manoeuvre and the ship being settled on her new course or speed. Once again, strictly speaking, knowledge of the manoeuvring characteristics of the ship is necessary. As stated above for headreach the actual manoeuvring characteristics can be very variable and for that reason again we make approximations and assumptions in allowing for alterations of own ship's course and speed in our plots. Usually the question will state something like the course was altered at 1218 and at 1222 the ship had settled on her new course.

In illustrating this on our plot we allow half at the old target relative track and rate, and half at the new. In Figure 10.21, instead of marking off the line parallel to O_1 A from the 1218 position, it is marked off at the 1220 position (marked B in the figure), half the steadying time. The other two minutes is laid off from the 1220 position along this new relative track at the new (O_1A) rate. This then gives the target position at 1222 (marked C in the figure) when own ship is steady on the new course.

An alteration in speed is treated exactly the same.

Good plotting practice requires that after an alteration in course or speed by own ship, that plotting be continued. This is reflected in some plotting exercise questions by being given a plot shortly after the alteration. For example the question might state that at 1218 own ship alters course 42° to starboard and when steady on her new course at 1222 the target now bears 334° at 4.6 miles (marked B in Figure 10.22). When given this information the plotting approximations allowing for the time taken to settle on new course/speed mentioned above should be ignored; the new plotted target supersedes any such approximations and the new relative track of the target can be laid off from this position to continue plotting to ensure the target is going clear, or to detect any change in the target's course and/or speed.

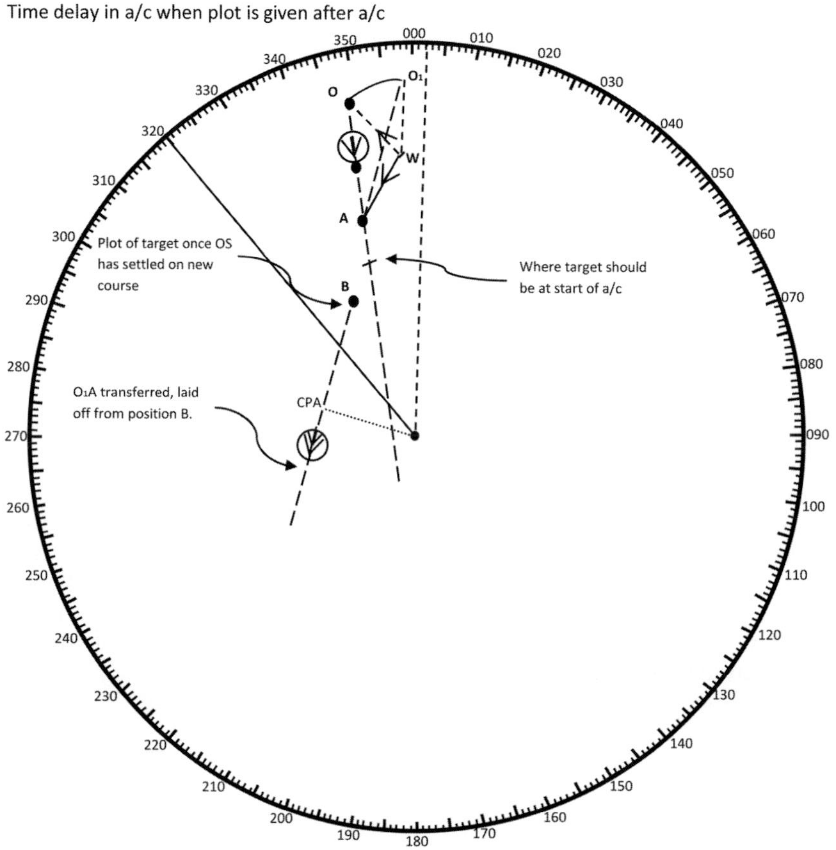

Time delay in a/c when plot is given after a/c

Plot of target once OS has settled on new course

Where target should be at start of a/c

O₁A transferred, laid off from position B.

CPA

FIGURE 10.22 Alteration of course with new plot.

These then are the techniques in manual plotting. Watchkeeping officers at all grades should be fully conversant with them. The significance of the OWA triangle should be remembered when observing the ARPA or TT display, and the change from relative vectors to true examined for consistency. For example if the true vectors of targets do not show the anticipated change of direction or length when changed to relative vectors it may mean a faulty speed input. Only familiarity with manual plotting will develop the 'eye' for such possibly dangerous inconsistencies.

Conclusions

Radar plotting is a skill and only practice can ensure proficiency and reliability. The basic OWA triangle and determination of the basic data is fundamental. Everything else in subsequent plotting depends on this.

Continuation of the plotting is essential to detect any changes in a target's motion.

The effect an alteration of course or speed is calculated on the original OWA triangle. This only has validity on the assumption that the target continues on its original course and speed. This emphasises the need to continue the plot until the target is passed and clear.

Manual 'true' plotting has little to recommend it. It does give an idea of how certain ARPAs calculate target data.

If PCPs (Potential collision points) and PADs (potential areas of danger) are used, their significance should be fully understood.

The techniques in allowing for the delay in reducing speed, stopping and altering course in radar plotting are merely approximations. They give a more realistic result to the plot but it must be understood that there are so many variables in a ship's manoeuvring characteristics that it is impossible to be more accurate. In any case the inherent inaccuracies in radar and in radar plotting make any attempt to be more precise futile. Once again the need to continue the plot after a manoeuvre is emphasised.

Chapter 11

Automatic Radar Plotting Setup

Introduction

In this chapter the set up of the automatic plotting functions are discussed. There is a certain necessary overlap with the contents of Chapter 9 on radar set-up and the functions discussed in the chapters on Radar for Collision Avoidance and Radar for Navigation. In this chapter the importance of a disciplined approach to the use of automatic plotting is particularly emphasised. This is discussed under the following topics:

- Historical development of modern ARPA.
- Setting up and checking inputs.
- Basic operation for collision avoidance.
- Plotting 'drill', the six 'musts'.
- Basic operation for navigation.

It is rare at this time that modern radar sets are not fitted with some kind of plotting aid. There is still a tendency sometimes to treat ARPA or other such plotting aids as an 'added extra' to the basic radar display whereas in modern terms it is an essential tool in the bridge team's or OOW's equipment. The new IMO regulations regarding the specifications for radars fitted after 2008 recognise this and all such radars must have some form of automatic plotting system.

Such plotting aids are very sophisticated computerised systems. Like any such complex system, it might seem to present a fairly simple operating procedure and output. However, in the hands of untrained or poorly trained people it can be positively dangerous. The old saying with regard to ARPA is 'ARPA makes good navigators better and poor navigators worse'.

Several factors brought about the development of ARPA. In the use of radar for anti-collision, it soon became obvious that plotting was a problem. There was a reluctance to plot on the part of ship's officers for various reasons; plotting is prone to serious errors; with multiple targets the plot soon becomes overloaded, and it was unrealistic to expect ships to slow down or stop, no matter what the regulations or marine administrations might say, to allow plotting to be carried out. Various plotting aids were developed over the years with more or less success. 'True motion' was an early development, 'situation displays' which were really a variation of true motion were another. 'Match-sticks' were another where the operator had a number of short graphic symbols or lines which could be moved across the screen. These gave relative bearings, if the target stayed on the 'match-stick' then it was on a collision course with own ship. The reflection plotter, which allowed the plot to be carried out on the face of the radar screen was a particularly useful aid.

None of these solved the plotting problem, they merely assisted the OOW/bridge team in assessing the situation and there was still a need to actually carry out a complete plot where the full information was required.

The big breakthrough came with the advent of the micro-processor. Computer technology quickly solved the problem and developments in data storage and graphics allowed the presentation of data in the now familiar form.

The old IMO requirements for ARPA state:

ARPA should '...reduce the workload of the observers by enabling them to automatically obtain information so that they can perform as well with multiple targets as they can by manually plotting a single target'.

Setting Up

This presumes that the normal radar settings have been carried out correctly, and the following deals only with those settings and adjustments which affect the plotting functions of the equipment.

Like all computers and computerised systems, operators are enjoined to 'RTFM' (read the friendly manual). This is something that people tend to be reluctant to do, and too often the manuals are poorly written, obscure and badly presented. However, no matter how tedious, there are details which the OOW needs to know about a particular set which can only be found in the manual.

While the heading in this section is entitles 'set-up', many of the details should be checked by an OOW on taking over the watch.

Gain. In most sets the tracking function works independent of the 'gain' setting by the operator. However there are sets on the market where the gain setting affects the signal level to the tracker and therefore a target can be lost if the gain is set too low. This may be detailed in the manual, but OOWs should check it out by testing on an acquired small target when opportunity offers.

Gyro/compass. Obviously of prime importance that the heading input is correct. While this input is usually automatic, and on switch-on the correct heading is usually applied without the need for the operator to do anything, he/she should be familiar with the function to adjust the heading to the radar/ARPA. Also checking the heading setting on the radar from time to time should be part of the normal routine of the OOW.

Speed. Accurate speed input is essential for the ARPA to carry out correct calculations. Once again the operator must be familiar with the functions for changing speed inputs. Speed can be input from a variety of sources:

1. Electro-magnetic log. This gives speed through the water in the direction which the ship is heading. There may be a percentage error, and it should be checked from time to time.
2. Doppler log. This can give speed through the water (water track) or 'over the ground' (ground track). For radar/ARPA input, water track should normally be selected.
3. Manual input. This is really an option in the event of failure of log input, although there may be other circumstances where it is necessary to input speed manually.
4. GPS or other electronic fixing system. This gives ground track (speed and direction). When this is being used the plot is 'ground stabilised'. It can be used in navigational mode, but should not be used in anti-collision situations.
5. Geographical reference. This is where a known fixed object is tracked and referenced by the computer. This also ground stabilises the display and the same provisos as for GPS apply.

In general, and certainly during a set-up procedure, the display/plot should be sea-stabilised.

GPS input. This is usually also a default position, and comes in automatically. However, the operator should be familiar with the function of disabling it, or switching between different positioning systems (such as LORAN) if fitted.

The CPA and TCPA warning limits should be checked and set during set-up. The actual settings will depend on circumstances, in open waters such limits will be set large, say CPA for 3 miles or more, and TCPA for 15 or 20 minutes. In busy waters such limits would have the system alarming continuously, and so the settings will be reduced to possibly 2 miles or less for CPA and 10 minutes for TCPA. In any case the operator should be aware of what limits have been set.

Many modern sets come on in a default mode, such as on the six mile range, with range rings on, in north-up mode and so on (see Chapter 9 dealing with radar set-up). In the ARPA functions such sets may have a default mode, often relative motion, vector time/length, vector mode (true or relative), CPA and TCPA warning limits. Each of these should be checked not so much to verify that they are working correctly, but to impress the data in the observer's mind.

In some sets the vector mode can be selected by the operator, no matter what the motion mode of the display, whether 'true' or 'relative'. In other words, if the display is on true motion, the vectors can be set as 'true' or 'relative' and they will remain as such. In other sets, the vectors will default to the motion mode of the display; if the display is in true motion, the vectors will be 'true' vectors. The operator can change them to 'relative' vectors, but after a minute or two, they will automatically revert to 'true' vectors. The same applies if the display is in relative motion, the vectors can be changed to 'true', but will revert to 'relative' after a few minutes. This is an important feature, and the OOW/bridge team should be fully conversant with it.

The way the operator can work the equipment, the 'operator interface' varies quite widely from one make of ARPA/radar to another; some have a minimum of knobs and buttons, and work by a system of roller-ball and a few buttons, and others have quite an extensive key-board. The main functions and operations are usually fairly obvious and readily to hand, whereas the more complex functions are often accessed by a series of 'menus'. Needless to say, the OOW should be fully conversant with all the functions and adjustments needed for watchkeeping and navigation. Care needs to be taken however in this regard, and there are certain functions, usually buried in several 'menu' stages where various technical adjustments can be made. These are usually protected by a password, but, in the hands of an untrained person, certain settings could be altered which could have serious consequences. In one make of radar, for example, the range of the manual tuning adjustment can be altered by such a function. If this is incorrectly set, then the tuning available for the normal set-up might not have the optimum range each side of the fully 'in tune' position, and thus could have a situation where the set is operated in less than optimum tuning (see 'Tuning' in Chapter 9).

Basic Operation for Collision Avoidance

Because it is an automated system, the tracking system can function, track targets, produce CPAs and so on without the OOW having to do very much. But, unless the information goes into the OOW's head, then it is useless. If the OOW has to physically plot a target, the work involved ensures that information deduced enters his/her consciousness. The danger with ARPA/ATA is that the information does not enter the OOW's consciousness. This is a psychological phenomenon well recognised in people supervising automated systems. For this reason a form of systematic drill or procedure needs to be adopted to ensure that the information presented is in fact absorbed by the OOW. One form of such a procedure is the six 'musts'. These are as follows:

1. Acquire. The ARPA/ATA can do nothing until a target is 'acquired'; that is designated, either by manual selection by the screen cursor, or automatically by the guard-zone/s. The use of the guard zones is discussed elsewhere, but normally it is good practice for the OOW to acquire the targets manually. This helps in the process of absorbing the information and his/her situational awareness.

   It should be noted, and guarded against, that there is a tendency not to acquire all targets near own ship. Often, on questioning an OOW as to why a certain target, clearly visible on the screen, and within a few miles, has not been acquired, the answer is, 'oh that's a buoy!' or a fishing vessel or such. It may obviously be going clear, but unless it has been acquired then the full advantage of the ARPA/ATA is not being utilised. Obviously common sense applies here, and it is not being suggested that, for example in very confined waters, with a lot of navigational marks and possibly small craft well clear of own ship, that everything be acquired, but the author has noticed this trait among OOWs not to acquire all relevant targets.

   Another related problem that has often been noted, and a more serious one, is that targets close to own ship have not been noticed, and hence not acquired by the OOW. This often happens when several targets have been acquired and the OOW's attention is focused on these. It is a form of plotting overload and the only solution is good ARPA attention and discipline.
2. Threat posed/CPA/TCPA. Once the target/s have been acquired and vectors generated, an assessment of the threat posed by them can be viewed. In this the vectors should be 'relative'. If the set is such that the default mode of the vectors is that of the display mode, then it is probably better to put the display into 'relative'. But whatever the display mode, the relative vectors are used. The vector time in increased causing the vectors to lengthen until they pass the origin. How close they pass to the origin gives an indication of CPA, and the vector time, at the stage where the end of the extended vector is closest to the origin gives the TCPA. This of course will be different for different target vectors.

   Vector time must be returned to its original setting once this check has been done, usually six minutes.

3. Aspect. How is the target heading? To ascertain this true vectors are used. Whether the display mode is 'true' or 'relative' 'True' vectors show in which direction the targets are heading. 'Aspect' can usually be determined by eye, but if needs be, the bearing cursor can be put through the target, and the angle between it and the target's true vector gives the aspect. This is the only indication on most ARPA/ATAs of aspect.

4. Report. The report is presented in the readout on the side of the screen. This has all the usual information on each target; bearing, range, CPA, TCPA, target course, target speed. What most sets do not give is 'Aspect', but many give 'Bow crossing range'. Some sets may give a readout for more than one target at a time, but many sets will just give a readout for one target, selected by the operator and the one selected having a symbol indicating which one it is.

The OOW should read the report on each relevant target carefully, and compare it with the information gleaned graphically from stages 1 and 2 above. This is part of the process of assimilating the information on the part of the OOW.

5. Trial Manoeuvre. This should be used whenever an anti-collision manoeuvre is contemplated. It is not a tool for advising the OOW which way to go, rather a manoeuvre should be decided upon, say a 30° alteration of course to starboard, in 5 minutes, and the trial manoeuvre operated with these details applied. The effect of the manoeuvre on the main target for which the manoeuvre is being carried out should be well anticipated by the OOW (he/she really should not need a trail manoeuvre to advise on the effect of the manoeuvre), but its effect on other targets in the vicinity can be seen, and whether the intended course of action is likely to cause problems with such other targets.

Misuse of the trial manoeuvre often takes the form of rotating the trial course to get all the extended relative vectors to go clear, and then deciding that such is the course to steer to get out of trouble. This may be the optimum course to steer, but such use of the trail manoeuvre function ignores the Regulations and own ship's obligations under those regulations.

6. History dots. These indicate the recent movement of the target. These should be 'true', no matter what the display mode (whether relative motion or true motion), although some makes of ARPA have them as 'relative' in relative motion, and 'true' in true motion. To be of any use, they need to be true. In some sets the dots (at least four, sometimes six) will all appear together once the function has been operated; in other sets the dots will only start

to be generated once the function has been set, and it will therefore take a few minutes before the whole series is visible.

These should be examined to detected any recent manoeuvre of the target/s. An alteration of course by the target is usually most easily detected, but a reduction of speed, the closing together of the history dots, is less readily apparent.

Basic Navigation Functions

The ARPA functions which are used for navigation, as opposed to collision avoidance are navigation lines, 'floating' EBL and VRM, maps, radar underlay. The use of these are discussed more fully in the chapter on radar for navigation. The operators of the set should be fully familiar with manipulating the controls and settings, and should practice with them whenever possible. It is too late, and dangerous, in critical pilotage or other such situations for the person on the radar to be searching out functions for moving and setting navigation lines; he/she should be able to do this almost without thinking.

The Electronic Bearing Line (EBL) or lines and the Variable Range Marker (VRM) are two essential controls that the operator should be most skilled in manipulating. Apart from rotating knobs very often these can be 'dragged and dropped' by the screen cursor.

Navigation lines are a series of straight lines, of infinite length, that is they stretch across the radar screen and normally their length cannot be altered by the operator. They can be moved to any distance from the origin and rotated to any bearing. Usually there are at least two, but depending on the make of the radar set there may be more. When 'set' the navigation lines usually remain fixed in relation to the origin, but care needs to be taken in this regard as in some cases they may be fixed geographically. In one particular make of radar, the navigation lines rotate with the ships heading in 'default' setting; the operator has to set a function so that they will remain fixed in azimuth.

The 'floating' EBL and VRM is another feature that the operator should be well able to manipulate quickly and effectively. Often two such of each are available. The reference point can be selected by the screen cursor, and the range and bearing measured and read off from this point. Alternatively the read-out can give a geographical reference (presuming there is global positioning input to the radar). The EBL/VRM reference point, selected by the operator, usually stays fixed in relation to the origin, but can sometimes be ground referenced. Again, care needs to be taken in this regard, and the operator fully conversant with this fact.

The drawing up of maps is something that is done less frequently.

Again, no attempt should be made to prepare a map under pressure, there is too much scope for making errors. Often the procedure for drawing maps is less than 'user friendly' and frustrating. Once drawn, the maps can be stored and then produced for the passages required. Geographically referencing is critical with maps and ensuring that they are 'locked' or fixed in their correct position.

Chart underlay. Like maps, the correct geographical location of the chart needs to be checked periodically, and the operator should be quite conversant with the process. Vector charts are the only ones which can be used in this mode, and only certain selected data can be displayed on the radar. Watchkeeping officers should ensure that they understand the datum being used, the process for adjusting the position of the chart underlay, and particularly the control to switch the chart on and off. It should be remembered that chart underlay is intended to enhance the bridge team/OOW situational awareness for traffic and collision avoidance by indicating the area of safe water in which own ship can manoeuvre. It is not intended for use as a navigation system, that is the purpose of the chart whether electronic or paper. Officers should know the filing system in the radar's processor in access charts, determine the designed scale and other related data.

There are other functions that involve going deeper into the system that OOWs should be aware of. These are accessed rarely, but professional ships officers must know of them. Making adjustments in this area should be done only by those who understand fully what they are about. One is the adjustment of the rate-aiding filters. The length of time in which the filter operates can be adjusted. At one extreme one gets a stable vector and read-out and a good averaging of the target data, but a time lag in indicating changes of target course and/or speed. At the other extreme one gets a more responsive vector and target data, showing more immediately changes in target course/speed, but with a more unstable or erratic vector and readout.

Radar Housekeeping

It is important for OOWs to develop good screen discipline. A modern radar screen can become very cluttered with the consequent danger of confusion or targets not being detected. In congested waters it is easy to have a screen full of vectors all pointing in different directions, some PiLs (parallel index lines) or other graphics such as a map and radar underlay all on together making a very complex display. While a radar observer who has been observing continuously and has set up the graphics and acquired the targets may have a clear 'picture' of the situation, someone else arriving to look at it would probably be at a loss. Even the observer, who might be continuously on the radar can get confused in such a situation with the consequent danger of error.

In order to guard against such confusion, the screen should be kept as clear as possible. Some points in this are:

Vectors. Delete vectors which are no longer relevant as soon as possible. Check the vector mode ('True' or 'Relative') at frequent intervals. Keep vectors short (e.g. 6 minutes); if needed to lengthen them, such as to check CPA, as in the six 'musts' then restore them to short.

Trails and history dots. These are important for collision avoidance assessment, but in circumstances where they can be done without they should be 'Off'.

Navigation lines PiLs. These should be used as required, and when not needed, switched off.

EBL/VRM. One of each should normally be on for quick ranges and bearings. Where a second one is required it should be switched off as soon as not required.

Conclusions

In summarising this chapter the main points are:

- checking the data inputs, particularly speed (log) and direction (gyro);
- for collision avoidance, using the steps of the six 'musts' ensures that complete target data is absorbed by the OOW; and
- in the use of radar for navigation the the necessary skills to manipulate and set up navigation lines, maps, ground stabilisation and other functions must be fully acquired.

Chapter 12

ARPA and Target Tracking

In this chapter we will examine the functioning of the radar's tracking computer and the way that calculated information is provided to the observer. The use of guardzones and alarms is also examined. The following topics are discussed:

- Target acquisition.
- Filtering and rate aiding.
- Data storage.
- Target swop.
- Warnings and alarms including the use of guard zones.

Acquiring a target

The automatic plotting function cannot operate until a target is 'acquired'. This is done either manually, where the operator designates the target on the radar screen with the roller-ball/joystick cursor and pressing the 'acquire' button, or automatically where the target enters a pre-set guardzone/area (see below under 'Warnings and Alarms'). Acquisition can also be achieved where all targets are 'tested' by the system and then acquired under a set of parameters some of which are set by the operator. In such a system large areas of echo response, such as land echoes will be rejected and others may be rejected under a prioritising system such as their proximity to own ship, or fulfilling certain CPA and TCPA values. Under such a prioritising system, the first twenty target which meet or exceed the values set will be tracked in order of such values and those which have the least values will be rejected. For example, if the criteria set are CPA within 3 miles in 15 minutes with range closing, then the targets where the range is opening will not be tracked, and those with the closest CPAs in the least time will be tracked first and those with greater values of CPA and/or TCPA will be tracked in ascending order up to 20 targets. This could mean that if there are more than twenty targets fulfilling the parameters set, then several of the least critical will be rejected.

The process of acquiring a target, whether manual or automatic will tell the computer to start processing a target in that vicinity; it will identify an echo signal, or more likely a group of signals called a 'hit matrix' above threshold limits on that particular bearing and range, which is the echo return from the target. It will give that particular target an identification designation (a number) and continue to track it. When the computer has tracked the target for a time to achieve sufficient accuracy (like the manual plotter plotting at least three plots) it will display the resulting calculation in the form of a graphic line on the display, called a vector, originating at the position of the target's 'spot', in the direction of the target's movement and having a length dependant on the target's speed and the vector time set by the operator. The vector may be 'true' or 'relative' depending on the mode selected by the operator, in which case the vector will show the target's movement accordingly. In the IMO radar specifications the automatic tracking must give the target's trend within 1 minute and full data within 3 minutes. Many modern sets do better than this, although it should be recognised that time is needed to register sufficient plots to get an accurate estimate of the target's motion. It will often be noticed that immediately after a target is acquired there may be small fluctuations in its vector and data (particularly target course and speed) until some time has elapsed.

At the same time as the vector appears, a readout on the side of the display will give the data calculated, such as target's course, speed, CPA, TCPA, range and bearing.

Filtering

In order that the equipment presents usable data to the operator a process of smoothing or filtering, takes place. The computer can carry out the necessary calculations very rapidly, and without such filtering, taking all the radar errors into account (such as own ship's movement in a seaway (see Chapter 13 on radar and ARPA errors)), the data presented would be changing very rapidly and erratically, for example the vectors on the screen would be flicking rapidly from side to side and lengthening and shortening and be completely unusable. In much the same way as a human plotter assesses the 'best fit' track through a series of plots which are not quite in a straight line, the computer does the same thing, but far more rapidly. It is on the basis of this averaging or filtering process that the resultant vector will be produced, both for direction and for length. The disadvantage of this is that when the target alters course (or speed) the computer assumes initially that this deviation off the track is just one of the aberrations for which it is designed to filter out and it may take a few minutes before the alteration registers as a new direction (or speed) of the target. This is an important feature of which all watchkeepers should be fully aware. It is obvious

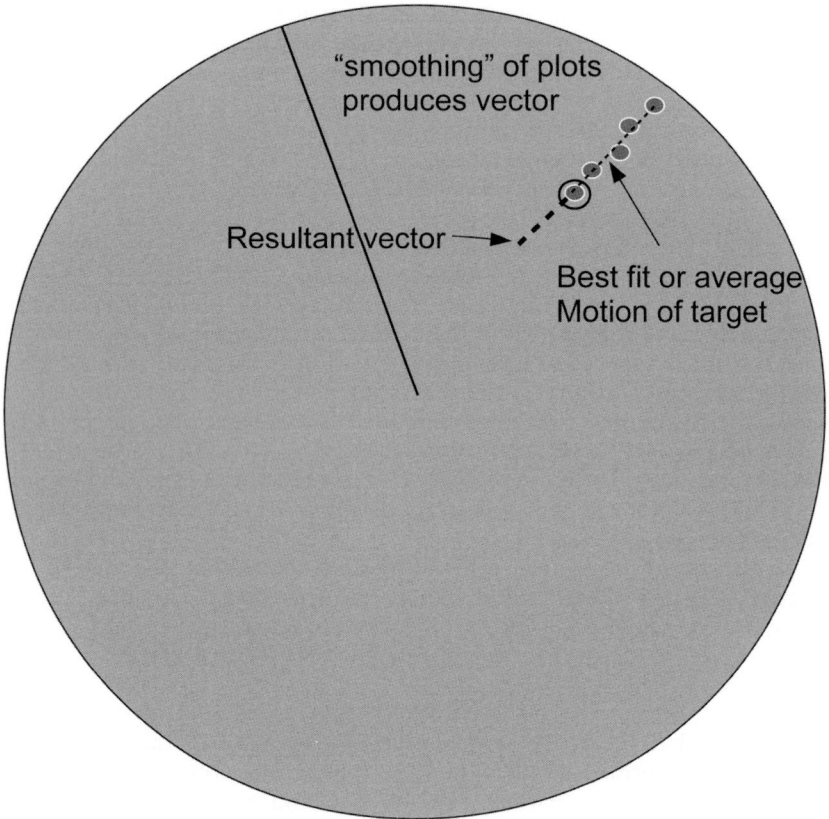

FIGURE 12.1   Filtering/smoothing.

from this that when the display shows that a target has altered course or speed that this has taken place some minutes ago. In some equipment the filtering can be adjusted to reduce the filtering effect, but this has to be used with caution because the 'trade-off' is more erratic behaviour of vectors and read-out data (see Figures 12.1 and 12.2).

ARPA data storage

Another important feature is the way the data is stored. Depending on the make of the equipment, it is either stored as 'relative' or 'true'.

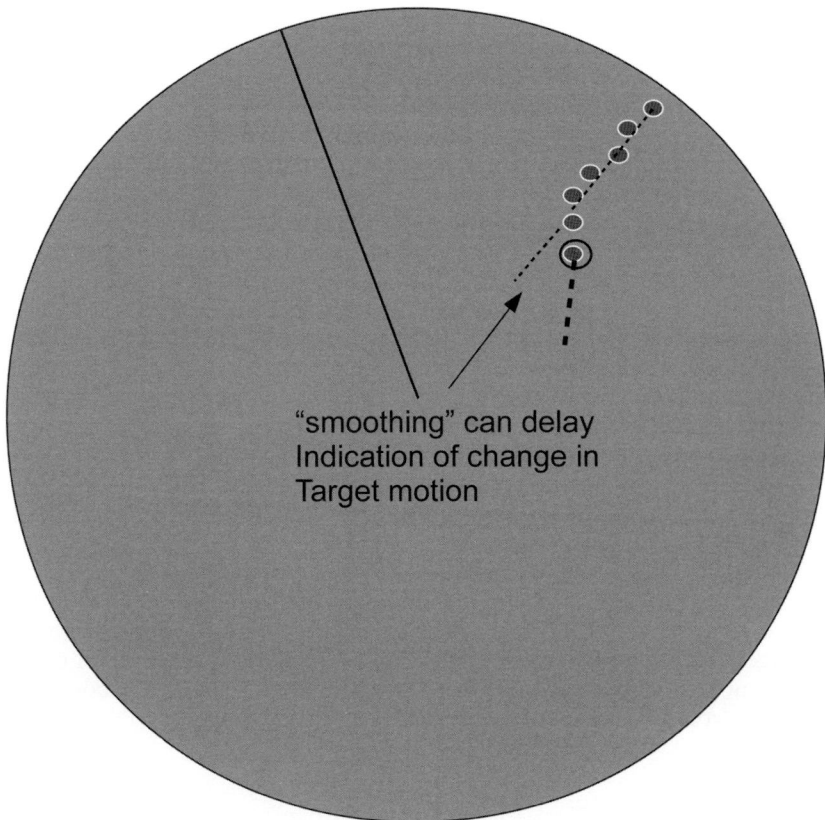

FIGURE 12.2   How smoothing can hide change in target motion.

In 'relative' storage, the bearing and range data is stored as such, i.e. a series of bearings and ranges. The filtering or smoothing process is applied to these and the target's relative motion is initially calculated, with of course CPA and TCPA. The 'Own ship' data (course and speed) is applied to this to calculate the target's true course and speed. This, as can be seen is much the same as a human plotter carrying out a manual plot.

In 'true' storage, the range and bearings are converted to a series of Cartesian co-ordinates, a series of x's an y's, in conjunction with own ship's position or movement associated with each plot. The smoothing process takes place on the series of these co-ordinates. This produces a 'true' track of the target. The 'Own ship' data is then applied to this track to calculate the target's relative track, CPA and TCPA.

The advantage of the 'relative' storage option is that it gives CPA and TCPA information independent of own ships course and speed data, in other words if there is an error in either or both of these, it will not affect the calculation of CPA and TCPA. The disadvantage is that if own ship manoeuvres, the smoothing filter will try to 'smooth' the changing range and bearing data of the target resulting from the manoeuvre of own ship. This will make the vectors (whether true or relative) unstable until own ship has settled on her new course and/or speed. To the uninitiated this might appear as if all the targets were also manoeuvring while, and for a short time after, own ship has manoeuvred. It can take about 1 minute for the vectors (and read-out data) to stabilise.

The 'True' storage system is not prone to this instability. As own ship's movement is applied to each range and bearing, the target's true motion plot is not affected by own ship manoeuvring. The vectors, both true and relative will remain stable, and a manoeuvre by the target, during own ship's manoeuvre, can be detected.

Obviously, if the own ships data (course and speed) is incorrect the target's true course and speed will be plotted incorrectly. However, where it might appear that a consequence of this, in a true motion storage, that the resultant relative motion calculated would be in error, this is not so. Provided the error/s in course and speed are constant, then the correct relative motion (at least for CPA and TCPA) will be calculated.

Rate Aiding

In order that the radar and computer can accommodate all the inherent errors in the system, such as own ship's movement, scanner parallax and so on, a process called 'Rate Aiding' is used. When the target is first acquired an area around the designated position, called a 'plotting gate' is applied. The processor searches for the target in this area. When it finds it, it starts tracking it and detects its movement. Over successive sweeps of the scanner it will search and identify the target in the 'plotting gate', but as the reliability of the plot increases, it will reduce the area of the plotting gate. It also moves the plotting gate ahead, once it has detected the target's movement, in anticipation of the target's position at the next scanner rotation.

In figure No. 12.3 below, the black dots are the actual successive plots of the target, the small white circles are the anticipated positions of the target and are central to the plotting gate circles. As can be seen, as

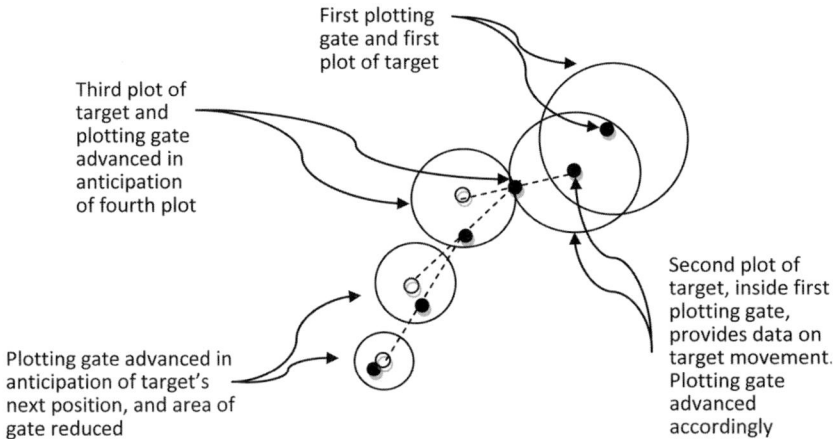

First plotting gate and first plot of target

Third plot of target and plotting gate advanced in anticipation of fourth plot

Second plot of target, inside first plotting gate, provides data on target movement. Plotting gate advanced accordingly

Plotting gate advanced in anticipation of target's next position, and area of gate reduced

FIGURE 12.3   Rate aiding.

the plot progresses the actual plots (black dots) and the anticipated positions tend to coincide. At the same time the area of the plotting gate reduces.

If on a number of sweeps of the scanner it does not detect the target within the plotting gate, it will start a process of increasing the plotting gate area until it picks up the target again. By this method it will find the target when it has altered course (or speed) and ensure that it continues tracking it. If the tracking gate was enlarged to find the target under such circumstances, once its movement has been determined, it will start reducing the size of the plotting gate again. If, after increasing the size of the plotting gate it still does not find the target (or another target, note the comments on target swop below) it will activate a 'lost target' Alarm.

Target Swop

It is this plotting gate which can lead to the phenomenon of 'target swop'. This is where another target comes within the area of a target's plotting gate; this can be an unacquired target or an acquired one. The computer may not distinguish which of the two targets within the plotting gate area is which. As a result of this the tracking can shift from one target to another, or even two acquired targets can swop vectors. Often a weak target can lose its vector to a stronger one passing close by; it can happen that an acquired strong target can pick up the vector of a weak target in addition to its own.

Certain makes of radar/ARPA have a system whereby if two tracked (acquired) targets' plotting gates overlap, it will stop tracking the targets while in the shared gate. It will also do this in cases where there is one tracked target and another untracked echo gets into the its plotting gate. In such cases it will continue calculating the motion previously calculated (and showing the vectors) until they are clear of each other, and then resume tracking, thus avoiding, or at least reducing the chance of target swop.

This phenomenon of target swop is less common that it used to be on older ARPA sets, but there is no way to prevent it entirely and watch-keeping officers need to be fully aware of it. In one case where a large container ship collided with a yacht, one possible conclusion was that the yacht's echo, which had been spotted and acquired by the container ship's OOW, had 'swopped target' with the sea clutter echoes and appeared to be passing clear, when in fact the yacht, the echo of which had lost its vector to the sea clutter, was actually on a collision course with the ship.

Vector mode, selection and adjustment

The 'vector' produced for an acquired target by the ARPA/ATA is in the form of a graphic line, emanating from the target representation (spot) on the screen. The vector can be 'true' or 'relative' indicating the predicted movement of the target in the sense of its true course and speed or its track and rate of movement relative to own ship as a fixed reference point. It should be noted that under the most recent IMO standards for ARPA/ATA, the vector symbol is a dotted line, whereas many older systems show it as a solid line.

The vectors must be adjustable for length as a result of the time selected for them. In other words the end of the vector indicated the predicted position of the target at the end of the set vector time; if the vector time is set at 12 minutes, then the target will be at the position indicated, the end of the vector, in 12 minutes, presuming there is no change in target's or own ship's courses and speeds.

The operator must be able to select 'true' or 'relative' vectors no matter what the display mode is, whether relative or true. It should be noted however that it seems to be increasingly the case that the vectors will revert to the mode of the display after a few minutes if they have been changed to the opposite mode; if the display mode is relative, and the operator has changed the vectors to true, after a few minutes the vectors will revert to relative automatically. This is an important point and watchkeepers should be very aware of this feature.

In addition to the vector the readout on the side of the screen gives

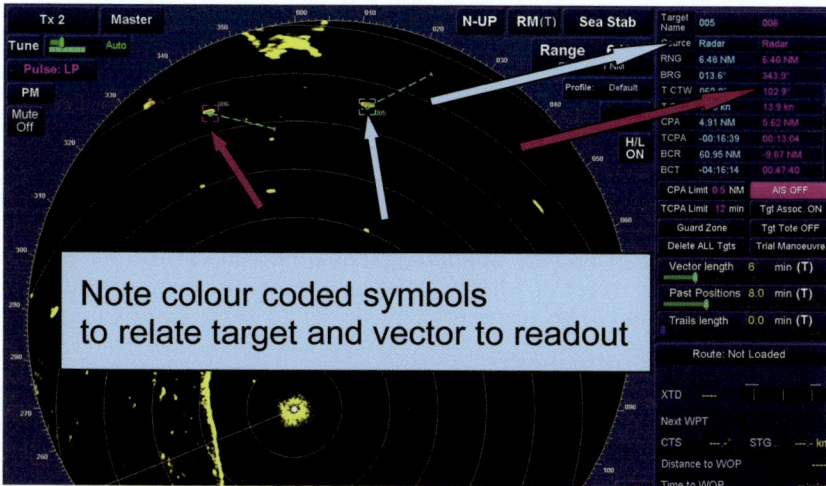

FIGURE 12.4   Radar screen showing vectors and readout.

the data on the target. Depending on the make of the ARPA/ATA, this readout will only show data on one or possibly two targets. The target who's data is required must be designated by the operator putting the cursor over it an 'clicking' it. The IMO specification is that the target who's information is being displayed in the readout will have a symbol in addition to the vector. There may be a colour coding where two targets' data is displayed to distinguish one from the other. On older sets where more than one target's data is displayed then the operator may have to refer to the target number or designation.

Warnings and Alarms

Lost Target
    As stated above, during the tracking process, if the tracker loses a target, an alarm, both audible and visual will be activated. The sound is usually a high pitched buzzer. The visual alarm is an alarm readout, for example 'Lost Target' appearing in the screen area around the radar display, in conjunction with a flashing symbol over the last position of the target and possibly a flashing vector symbol as well. The alarm can be acknowledged by the operator, which silences the audible alarm, but it usually requires the operator to actually mark the last plot of the

target and 'unacquire' it before the lost target symbol will disappear off the screen.

Guard Zones

These can take three main forms. One form is an area around the origin, the size of which can be set by the operator and when a target enters it the alarm is activated as well as the target being acquired, if it has not already been (see Figure 12.5a). Another is a system, of usually two circular segments, the angular extent of which can be set by the operator and the range of which can be also set by the operator (at least for the outer one). The segments can be set to form complete circles

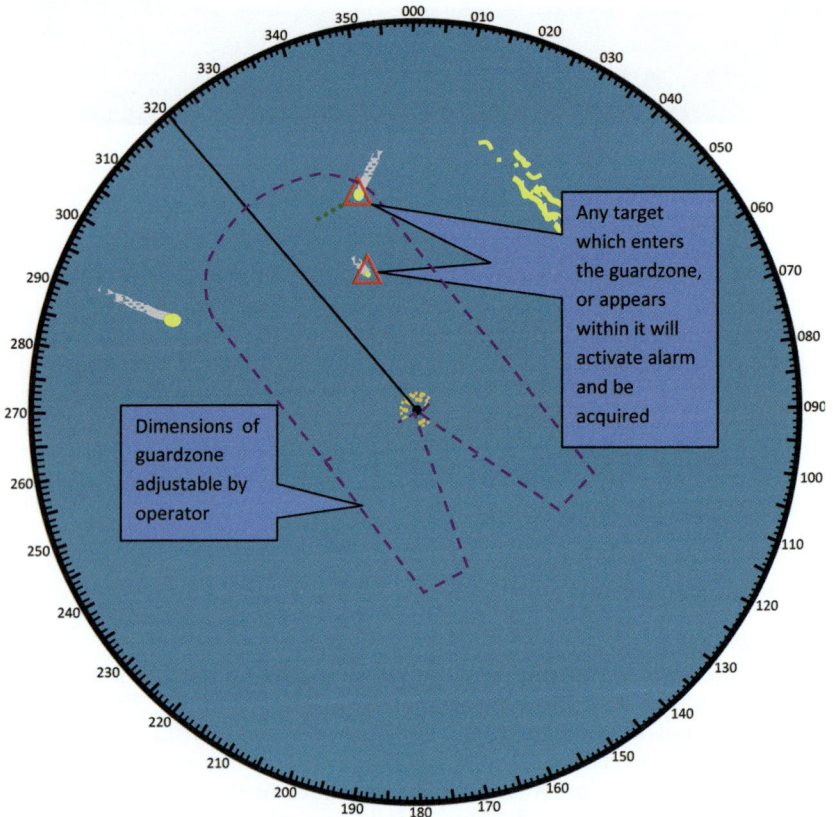

Any target which enters the guardzone, or appears within it will activate alarm and be acquired

Dimensions of guardzone adjustable by operator

FIGURE 12.5a   Guard zones, area.

around the origin, if required (see Figure 12.5b). The third system is two 'rings', at different ranges, which can be set by the operator. In this system, two lines are available which are called 'Area Rejection Boundaries' or 'Exclusion Lines' beyond which targets will not be picked up by the guard zones (see Figure 12.5c).

As stated, when a target enters the guard zone an alarm is activated. This will have an audible form, a readout around the edge of the screen, and a symbol, usually a flashing one, appearing over the target echo. The alarm must be cancelled by the operator. If the target has not already been acquired, the guardzone process will acquire it and soon a vector will appear for that target.

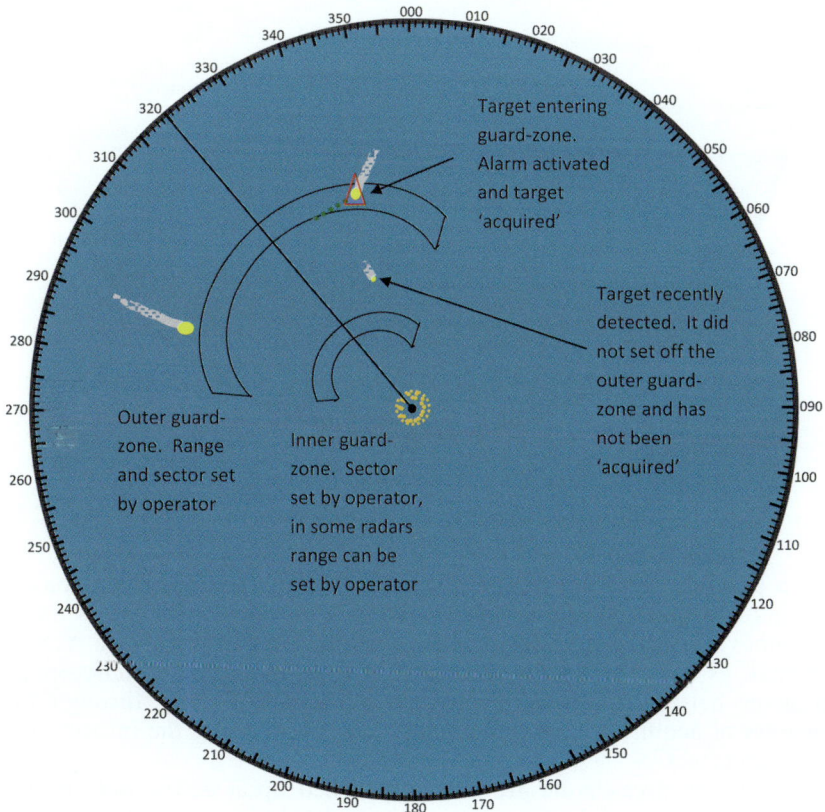

Target entering guard-zone. Alarm activated and target 'acquired'

Target recently detected. It did not set off the outer guard-zone and has not been 'acquired'

Outer guard-zone. Range and sector set by operator

Inner guard-zone. Sector set by operator, in some radars range can be set by operator

FIGURE 12.5b   Guard zones (segments).

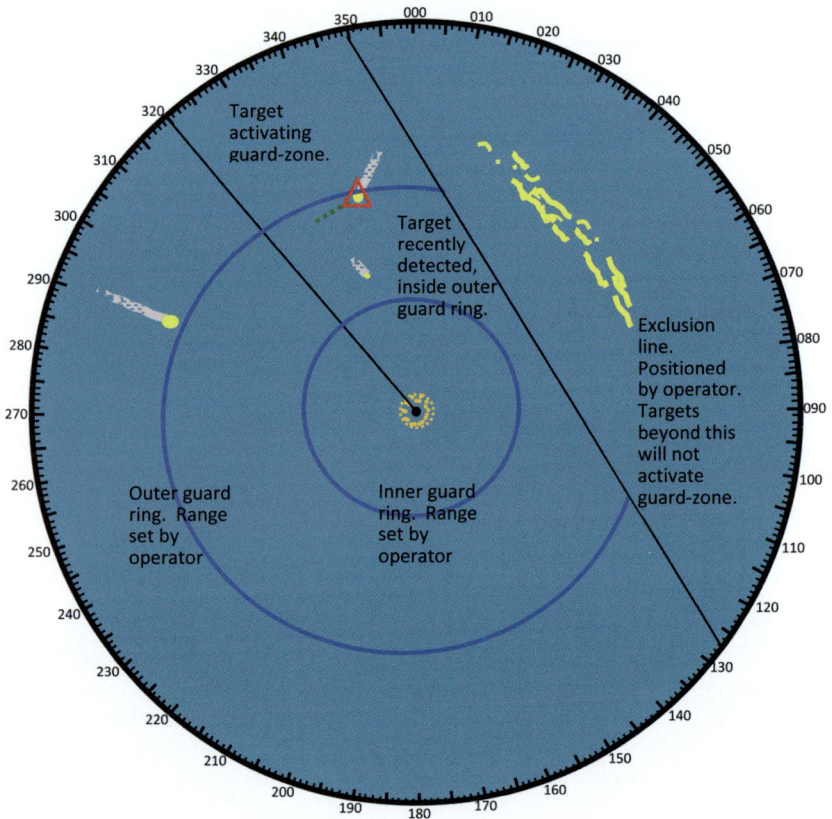

FIGURE 12.5c   Guard zones, rings.

In practice, guard zones tend not to be used very much. In coastal waters too many spurious echoes cause it to alarm constantly and it becomes a distraction. This includes sea and rain clutter echoes. As well as that, as discussed in the section on operation and on watchkeeping, it gives better situational awareness if the OOW goes through the process of acquiring targets manually and assimilating the information on each target.

The guard areas/zone are not to be relied upon as the only means of target acquiring/alarming. Using them should not allow the OOW to reduce vigilance in other aspects of good watchkeeping such as keeping a good visual lookout. They can also suggest a dangerous practice of depending on them to give warning of approaching danger while the OOW is engaged in some non watchkeeping function.

CPA TCPA limits

If an acquired target is going to pass own ship within a distance set by the operator, and within a time set by the operator, an alarm will be activated. This also will take the form of sound, a readout and a flashing symbol over the target echo. This also must be cancelled by the operator, but usually the flashing symbol will remain as long as the target fulfils the conditions set, that is, as long as it's CPA and TCPA are within the limits.

This is a useful function, but in situations of heavy traffic is too can become a distraction. Under those circumstances it may be necessary to set a very low CPA value or short TCPA time.

It should be remembered that this function only works for acquired targets; if a target has not been acquired it may well be within the limits and no alarm will be activated.

Trial manoeuvre

The operation of the 'trial manoeuvre' function is discussed elsewhere. The IMO requirement is for the ARPA equipment to be able to display predicted positions and vectors of acquired targets as a result of alterations of own ship's course and/or speed, either immediately or at some time in the future, usually within 10 or 15 minutes. It is mentioned here in so far as some sets have an alarm that sounds and shows a readout when own ship has arrived at a point where an alteration of course or speed was set in the trial manoeuvre function. This alarm will sound even when the trail manoeuvre display is not showing. The trail manoeuvre function must not interrupt the continuous tracking and updating of the plot which is immediately restored when the trail manoeuvre display is cancelled.

System alarms

There are various other alarms usually fitted such as failure of various inputs such as gyro, log, GPS, etc. Other internal tests are carried out automatically and any failures of there will cause alarms to be activated.

With regard to alarms for input failures, watchkeeping officers should be aware that these have limited capabilities. For example, a gyro can be malfunctioning and giving an incorrect heading; the radar computer system has no way of determining if this incorrect heading is not the correct one. Also, the gyro may be 'wandering' and if the rate of wander is not excessive, then it will appear to the ARPA as an alteration of course. Similar serious errors can occur with other inputs such as log and GPS with similar possible serious consequences.

Conclusions

In this chapter we have discussed the process of designating a radar target to be tracked by the computer, called 'acquiring' the target, and the different methods of achieving this. We have looked, in simple terms the process of filtering or smoothing the computed data presented to the operator. While the processing of the data may be 'true' or 'relative', this is not normally apparent to the operator, but the operator must be aware of the limitations resulting from each of these methods.

Target swop is less of a problem with modern tracking systems, but it cannot be entirely ruled out as a danger in relation to tracked targets and the dangerous effect this might have on the plot and decisions which ensue.

Warnings and alarms are useful functions, but need to be treated with caution. Some are more useful than others. If not correctly used they can become a distraction and this then diminishes their usefulness. Under such circumstances the tendency is to just cancel the irritating buzzer without taking in its significance. Other expected alarms may not give warning under certain circumstances such as gyro or other input failure.

Chapter 13

Radar and ARPA Error

This chapter examines an important aspect of radar observation and use. The watchkeeping officer must have an appreciation of the limitations of the equipment. This chapter examines the errors produced by:

- the size of the pulse,
- the random point of echo on targets,
- the positioning of the scanner and ship movement,
- target movement, and
- by the processing of the data.

Several points in this chapter are also mentioned in the chapter on ARPA target tracking.

Every watchkeeping officer must be fully aware of the degree of accuracy which radars and ARPAs are capable. While an ARPA display can produce data with a good deal of precision, such as giving a CPA to two places of decimals of a mile, this does not mean that this figure is accurate; the degree of accuracy might only be anything within a mile in certain circumstances. The ARPA computer is capable of very rapid and precise calculation, but the raw data on which this calculation is based is of limited precision and accuracy.

In the following discussion it is assumed that everything is working correctly and all input and settings are correct; the discussion is on the inherent accuracy of the system, not faults or mistakes.

Pulse dimensions

The first thing to consider is the size of the pulse. This defines the smallest theoretical accuracy of the measurements of which the equipment is capable. Any object returning an echo cannot return an echo smaller than the size of the pulse, no matter how small the object itself is. The size of the pulse is defined by the pulse length and beam width. Pulse length is controlled by the modulator, and this in turn is decided by the range scale in use and the pulse length selected by the operator (see Chapter 5, 'The Display' and Chapter 9, 'Setting up a Marine Radar'). Beam width is fixed by the design of the antenna and of course the pulse widens with distance from the scanner.

195

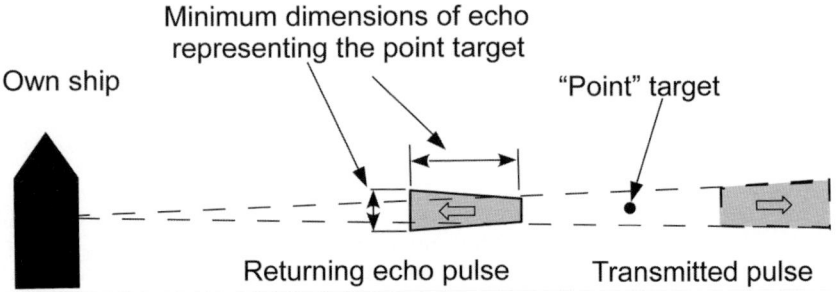

FIGURE 13.1a   Dimensions of pulse.

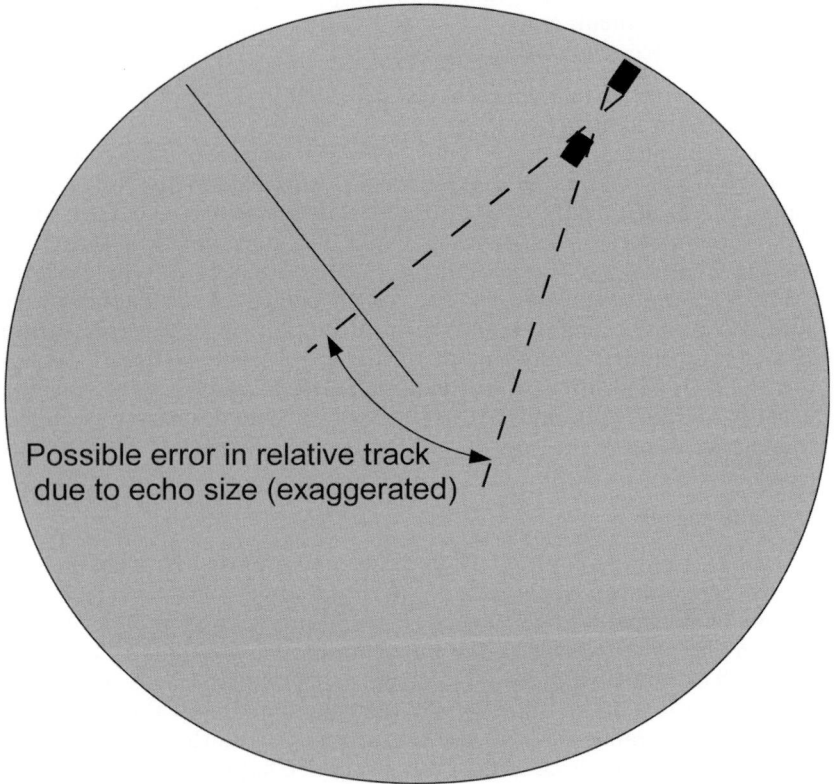

FIGURE 13.1b   Possible error in relative due to echo size.

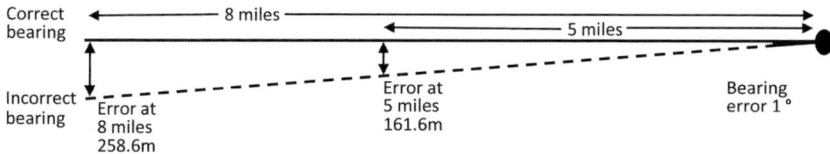

FIGURE 13.2   Bearing error.

Taking a theoretical pulse length of 0.5 microseconds, this gives a pulse length of 300 × 0.5 metres, which is 150 metres.

If the beamwidth is, for example 0.5°, then at 5 miles the beamwidth is 5 × 1852 × sin (0.5) metres, which is 81 metres.

Therefore, theoretically a 'point' target, no matter how small, at 5 miles from the ship cannot be displayed as anything smaller than an object 150 metres by 81 metres. Of course, there is no such thing as the theoretical point target, any object returning a receivable echo will have a dimension that will probably produce an echo larger than this, and it is from such an echo that range and bearing is measured.

The pulse size also has implications for the degree of precision which the radar can give in taking bearings and ranges. The echo returned from the theoretical 'point' target is the very limiting factor in such measurements. This of course is ignoring for the moment the actual size of any target and the size of echo or group of echoes which it returns. The IMO requirements are for the equipment to be capable of a maximum system error of 1% of maximum range of the range scale in use or 30 m whichever is the greater distance, and bearings with a maximum system error of 1° at the periphery of the display. Just taking the allowable bearing error into account Figure No. 13.2 gives some indication of the significance of this error.

Target response

The next thing to be considered is where the actual response comes from on a target object. A ship, even quite a small one, is a complex echoing surface. Where, precisely on the ship is the echo return actually coming from? Is it from the bows?, possibly the bridge-front, the ship's side. One pulse might be from one part of the ship, the next from another part; and this is ignoring, for the moment, the movement of the ship which further complicates the situation. This phenomenon is called target 'Glint', and it means that the response or echo representing the target is a series of random echoes around a small mean or average point (see the comments below on 'centre of gravity'). Obviously the movement of the ship (or other object) in a seaway, its rolling and pitching, increases the random error produced by 'glint' but it should be noted that 'glint' is present in stationary objects, albeit to a lesser degree.

Antenna (scanner) movement

Now we must consider the antenna of our own ship. Placed high above the waterline, in any kind of sea-state, it is describing arcs or ellipses in the air of a fair degree of magnitude. If, for example the antenna is 20 metres above the waterline, and the ship is rolling through 10°, that is 5° each side of vertical, then the scanner is moving 3.5 metres from side to side with each roll. This means that the echo from a target when the scanner is at one extremity of the roll can be 3.5 metres away from an echo from the same target when the scanner is at the other extremity of the roll.

Displacement of antenna during rolling

FIGURE 13.3   Error due to scanner movement.

Quadrantal error

This an error produced by a bearing measuring device such as a pelorus, gyro repeater azimuth ring, or in our case the scanner when the plane of the device is not horizontal. The effect of this will produce bearing errors, depending on the degree to which the device is out of the horizontal. Such error will be zero on the axis of the rotation causing the device being not horizontal, and zero at right angles to that axis. Say for example the ship is rolling from side to side the error will be zero ahead and astern and on the beams but maximum on the bows and quarters.

All of these are random errors and as such can be reduced by repeated observation, but only to a certain limiting point. No amount of repeated observation can reduce the target size, and changing circumstances can change the degree of random error in various ways. For example the changing aspect of an approaching ships can change the apparent target size (e.g. it was initially 'End-on' showing a small profile, but is now showing more of a side view presenting a larger profile) with the implication this has for range and bearing measurement and this will also change the degree of 'Glint' error value in the estimation of the mean position of the target.

In many ways, in the era of manual plotting, most of these random errors were hidden in the basic inaccuracy of the plot. As discussed in the chapter on manual plotting the degree of accuracy is very poor, even not allowing for plotter error and mistakes. The use of a chinagraph pencil on a reflection plotter reduced the error in transferring data from the PPI to a plotting sheet, but anyone who has looked at the thickness of the chinagraph drawings and the small size of the subsequent triangles to calculate target data could not but be impressed by the lack of precision of the system. This, in turn prompted caution in analysing the data presented by such plotting techniques.

Now we are in the computer age. The machine can rapidly and very precisely carry out the calculations and eliminate certain aspects of error problems. But, like everything in life, the new technology brings new problems, not least those brought about by the benefits of the technology, both in is functioning and presentation of data and in its use and interpretation by bridge personnel.

Processing errors

In considering ARPA errors, as opposed to radar errors there are several points to be considered:

- The first is the process by which the computer locates an echo, and then decides what point of the echo represents its actual position.

- In processing the data in range and bearing, an error called 'quantization' becomes involved. This is the 'rounding-off' of data to the nearest value that the digitising process of the computer can handle. If, for example the range of an echo is 5.436 miles, and the range register only allows for two places of decimals of a mile, then the range will be recorded as 5.44 miles.
- Having located or 'acquired' a target, the computer then must calculate the necessary data.
- In the process of carrying out and presenting the data in a usable form, filtering, or averaging is required and used. This reduces accuracy.
- When the circumstances change, such as the target altering course, the necessary filtering can delay the presentation of the change.
- The filtering process also produces errors in the plot when own ship alters course.
- Random gyro and log errors will be added to the other random errors.

Locating the target

Different makes of automatic plotting systems, generally called TT (target tracking) or ATA (automatic tracking aids) including automatic plotting aids which do not meet the full IMO specifications for ARPA, use different methods for acquiring and locating a target. Two methods are invariably used to set the computer process in motion to lock-on to a target; the first is the action by the observer in marking or 'acquiring' the target by joystick or roller ball and cursor on the screen. The second method is by the guard zones. (See previous Chapter 12, 'ARPA and Target Tracking'.)

Whichever of the two methods have set the computer working, it now has to work on the data indicated, a series of return range echoes on a series of bearings which compose the target (called a 'hit matrix'). These can be represented by a close and overlapping series of echoes. The computer must now calculate what point represents the actual position of the target. Some makes of equipment will work on a 'centre of gravity', i.e. the mean or average position of the series of echoes; others will work on that part of the group of echoes closest to own ship. In either case it can be seen that these points can change as the situation develops; more or less echoes composing the target due to the various factors mentioned above can change the 'centre of gravity' or the closest point, and the closest point can be changed by changing aspect. All in all, within the 'blob' which represents the target, the point of measurement is constantly changing.

Filtering

Such an erratic behaviour of the supposed position of the target is unusable, and therefore the first stage of filtering occurs. This process takes all the positions over a short period into account and produces a mean position which is considered 'the' position of the target, and on which further calculations will be based.

(This section should be read in conjunction with the previous chapter on the functioning of ARPA and target tracking, particularly the part on 'rate aiding' and 'target swop' which are not really errors within the meaning of the word and the theme of this chapter.)

Having refined the target's position into a usable point, the next consideration is the target's movement. This also is subjected to a certain amount of error, due to factors mentioned above. If plotted on a sufficiently large scale, the target would seem to zig-zag erratically, even though it is in fact on a constant course and speed. If, at this stage a vector was displayed without any further refinement, it would be constantly flicking from side to side and lengthening and shortening very rapidly, and be completely unusable. Once more a filtering process calculates the mean course and speed, and displays a reasonably stable vector for the target and the readout of the target's data (CPA, TCPA, target course/speed, etc.).

The filtering process of course reduces the accuracy of the data and can obscure certain information. The most critical of these is the situation where the target makes an alteration of course. Initially the filtering process will assume that the displacement of the target off the mean course is just one of the aberrations which it is designed to filter out, and therefore does not display the change in course. Also the computer works on a process that anticipates the next position of the target at next rotation of the antenna (rate aiding). This is designed to reduce the occasions of lost targets. Between the two of these processes it can take quite a few seconds, and possibly a few minutes of time before it becomes apparent that the target has in fact altered course.

Own ship making a substantial alteration of course, particularly at normal speed, can upset the filtering process and hence the plot quite noticeably. Immediately after the alteration it can often be noticed that the vectors on most, or all targets have changed, maybe not by a huge amount, but certainly by an amount that is quite obvious. After a few minutes these should settle back into their previous condition, assuming the targets have not, in fact altered course or speed.

What does all this mean, in practical terms, to the officer of the watch observing the radar in actual watchkeeping situations? In the first instance it means that the apparent precise data presented by the

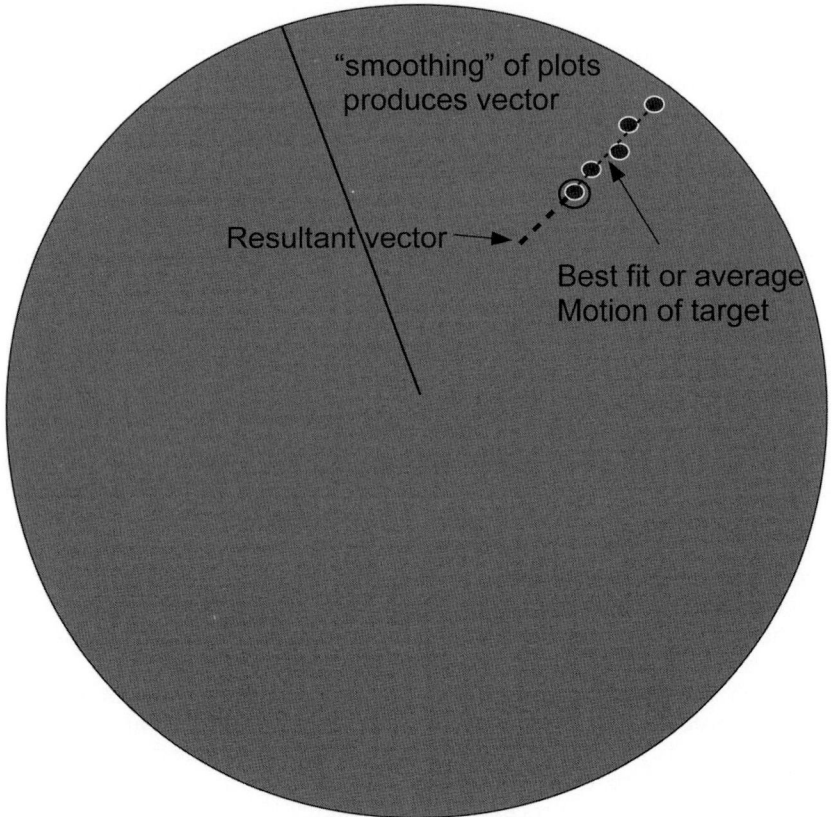

FIGURE 13.4   Target filtering.

equipment must be treated with caution, an indicated CPA of 0.2 miles might be anything from zero to 1 mile or more. Targets on slow <u>relative</u> rates of approach are likely to have greater error than fast ones.

By the time the display shows that a target is altering course, it probably has already completed the alteration. A useful exercise is to observe situations in clear weather when a ship is seen to alter course, and compare it to the ARPA vector and read-out.

When own ship makes a substantial alteration of course the data, whether vectors on the display or read-out must be treated with great caution until the ship has settled on her new course.

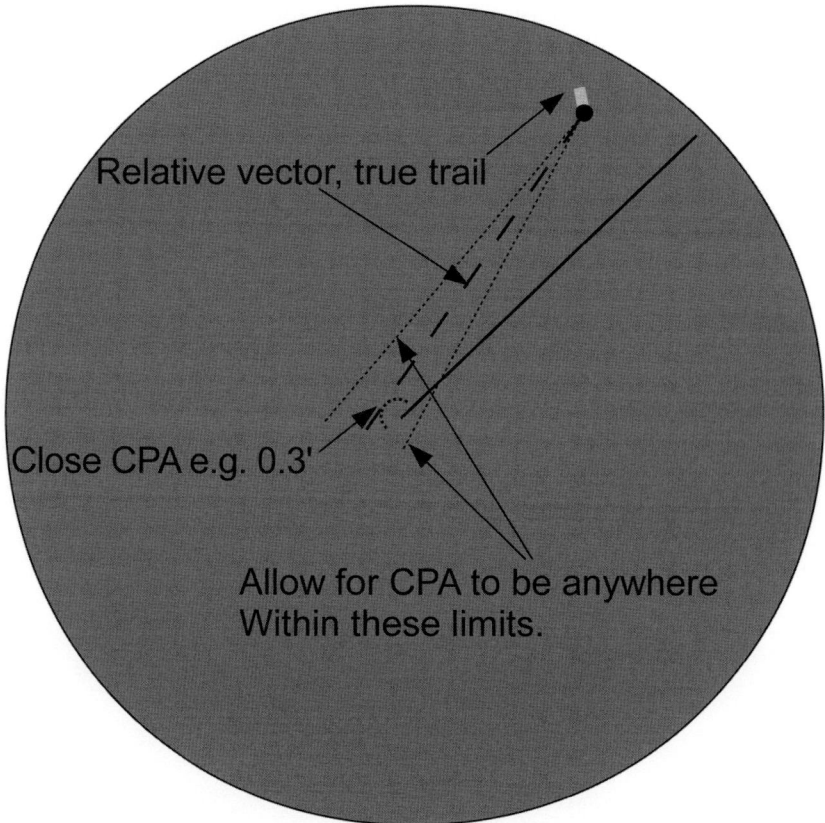

FIGURE 13.5   Allowance to CPA.

Where the ARPA gives a CPA of less than 1.0 mile, add at least another mile, but in this recognise that the CPA could be on the other side of own ship than indicated.

New Technology (NT) radar

New Technology radars claim greater accuracy than the traditional radars. However, much of what has been discussed above remains valid. While the detection of a target within the returning very long pulse can give greater range accuracy than purely detecting the leading edge of the much shorter pulse of traditional radar, beamwidth is still a factor as is the build up of a series of random echoes to produce a 'target'. All the variables such as scanner height, ship movement, glint and so on are still present in the basic data on which the computer works, and the errors produced by the filtering process are still to be taken into account by the radar observer.

Conclusions

The size of the pulse provides some idea of the theoretical accuracy of the radar.

The point of echo from any target produces an effect called 'glint' which is a random movement of the point of echo.

The movement of the scanner produces substantial error including quadrantal error.

The point on a series of echoes representing a target from which the target tracking calculates its movement varies randomly.

Filtering produces errors in target vectors and data.

The degree of accuracy must be kept in mind, particularly with regard to close CPAs.

Chapter 14

Integration of Radar and Electronic Chart Displays

This chapter must be read in conjunction with Chapter 16, 'Radar for Navigation'.

In this chapter the following points are discussed:

- Advantages and disadvantages of radar overlay on an electronic chart display.
- Radar as a positional back-up on the electronic chart.
- Chart underlay in chart radar.
- Collision avoidance data displayed on an electronic chart display.
- Separate work-stations.

Today's radar operation includes a large variety of user functions. Over the past decades, new functions have been added to improve radar performance, and also to integrate radar with other navigational systems. In the opinion of some experts, some functions have only historic and 'we-always-used-radar-this-way' meaning, but such attitudes need to be treated with caution by practicing mariners. There is a tendency by the developers of such systems to tell mariners what they need, rather than ask them for their opinions. This has the result of making it increasingly difficult to retain traditional navigational and collision avoidance methods and skills as a 'fall back' in the case of equipment failure.

By the introduction of satellite position fixing, ECDIS and AIS, the role of radar has changed, and there are two basic consequences for the user:

- All relevant data of own ship and targets must be correlated to allow an unambiguous and easy situation analysis.
- A number of conventional functions and display modes may no longer be available with the intention of 'simplifying' the system.

Radar overlay

Transferred radar information can contain both radar image and ARPA (or tracker) information. If ARPA information only is transferred then only the acquired targets will appear on the SENC (electronic chart) display. These will appear as a symbols, usually with the ARPAs target identification, and a vector. The vector mode and length can usually be adjusted separately on the electronic chart display from that on the radar (the radar display could have relative vectors while the chart display has true).

If radar video is transferred, the radar 'picture' is superimposed on the electronic chart. This can be done in addition to, or instead of the transferred ARPA data. In this, all the radar picture, including unwanted echoes such as sea clutter, rain and spurious echoes will appear on the electronic chart.

Obviously, if the radar image is added to the ECDIS display, the chart and the radar image must match in scale and orientation.

The radar image and the position from the position sensor should both be adjusted automatically for antenna offset from the conning position (CCRP).

It must be possible to adjust the display position of the ship manually so that the radar image matches the SENC display. This needs to be treated with great caution, as one of the benefits of the radar overlay is a confirmation of ship's position (see below).

It should be possible to remove the radar information from the electronic chart by single operator action for ARPA data and for radar video.

Radar as position back-up by ENC object correlation

It is essential that not only ARPA data, but raw radar data are integrated with chart data. Integrating an electronic chart with radar, either overlaying radar data on the electronic chart or at least enabling the user to underlay some selected SENC information on the radar display (chart radar), provides a system which can bring the two principle watchkeeping objectives together, keeping the ship in safe water and collision avoidance. The mariner can receive the following relevant information in one display:

own ship's position,
all other vessels in the sea area (or at least those detected by radar) and
all charted objects and obstacles.

The radar-ENC overlay and object correlation are of particular value:

- For position fixing because, in essence, the overlay provides a permanent and continuous radar fix without any action on the part of the watchkeeper provided there are suitable and correctly identified radar targets in the vicinity.
- For real time monitoring of the ship's Electronic Position Fixing System (EPSF) e.g. GPS, again provided there are suitable and correctly identified radar targets in the vicinity.

As long as the radar echos and corresponding ENC object match, own ship's EPFS (GPS) is reliable. However, if the radar echoes of fixed objects do not match the ENC objects, but are shifted for some distance, there is probably an error in the EPSF, potentially in the geodetic datum. Moreover, if the radar picture and the ENC picture are rotated against each other, there is obviously a directional (gyro or compass) malfunction.

Thus the radar may serve as a second and independent position fixing system, in its own right and as a check on the EPFS provided there are sufficient suitable radar targets in the vicinity. It has been proposed that radar is the only secondary position fixing system required; this is currently a contentious point as GPS is the only global positioning system and the debate on the need for a secondary automatic positioning system is ongoing.

The ECDIS-radar overlay is particularly relevant in automatic track control (TC) mode where the current GPS position may be always displayed on the planned track, on the ENC, independent of the actual GPS error. This might seduce the mariner to over-rely on the system. A radar overlay showing discrepancies between radar and electronic chart objects will immediately indicate danger if there are suitable and correctly identified radar targets in the vicinity.

Accuracy limitations

Particularly for large commercial vessels (and also for smaller ones) a scaled outline of the vessel will appear on the ECDIS screen when the scale of the chart can show it. The vessel's display size will adjust with the operator's changes in display scale. This helps to maintain the navigator's perspective of his/her vessel with respect to the surrounding navigational features (e.g. width of channel in port approaches). Such a feature will also apply to many radar displays, particularly when chart underlay is being used.

The overlaid radar display must automatically adjust to the scale of the ECDIS display in order to prevent misinterpretation.

Radar beam width and pulse limitations will limit the correlation accuracy. For example, the radar echoes of navigational buoys can appear as comparatively large 'blobs' in relation to the chart symbols for the same objects. This is one of the reasons why radar as a positional back-up has limitations which need to be taken into account.

In addition to the accuracy limitations, the presence of clutter, false echoes, and possible mistaking radar objects for the wrong navigational feature, reduce the effectiveness of radar providing the positional back-up to the EPFS (GPS). It points to the necessity of watchkeepers maintaining and using traditional and all methods to confirm the ship's position.

Radar with chart display (chart radar)

According to current radar performance standards (2004) the radar system may provide the means to display vector chart information (ENC) within the effective display area to provide continuous and real time position monitoring. It must be possible to remove the display of ENC data by a single operator action. The ENC information should be the primary source of information supplied in S-57 format. Unofficial electronic chart information must be identified with a prominent indication. There must be a minimum set of electronic chart objects that should be displayed if chart data is available:

- A user selected own ship safety contour.
- An indication of isolated underwater dangers of depths less than the safety contour.
- An indication of isolated dangers which lie in safe water defined by the safety contour such as bridges, overhead wires, etc. and including buoys and beacons, whether or not these are being used as aids to navigation.
- coast lines must be shown.

Means must be provided to enable the user to select parts of the available electronic chart information. Such information may only be selected on a class or layer basis, not as individual objects (e.g. soundings).

There must be an indication of the electronic chart status in terms of source, authorisation and update information.

The chart information must use the same reference and co-ordinate criteria as the radar/AIS system. Means to align the radar and chart information is permitted. The application of such alignment must be clearly indicated. A simple alignment reset facility should be available.

The display of radar information should have priority. Chart information should be displayed in such a way that radar information is not substantially masked, obscured or degraded. Electronic chart information should be clearly perceptible as such.

A malfunction of the source of electronic chart data should not affect the operation of the radar/AIS system. Symbols and colours should comply with the IMO and IHO standards.

ARPA information

ARPA features are generally not fully integrated into ECDIS displays. Watchkeepers must maintain caution that ARPA vectors are usually based on relative motion and speed through the water, while ECDIS is operating in a speed and direction over the ground mode. For this reason, while ARPA or tracker derived targets can be displayed on the ECDIS display, the anti-collision plot should remain the radar display, and anti-collision manoeuvres should not be assessed on the ECDIS.

The use of the ARPA derived targets on the ECDIS can improve situational awareness, the ability to see where other ships are in relation to the navigational situation obviously improves the ability to assess the anti-collision situation. The assessment of the anti-collision situation, and the possible threat posed by other ships can generally only be done safely on the radar/ARPA display.

Route and track information

In many systems the route entered into the electronic chart system can be displayed on the radar display. This is usually independent from the chart underlay function. Where available it must be displayed or hidden by a single operator function.

Separate workstations

The tendency among developers of integrated chart and radar systems has been to emphasise the possibility of one display, or workstation, providing all the required navigational and collision information to the watchkeeper. The IMO regulations are ambiguous on this point but seem to require two displays, one in which radar is the predominant source of information, in other words a radar display, and the other which is predominantly a navigational chart display. For example in the regulations relating to a radar display, chart data may be displayed, but must be subordinate to the radar picture, and limited in the amount of data it may display on a radar screen. Similarly radar overlay on a chart display must not inhibit or obscure chart data. In an integrated system, where the function of a monitoring screen or station can be changed, for example, under certain circumstances it can be an electronic chart display and under other circumstances it can be a radar display, it must fulfil all the requirements laid down by IMO for a radar display in its functioning, controls and presentation of display and data.

The development of bridge layout and equipment, and bridge watchkeeping practices are evolving very rapidly. It means that different shipping companies are putting different bridge configurations and practices into place. While developments into 'e-navigation' and 'S-mode' may serve to bring about some kind of universal bridge layout and watchkeeping practices, for the moment mariners must adapt the traditional practices as best as they can to the modern and changing situations.

Conclusions

- The integration of radar and ECDIS provides enhanced situational awareness.
- Radar overlay on an electronic chart can provide continuous, 'at a glance' positional confirmation in certain circumstances.
- Radar overlay on an electronic chart as a positional back-up has serious limitations and must be treated with caution. Traditional methods must not be neglected.
- Chart underlay on a radar display can provide improved anti-collision assessment.
- Similarly ARPA target display on the ECDIS can improve anti-collision assessment, but must not be used as the main anti-collision plot.
- There is a need for two distinct displays, one for the navigation of the ship and keeping it in safe water, the other for tracking and monitoring traffic and avoiding close quarter situations with such traffic.

Chapter 15

AIS

One of the most significant developments in recent years is AIS. Its effect on navigation, collision avoidance and marine traffic is still evolving. The fact that so much information about other ships in the vicinity is so readily available has changed the shipping environment significantly. Like the impact which radar had when introduced aboard the bridges of most ships, many years ago, it will take time for the procedures and safe usage to evolve and the training needs to clarify. There has already been AIS assisted collisions. However, unlike the early days of radar, the need for training has been recognised from the start although without its existence being recognised in the IRPCS, this training must remain somewhat curtailed and speculative.

In this chapter the following aspects are examined:

- The regulatory requirements for AIS.
- The information handled by AIS.
- The technical features of the system.
- AIS and Aids to Navigation (AtoNs).

Regulatory Requirements

AIS is an abbreviation for Automatic Identification System. It is a based on the marine very high frequency (VHF) radio system and it automatically communicates between units using a transponder system. That is, it is an automatic device that transmits a predetermined message and if required, will respond to a predefined received signal. The International Maritime Organisation (IMO) requires that all vessels complying with the SOLAS convention must have this system fitted. Other vessels such as fishing boats and leisure craft are not fully required to have this system fitted as yet. Although primarily used for the safety on navigation it is capable of including information in a wide marine environment, including VTS centres, Coast Radio Stations (CRS), Port Authorities and Lighthouse Authorities.

AIS units fitted to ships automatically and continuously transmit up-to-date navigational information. This data is taken from the ship's own information systems and is transmitted to all stations in range.

The SOLAS Regulations, Regulation 19 of SOLAS Chapter V – Carriage requirements for shipborne navigational systems and equipment-state that AIS must be fitted to:

all ships of 300 gt and upwards on international voyages,
cargo ships of 500 gt and upwards on non international voyages, and
all passenger ships irrespective of size.

This can be found at:

http://www.imo.org/Safety/mainframe.asp?topic.id = 754

The AIS system is intended to enhance the safety of navigation by assisting in the efficient navigation of ships, protection of the environment and the operation of Vessel Traffic Services (VTS), by satisfying the following IMO requirements:

- As a means for coastal States to obtain information about a ship and its cargo.
- As a VTS tool, i.e. ship-to-shore and shore-to-ship (traffic management).

In addition AIS can:

Help identify vessels.
Assist in target tracking.
Simplify information exchange (e.g. reduce verbal mandatory ship reporting).
Provide additional information to assist situation awareness.
Assist in protection of environment through monitoring of vessels.
Be used in fisheries management and other offshore activities.
Integration with voyage data recorders (VDR) ensures a more complete record for analysis in the event of an investigation.

Technical Features

As AIS is a VHF radio system, operating on marine-band frequencies, it therefore has a similar range, depending on atmospheric and other conditions to normal voice communications VHF. Its transmission power is about half that of 'normal' VHF being 2 watts (low power) and 12.5 watts (high power). Typically, the range of AIS is around 30 nautical miles or more and is not normally affected by weather conditions. If the antenna is shielded by the ship's superstructure its range will be compromised.

Transmissions are on two frequencies, VHF channel 87B (161.975 MHz) and 88B (162.025 MHz). These are designated AIS 1 and AIS 2, respectively.

Each AIS transponder consists of one VHF transmitter, two VHF receivers, one VHF Digital Selective Calling (DSC) receiver, and links to shipboard display and sensor systems (GPS, Radar, ECDIS, etc.) via standard marine electronic communications (see Figure 15.2).

Ships using AIS automatically and continually transmit up-to-date navigational data. This includes 'Static, 'Dynamic' and 'Voyage' related data (see below). This data can be displayed on other ships and stations in a variety of ways and allows instant information about the transmitting ship to the operator. This includes ship's name, present course and speed.

The ITU (International Telecommunications Union) is the UN agency for information and communication technology and it co-ordinates global use of the radio spectrum and establishes worldwide standards for communications systems; this includes AIS. The technical characteristics of the AIS such as variable transmitter output power, operating frequencies (dedicated internationally and selected regionally), modulation, and antenna system must comply with these recommendations

The IEC (International Electrotechnical Commission) is the world's leading organization that prepares and publishes International Standards for all electrical, electronic and related technologies.

The specification for AIS therefore is governed by three international agencies under the following:

IMO Regulation 19 SOLAS+Annex of MSC.74(69).
Physical characteristics ITU-R-M1371.
Test spec IEC 61993-2.

Information Transmitted and Received by AIS

The AIS equipment:

- provides information automatically and continuously to a competent authority and other ships, without involvement of ship's personnel;
- receives and processes information from other sources, including that from a competent authority and from other ships;
- responds to high priority and safety related calls with a minimum of delay; and
- provides positional and manoeuvring information at a data rate adequate to facilitate accurate tracking by a competent authority and other ships.

Much of the data exchange does not concern the ship-board user. These exchanges have to do with the control by shore stations and SAR under certain conditions and situations and are concerned with frequency allocation and power, management of the 'assigned' and 'polled' modes for such purposes as VTS and SAR. These functions are normally 'invisible' to the user aboard ship and require no action.

For the shipboard user, information is managed under five headings:

1. Static data (ship-related information which rarely changes, such as ship's name, length, etc.).
2. Dynamic data (data which is continually changing, such as position and speed over the ground).
3. Voyage-related data (data which is specific to the voyage, such as destination port and estimated time of arrival).
4. Safety and security related messages.
5. Other user-relevant AIS messages (see Binary Messages).

Static data

This consists of data which normally needs to be entered only once into the ship's AIS system as it remains generally constant (i.e. static). It consists of the following entries:

- Maritime Mobile Services Identity Number (MMSI).
- Ship's name.
- Call sign.
- IMO number.
- Type of vessel.
- Length and beam.
- Location of the position fixing antenna (referred to bow and centreline).

This data is usually broadcast once every six minutes by each ship (unlike 'dynamic' data which is more frequent).

The MMSI is the unique identification of a particular ship and it is this, rather than the ship's name which appears at the transmitting station identifier for all transmissions. As the ship's name is normally transmitted every six minutes and the dynamic data can be transmitted every few seconds it can happen that a name may not be assigned to the transmission of such dynamic data, whereas the MMSI will always identify a transmission. Users should therefore be aware that the MMSI may sometimes appear in place of a ship's name.

The location of the position-fixing (GPS usually) antenna on board the ship is needed to ensure accuracy where large scale electronic charts or radars are being used to the extent that the outline of the ship

appears on other ship's or VTS displays. In the event of the external
position-fix system failing the AIS will use its internal system (e.g. its
own GPS receiver) to find the ship's position. It is therefore necessary
to provide the position of both the AIS and external position fix
antennas.

Dynamic data

This is the ship's current position and movement information. Most
of this is automatically supplied directly to the AIS unit by the ship's
navigation equipment. It consists of:

- Ship's position (with accuracy indication and integrity status, if
  available).
- Time in UTC.
- Course over ground (COG).
- Speed over ground (SOG).
- Heading.
- Rate of turn (where available).

There is a manual input under this heading which has to be made
concerning the navigational status of the ship, such as at anchor,
underway using engine, etc. (see table of such definitions below).

The accuracy of the positional data is indicated as either high or
low. High means that the positional accuracy is better than 10 metres.
The better accuracy is usually achieved by a differential system on the
position fixing system such a DGPS or differential E LORAN or other
such system.

Dynamic data is transmitted at regular time intervals. These time
intervals depend on the dynamics of the vessel, as below.

Ship's Dynamic Definitions	Transmitting Interval in seconds
At anchor or moored and not moving faster than 3 knots	180
At anchor or moored and moving faster than 3 knots	10
With a speed of 0 – 14 knots	10
With a speed of 0 – 14 knots and changing course	3.33
With a speed of 14 – 23 knots	6
With a speed of 14 – 23 knots and changing course	2
With a speed over 23 knots	2
With a speed over 23 knots and changing course	2

Note:

Experience has shown that many users are negligent in correctly entering the navigational status of the vessel. This is why the peculiar definition of 'at anchor or moored and moving faster that 3 knots' is provided. It should be part of normal arrival and departure routine to change the navigational status on the AIS.

Position Accuracy and Data Terminal Equipment (DTE):

PosAcc: High >10 m.
DTE: Available.

This information indicates that the vessel's AIS transponder is connected with a user interface and can show AIS Data. This function basically ensures, that the current transponder being used is fitted with a display and can therefore send and receive messages.

Voyage Related Data

This data is entered manually. This, as the name suggests changes with each voyage. It is normally entered at the start of a voyage, but may need to be updated in the course of that voyage, for example the 'destination' and consequently the 'estimated time of arrival' may change.

- Ship's (static) draught.
- Destination.
- Estimated time of arrival.
- Type of ship.
- Hazardous cargo type.

Number of persons on-board (not a requirement of the IMO and is therefore not included on all systems).

Sometimes the equipment has the facility to input a route plan (textual or with lat/long waypoints) as part of voyage-related data. This is an International Function Message (IFM) which is explained below.

The type of ship and hazardous cargo type are entered from a list of options and will be transmitted by the AIS system.

Safety and Security Related messages

An important part of the AIS shipborne equipment is receiving and transmitting safety and security related messages. These may be sent to an individual station (e.g. another ship), or broadcast to all stations. The message length is restricted to about 160 characters.

For a message to an individual station (addressed message) there is an automatic response given from that station stating whether the message was fully received or if there was a problem. This is usually made evident to the sender, although not all AIS equipment does this. (**Important note:** This is no guarantee that the message has been read and understood in the receiving ship/station.)

For a broadcast message (to all stations) there is no automatic response.

Messages can also be broadcast from shore base stations, or a selected group of base stations if there is a network. These again can be to selected ships or groups of ships to broadcast generally.

AIS and Aids to Navigation

An interesting development of AIS is its use to mark the positions of navigational aids or dangers such as buoys, beacons or isolated rocks. In conjunction with this it can provide additional information about the AtoN (Aid to Navigation). RACONs have traditionally been used to help identify certain AtoNs. AIS enhances this, but in some cases may in time replace RACONs particularly in the light of developments in new technology (NT) radar.

The advantages of an AIS RACON are:

It gives automatic precise position of the aid (e.g. buoy).
It can also transmit/display other information such as tidal or meteorological data.
It does not rely on radar and thus is not subject to typical radar characteristics.
After 1st July 2008 new fit radars are not required to receive S-band RACONs.

218

However, a conventional RACON's position is entirely related to the ship and its immediate environment and does not depend a position-fix system (GPS) unlike AIS. In this regard an interesting and important warning is contained in the Volpe Report outlining the vulnerability of GPS. It can be found at:

http://www.navcen.uscg.gov/archive/2001/Oct/FinalReport-v4.6.pdf

While the AIS transmitter may be located on the AtoN itself, AIS gives the facility of locating it elsewhere but still giving the location and other information about the AtoN. For the mariner this appears as if the transmission is from the aid itself; the symbol on his electronic chart display appears in the correct position for the aid. An on-shore station could broadcast the positions of several buoys in its locality each of which is too small to carry a transmitter itself.

The term *Virtual AIS*, also known as *Synthetic AIS* or *Artificial AIS* can be applied to the above application, It is more correctly applied where AIS is used to indicate a marker which does not physically exist. It can, for example be used to mark an invisible danger such as a wreck or submerged rock. Such transmission is of course also from some other location such as a shore station. Such a marker would be only 'visible' to AIS equipped ships.

The low cost and flexibility of the use of AIS for AtoNs will see its increasing use in this regard. Information on the aid can be easily updated; virtual aids can be located quickly in critical circumstances, for example to mark a new wreck; virtual aids can be located where it is difficult or expensive to locate physical ones. AIS provides an opportunity to improve the functionality of existing aids and develop new aids to navigation services. With an AIS AtoN system the user automatically receives, at the conning position, up to date information on the status of the aid to navigation.

As stated above AIS units can be fitted to any existing physical AtoN to provide real time status information direct to the mariner. Such information can include confirmation of correct operation of AtoN systems and of position in the case of floating AtoN. Additional information relating to tide and weather can also be provided where appropriate sensors are fitted.

Where AIS is provided as an overlay on ECDIS, Radar or other display systems, AIS AtoN are indicated by a Diamond shape with crossed lines at the reported position of the AtoN.

Where the AtoN is on station the diamond will be Blue and where the AtoN 'Off Station' flag has been activated the diamond will be Red.

In the case of a Virtual AIS AtoN there will be a V below the crossed lines.

AIS for Search and Rescue and Protection of the Environment

Ships governed by the terms of SOLAS are required to carry one or more Search and Rescue Transponders (SARTs). These are robust simple devices used to locate vessels in distress or survival craft. A SART will cause a series of 12 dots to appear on a searching ship's X-band (3 cm) radar display, in the direction of the SART. The range at which such a signal can be received will depend on the height of the searching craft's scanner and the height of the SART. About 7 miles is normal but ranges considerably more that this and even much less have been encountered. See Chapter 8, 'Target Response Characteristics'.

From 1st January 2010 an AIS SART (AIS – Search and Rescue Transmitter) or a conventional Radar SART may be fitted. The AIS SART will be able to send data using a standard AIS class A (see below) data.

The AIS-SART transmits data including position and time synchronization from a built in GNSS receiver and transmits its position with an update rate of 1 minute and will send a series of 8 reports within each minute.

AIS for VTS and Ship Reporting

While the most obvious and useful application of AIS is the information it provides on other vessels and on aids to navigation it has also been developed to assist shore authorities to monitor or control shipping. Some would argue that this was the primary intention of the development of AIS and its rapid adoption by the IMO.

Shore stations may simply monitor the AIS transmissions of vessels within their area of operation, such as a VTS or a state's coastal waters, or it may actively 'poll' vessels by means of the AIS system requesting data such as identification, destination, ETA, type of cargo and other information. Either of these activities are 'invisible' to the mariners in the ships. Coast stations can also use the AIS channels for shore-to-ship transmissions, to send information on tides, notices to mariners and local weather forecasts. Multiple AIS coast stations and repeaters may be tied together into Wide Area Networks (WAN) for extended coverage. This is the basis of the EU directive Safe Sea Net (SSN).

Coastal states may also use AIS to monitor the movement of hazardous cargoes and control commercial fishing operations in their territorial waters. It can also be used as a valuable tool to record data that can be logged automatically for playback in investigating an accident, oil spill or other event. It can also be used to gather statistical data, i.e. a means of analysing traffic of AIS equipped vessels for coastal states or international bodies like the IMO, or EU.

Similarly AIS can also be a useful tool in search and rescue (SAR) operations, allowing SAR coordinators to monitor the movements of all surface ships, aircraft and helicopters involved in the rescue effort.

AIS and VTS

AIS provides an ideal system for monitoring vessels through restricted harbours and waterways. Where traditionally such monitoring is done by radar and by ships reporting their positions and passage plans at various stages, AIS offers obvious advantages.

Compared to radar and visual coverage AIS can offer:

- Accurate information of vessels in area (course, speed, position, name, size, nature of cargo, etc.).
- Greater geographical cover than offered by radar and not requiring extensive outlay in radar installation/s.
- Greater positional accuracy than shown by radar echoes of vessels.
- No radar shadow, ships which are in parts of the waterway or harbour that may be behind high land for a radar installation are 'visible' to AIS.
- Reduced 'target swap'. The vector of one ship's AIS transmission will not transfer to another as readily as from one radar echo to another. It has been known however to happen but this apparently is a fault of certain earlier AIS units.
- Near real-time data.

Examining some of the above points in more detail:

Accurate Information

When a VTS is equipped with AIS it receives both the identification details, position and movement and other details of all AIS equipped ships in its area. This allows for a more complete overview of the port or waterway traffic. A major advantage is that it reduces the amount of voice radio traffic and thus takes some of the workload from ships' staff and pilots.

Geographical coverage

While VHF radio in itself does not give much greater range than radar, it is easier to place a VHF antenna in a high location than a radar scanner thus giving increased range. Also a network of stations can give great coverage over a long waterway of a large coastal area or estuary.

Greater positional accuracy

If AIS is taking its position from differential GPS (DGPS) or other differential systems a positional accuracy of better than 10 m can be expected. Radar, being subject to beam-width and pulse length distortion can often only achieve an accuracy of 30 to 50 metres.

Radar shadow

Depending on the position and number of shore based radar scanners, certain areas of a VTS's coverage may be in the radar shadow of high land, or buildings. This may cause traffic to be obscured to the VTS in these areas. AIS transmissions are not normally obscured to such an extent by land or other obstruction between the AIS transmitting ship and the receiving antenna.

Target Swap

While the phenomenon of 'target swop' has been much reduced in modern ARPA radars, it can still occur. This is where the vector of a weaker radar target is picked up by a strong target when it passes close to the weaker one. Such a phenomenon has occurred with AIS, but this is rare and on investigation it seemed that this only happened with one particular make of AIS ship-board installation.

Radar and AIS overlay

The optimum arrangement is to have both radar return and AIS appearing on the one display. AIS being not affected by clutter can track targets where radar might lose them. Radar can display targets which are not, for whatever reason, transmitting AIS signals such as small leisure craft. By using both, where possible, a more complete and safer picture of the traffic in the VTS area is achieved.

The Big Picture

The AIS channels can also be used to transmit port data, pilotage, berth assignments, shipping agency information, tides and currents, notices to mariners and other information from shore to ship, as well as ship-to-ship and ship-to-shore AIS reports. Apart from improving safety it enhances the efficient operation of the port or waterway and even adjacent ports. For example the ETA of a vessel, being accurately known can be passed on to pilots, tugs and terminal or berths, often automatically. It is also possible for the VTS to broadcast the complete harbour picture to all ships in the area so the masters and pilots all share the same big picture.

Special dedicated channels may be designated for local-area AIS operations

Precise navigational advice

In situations where precise navigation is required, such as very large ships in a dredged channel, or following a bend, AIS providing almost real time information can be used to monitor the progress in such cases. This is much better than the radar monitoring of such navigation at there are inherent delays in the presentation of tracked target data in radar tracking (see chapters on ARPA tracking and radar and ARPA errors).

AIS Channel control

As the AIS equipment has the ability to shift to different channels automatically when directed the VTS centre can assume control over the assignment of timeslots for AIS messages. This would ensure optimum data exchange within the coverage area. As a result special dedicated channels may be designated for local-area AIS operations.

A summary then of the practical benefits of AIS for VTS and Pilotage areas are:

- Detect a change in a ship's heading almost in real time.
- Automatic identification of radar targets.
- Constant coverage even when radar picture is degraded by weather and interference.
- Tracking of vessels behind islands and obstacles.
- Ability to send port data, weather forecasts and safety messages automatically to all ships in the area.
- Ability to send ship data to other port or state agencies.
- Automatic logging of all data for analysis and research purposes.

Message	Description
1, 2 & 3	Vessel position reports (scheduled, assigned & special)
4	Base station report
5	Vessel static and voyage related data
6, 7 & 8	Binary messages. Addressed, acknowledgement & broadcast
9	Search and Rescue aircraft position report
10 & 11	UTC inquiry & response
12, 13 & 14	Safety Related Messages (SRM). Addressed, acknowledgement & broadcast
15	Interrogation message to request a message from another station
16	Assignment command message used by a Base Station to control traffic
17	Differential GNSS broadcast
18 & 19	Class B position reports
20	Data link management for base station slot reservations
21	Aids to Navigation
22	Channel management by base station

AIS messages

The specification for AIS Messages is set out in ITU M1371-1. The table above sets out a summary of these message types and their purpose. An understanding of message types can assist in evaluating AIS data.

A ship's Minimum Keyboard Display (MKD) is capable of receiving all of these messages. However, some receivers may not decode the full detail due to uncertainty about message content at the time AIS became an IMO carriage requirement. This will not affect ship-to-ship messages but could affect Met Hydro information in Type 8 Messages and Aids to Navigation Messages.

The AIS system design allows many additional uses. Some of these are already built into the system (such as AIS AtoN messages, as mentioned previously) but also there is a flexible message facility, known as AIS Binary Messages.

The binary messages are designed to be in two categories – international applications and regional applications.

Minimum Keyboard Display or MKD

The original carriage requirements for AIS only specified what is referred to as a Minimum Keyboard Display (MKD). The basic MKD displays the AIS messages in alphanumeric format on a scrollable 3 line display. All of the information is there but the format is not particularly useful to the mariner particularly in an area of heavy traffic. Some manufactures provide a small situation display in the MKD but it has been recognised that the overlay of AIS data on Radar and ECDIS offers the best means of displaying AIS information. As it is now a requirement for radars fitted after January 2010 to be capable of showing AIS data, and as many radars before this date already had such a capability, it follows that it is increasingly to be expected that the graphical display will be the norm.

How the System Works

While from the practical point of view it seems that all AIS stations are transmitting and receiving simultaneously a process called Time Division Multiple Access (TDMA) is used where each station uses defined time 'slots' for its transmissions. This ensures that the transmissions of different transponders do not occur at the same time. This also allows for large numbers of transmitters to share one single narrow band VHF radio channel by synchronizing their data transmission to an exact timing standard.

The precise time signals of Global Navigation Satellite systems (GNSS) is used. This provides the time slots for AIS transmissions which are all precisely aligned to Co-ordinated Universal Time (UTC). Each minute is divided up into 2250 equal slots, allowing 2250 transmission bursts to be made on each channel per minute.

Each station automatically determines its own transmission slot or slots per minute. This is based upon the history of transmissions from other stations and of future actions by these other stations. A position report from one AIS station occupies one of the 2250 time slots for every 60 seconds on each of the two AIS VHF frequencies. Initially a station (a ship's AIS unit) will 'listen' to existing traffic to establish the 'history' and which slots are free. It then uses its computerised program to select suitable slots. This process of selecting transmission slots is called Self organising Time Division Multiple Access (SOTDMA). Having established which slots are free the AIS station then registers into one or several slots where it will transmit again on whichever of the two frequencies.

As each minute is divided into 2250 slots it allows that number of transmission on each of the two channels. Theoretically each ship transmits on average every 6 seconds, that is 10 times per minute. That allows the system to handle 225 ships for each channel which is a total of 450 ships in any one area. This is in excess of the IMO performance standard which requires a minimum of 2000 time slots per minute. The frequency of a ship's transmissions will vary according to the rate at which it is moving or changing direction and can be from every three minutes for ships at anchor or 'moored' to every 2 seconds for fast moving vessels.

When the system becomes very busy for any one ship, preference is given to the transmissions from the closer ships and those further away may be dropped. This ensures that transmitting ships within 8 to 10 miles are received. This means that the range at which AIS transmissions are received will vary depending on the number of transmitting ships in the area, but that those of concern to the OOW, those closest, are sure to be received. The term 'degrade gracefully' is used to describe this process.

Since transmission occurs at a rate of 9600 bps, and each slot has an equal-length duration of 26.67 ms, each data packet (message) fitting in a single slot has a maximum length of 256 bits, including a guard time for the propagation delay. However messages (i.e. text messages) can cross the slot boundaries where messages are bigger than 256 bits.

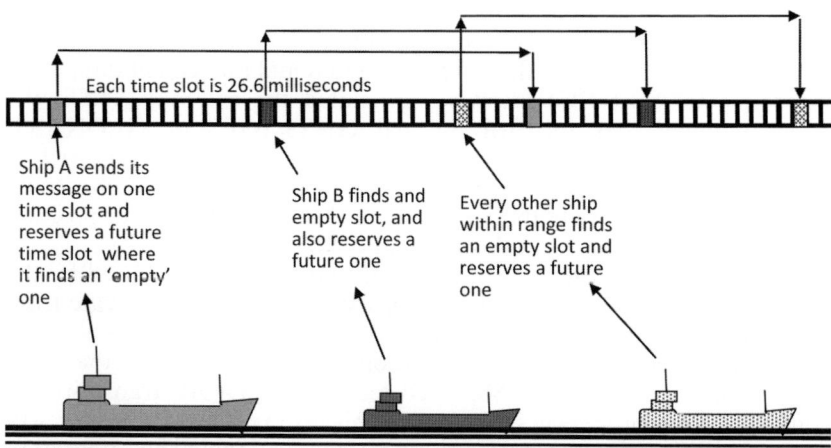

FIGURE 15.1   AIS time slots.

The station that is transmitting alternates between channels using SOTDMA. When the station is transmitting in one channel it will not be able to receive in the other channel.

By that reason both channels are evaluated before selecting an appropriate slot.

While the SOTDMA process works for ships at sea in any part of the world, shore stations or network of stations can control the assignment of time slots. This is 'assigned mode'. This would typically be used in a VTS network, or within range of a coast radio station which might be controlling a SAR operation. There is no apparent change to the people on the bridge of a ship or any action required by them when their AIS transmissions are assigned slots. Similarly a shore station can 'poll' a ship's AIS transmission, that is interrogate it at a particular time. This also would be 'invisible' to the ship's people.

Ship's AIS Installation

The installation consists first of all of three VHF receivers, two VHF receivers to be able to receive data on the AIS1 and AIS2 channels simultaneously. In addition, one DSC receiver, capable of operating on all marine VHF channels, is required in order to be able to receive polling requests and other data sent over Channel 70. Such DSC operation is automatic and requires no operator intervention.

There is one VHF transmitter which is capable of operating on all marine VHF channels and responding to Channel 70 DSC interrogations. At any one time AIS data is only being transmitted on one channel; the system automatically selects the correct transmission channel (AIS1, AIS2 or DSC Channel 70).

While the system normally takes positional and movement data from the ship's GNSS (GPS) receiver, designated the external GNSS, the AIS equipment has its own internal GNSS (GPS or GLONASS) receiver, with its own separate GNSS antenna, which is required to determine UTC accurately. If a loss of positional information occurs from the ship's external GNSS equipment the AIS will transmit position and other GNSS data from this internal GNSS.

The AIS installation has a single VHF antenna which is used by the transmitter and the three receivers. This VHF antenna is separate from the other VHF antennas used on the ship.

All the functions in the operation and handling of the data concerning the internal GNSS, VHF receivers and VHF transmitter is controlled by a processor sub-system. This also performs all other AIS data handling and computational operations, including the integration to and from the navigational equipment on the ship such as external

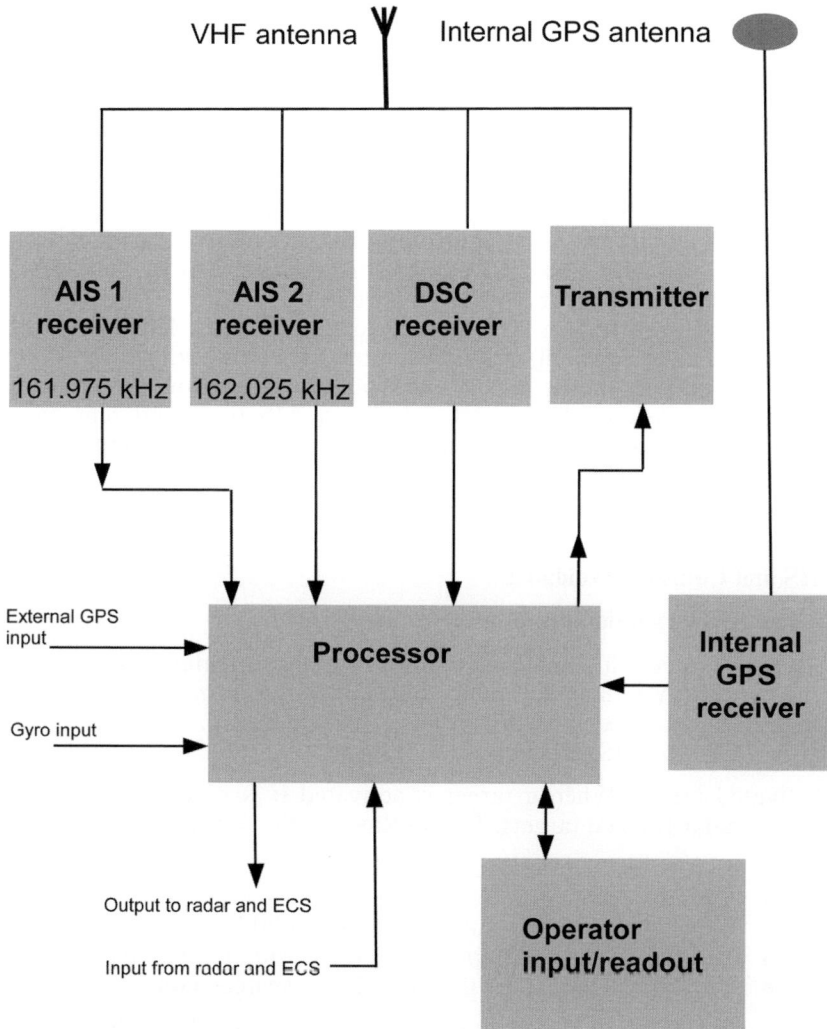

FIGURE 15.2   Diagram of AIS equipment.

GNSS, gyro, ECDIS and radar, the user input of information by the keyboard and the display of information. The processor sub-system also controls the 'built-in integrity tests' of the unit. These provide monitoring of any failures in the equipment, detecting the availability of data and error checking of both the transmitted and received data.

The shipborne unit includes a display and keyboard. The keyboard is used to allow the user to control the operation of the unit, and input necessary data such as voyage-related data and safety-related messages. This may be the Minimum Keyboard Display (MKD) described above, but it is also the means of controlling the system even when data is displayed on radar and ECDIS. Increasingly control of the AIS is incorporated in the radar installation (see Fig. 15.3).

The ideal interface for AIS is the Radar or ECDIS. This helps with situational awareness but, does not remove concerns outlined elsewhere in this book. It should be treated with great caution for collision avoidance decisions. The target data from AIS is ground referenced (targets show ground track) with the danger that this can imply when making collision avoidance decisions.

AIS and Collision Avoidance

The AIS target definitions are:

Sleeping target: This indicates the position and orientation of a vessel transmitting AIS data. To reduce display 'clutter' a means to filter the presentation of sleeping AIS targets is provided, e.g. by target range, CPA/TCPA or AIS target class (A or B).

Activated target: When a target is activated it is treated in a similar way to radar tracked targets. The display additionally shows:

- A speed and course over ground vector.
- A heading indicator.
- A turn direction 'flag', if the data is available.
- An optional turn path predictor, (providing ROT is available).
- The target's past track if this facility is switched on.

Note: Activation/deactivation of targets may be achieved by individual selection or optionally by user defined zones or when meeting user defined parameters such as target range, CPA/TCPA limits or AIS target class. On some systems it may be possible to select more than one target. If so there will be a clear identification of the data which is applicable to each selected target.

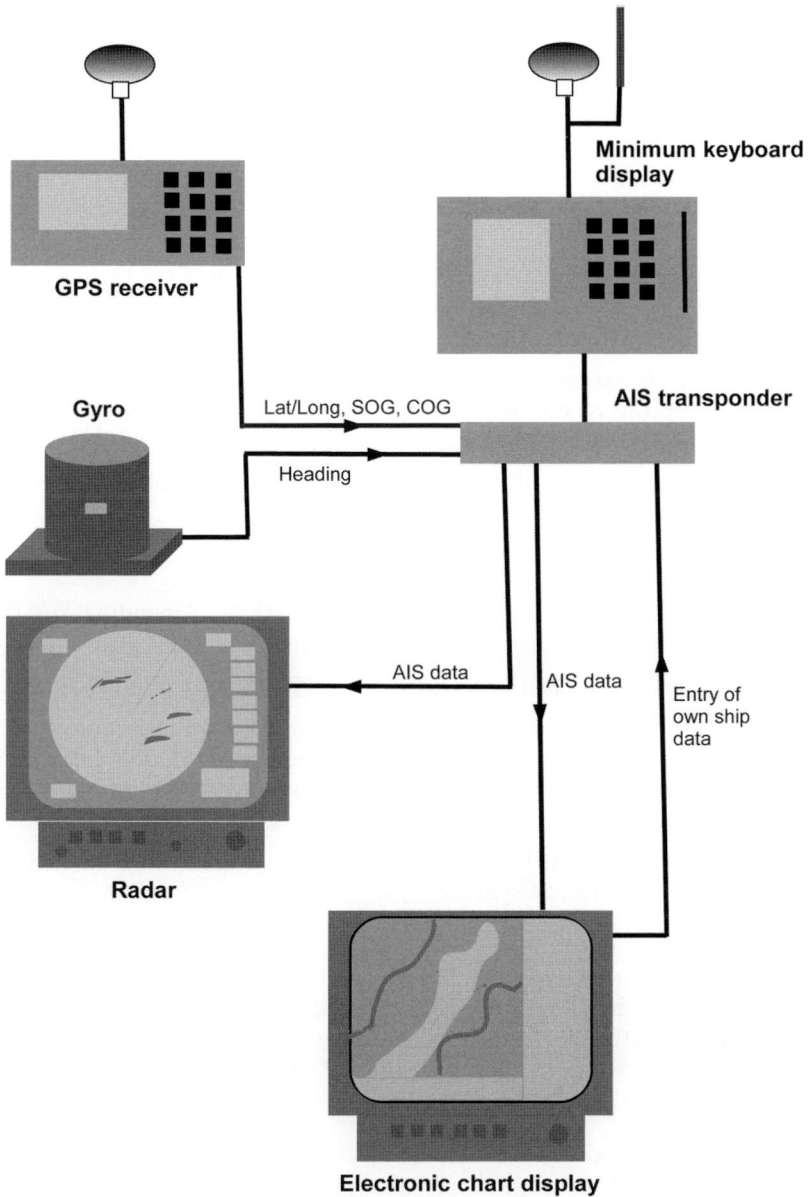

GPS receiver

Gyro

Lat/Long, SOG, COG

Heading

Minimum keyboard display

AIS transponder

AIS data

AIS data

Entry of own ship data

Radar

Electronic chart display

FIGURE 15.3   Diagram of AIS installation.

Dangerous target: A target (activated or not) that is calculated to violate user-set CPA and TCPA limits.

Note: If the calculated CPA or TCPA of an activated AIS target is less than the set limit a CPA/TCPA alarm will be given and the particular target will be indicated with a dangerous target symbol. The set CPA/TCPA limits may apply both for radar and AIS targets. The user can select whether the limits apply just to activated AIS targets or to both activated and sleeping targets.

Selected target: When a target is selected its detailed information is displayed in a window on the display. This will include CPA and TCPA data as well as AIS target data.

Lost target: A target representing the last valid position of an AIS target before the reception of data was lost. It will be shown as such until the alarm is acknowledged or the target signal reacquired.

Note: There is a lost AIS target alarm that can be enabled or disabled if any of the following events occur:

The AIS lost target alarm function is enabled.
The target meets the user-set lost target filter criteria.
A message is not received after a set time, depending on the nominal reporting time of the AIS target.

In these events the last known position will be clearly indicated by a lost target symbol and an alarm given.

The lost target symbol will be extinguished if the alarm is acknowledged (or replaced by the previous symbol if the signal is received again).

Note: When a target first enters the display area/zone it will be identified by the relevant symbol.

AIS Target (sleeping)

An isosceles, acute-angled triangle symbol is displayed. The triangle is oriented by heading, or COG if heading missing. The reported position is located at centre and half the height of the triangle. The symbol of the sleeping target is smaller than that of the activated target.

Activated AIS Target

An isosceles, acute-angled triangle symbol. The triangle is oriented by heading, or COG if heading missing. The reported position is located at centre and half the height of the triangle. The COG/SOG vector is displayed as a dashed line with short dashes with spaces approximately twice the line width. Optionally, time increments may be marked along the vector. The heading is displayed as a solid line thinner than speed vector line style, length twice the length of the triangle symbol. Origin of the heading line is the apex of the triangle.

The turn is indicated by a flag of fixed length added to the heading line. A path predictor may be provided as curved vector.

For a 'Dangerous AIS Target, bold, red (on colour display) solid triangle with course and speed vector, flashing until acknowledged.

AIS Target – scale outline

A true scale outline may be added to the triangle symbol. It should be located relative to the reported position and according to antenna position offsets, beam, length and oriented along target's heading.

Selected target

A square indicated by its corners should be drawn around the activated target symbol.

Lost target

A triangle with bold solid cross. The triangle should be oriented per last known heading value. The cross has a fixed orientation. The symbol should flash until acknowledged. The target is displayed without vector, heading and rate of turn indication.

Target past positions

Dots, equally spaced by time.

Assessment of Radar and AIS

Both AIS and Radar are designed to assist the navigator to become aware of targets in his vicinity.

The advantages of Radar:

1. Radar can detect a vast range of objects, most of which need no electronic device to send a 'return' to the radar.
2. All received data is relative to own-ship and its effective use does not fundamentally require any positional input.
3. AIS requires the co-operation of other targets. The target must be fitted with an AIS transponder and it must be switched on.
4. AIS relies on the target to provide accurate data.
5. AIS relies on all targets and own-ship continuing to have accurate positional date referenced to a common geodetic datum (WGS84).

6. Many objects which radar can detect are not AIS targets and are never likely to be.

Advantages of AIS:

1. Radar targets are easily obscured in clutter and by line-of-sight shielding.
2. Radar target information is limited. It gives only its present position (in relation to own ship) and past (relative) track and is open to confusion with other targets.
3. In radar, the movement of targets has to be calculated (even by a tracking radar such as ARPA display) from consecutive radar returns. When a target alters course it can take an appreciable time for this to become apparent.
4. The effective centre of the radar return from large target can vary for a variety of reasons thus causing speed and course errors particularly during target alterations of course (see Chapters 12 and 13 on 'ARPA Target Tracking' and 'Radar and ARPA Errors').
5. Target heading is not given by radar. It can only give the direction of the track which the target has been making and in some situations the displayed course of a target may not be reflected by its actual heading. This may give different interpretations of the collision regulations depending on whether radar or visual data is being used to assess heading.
6. AIS targets can give a vast amount of useful data about themselves including absolute positional, velocity and heading data potentially far superior to that obtainable by radar, also size and type of vessel.
7. AIS does not suffer from clutter problems and is relatively capable under conditions where the radar 'line-of-sight' is compromised (shielding by headlands and islands, etc.).
8. AIS provides better vessel tracking in that there is less likelihood of 'target swap'.
9. AIS generally has greater range, although in some circumstances e.g. mountainous areas, this may only be achieved with the provision of shore-based repeater stations.
10. AIS can also be used as an aid to navigation and possibly replace RACONs. It has greater positional accuracy, but is dependent on the position input sensor.
11. AIS provides near real time manoeuvring data.

> **It must always be remembered that neither radar nor AIS will detect all objects of interest or concern to the mariner.**

Ground versus Sea Stabilised Displays

In the chapters on radar and particularly on ARPA, the relative merits of ground and sea stabilisation are discussed. The importance of the distinction between ground and sea stabilised displays and the importance of using sea stabilised display for collision avoidance should be well understood by watchkeeping officers. AIS information on a ship's course and speed is based on its onboard equipment (usually the GNSS) and will be therefore be ground stabilised. Consequently AIS information displayed on the Radar screen or ECDIS will show target's course and speed over the ground.

The added complication of overlaying ground stabilised AIS data on a sea stabilised display is shown in Figure 15.4. Here an AIS 'over-the-

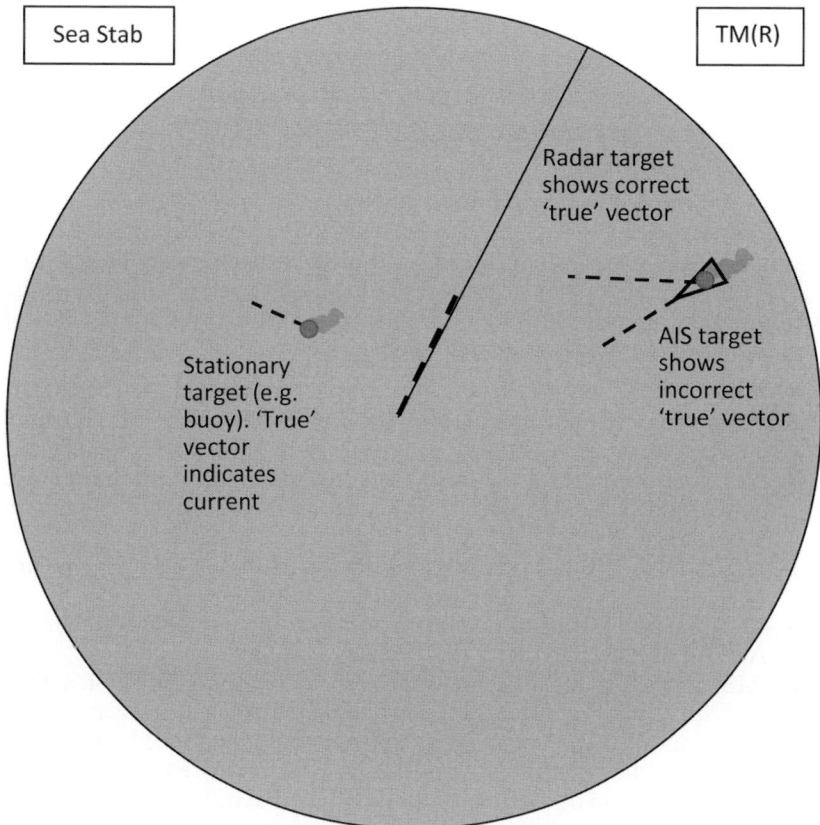

FIGURE 15.4   Showing AIS v radar in sea stabilised mode.

ground' ship shape and vector is introduced incorrectly suggesting a Green to Green situation.

Similar confusion can exist with AIS overlay on a non stabilised Relative Motion display.

These pitfalls can be overcome by operator awareness and training. The data is not incorrect, it is labelled correctly, but can be displayed inappropriately,

Separate displays optimised for collision avoidance and navigation is a good practice and is recommended by current IMO circulars. In collision avoidance (and navigation generally), the OOW must use all available information. AIS will give valuable additional information but awareness of the data and its reliability is critical and of course what can be seen in clear visibility by looking out the windows completes the picture.

AIS and the International Regulations for the Prevention of Collisions

Currently the IRPCS does not mention AIS as it does radar ('a vessel which detects the presence of another by radar alone ...'). Therefore the information derived from it has to be used in the context of using all available means of assessing risk of collision. Great care needs to be taken to avoid cases where a collision avoidance manoeuvre is based solely on the information from AIS. For this reason officers of the watch should be fully conversant with the capabilities and limitations of AIS. Some of the points to remember are:

- The IRPCS are based on movement through water, the dispositions of vessels in relation to each other – AIS is a ground based system.
- AIS does not replace the need for good seamanship or compliance with IRPCS or keeping a lookout by all available means in accordance with STCW 95.
- AIS data should not be used in isolation and never as a sole means of assessing the close quarter situation of any targets.
- The data being transmitted from another ship can be erroneous.
- AIS is not mentioned in the IRPCS as is radar.
- AIS does help in situational awareness on the bridge of a ship and the tracking of target, particularly with the near real-time updating of a targets change of status, i.e. course/speed, etc.
- Knowing the identity of another ship tends to encourage radio communication and discussion on collision avoidance manoeuvres. Such discussions are potentially very dangerous and should be treated with great caution.

Tx 2 Master   N-UP RM(T) Sea Stab

Tune Auto
Pulse: LP
PM
Mute Off

Range 6 NM
Rings 1 NM
Profile: Default

H/L ON

Target Name	005	006
Source	Radar	Radar
RNG	6.48 NM	6.40 NM
BRG	013.6°	343.9°
T CTW	062.8°	102.9°
T STW	16.0 kn	13.9 kn
CPA	4.91 NM	5.62 NM
TCPA	-00:16:39	00:13:04
BCR	60.95 NM	-9.67 NM
BCT	-04:16:14	00:47:40

CPA Limit 0.5 NM   AIS OFF
TCPA Limit 12 min   Tgt Assoc. ON
Guard Zone   Tgt Tote OFF
Delete ALL Tgts   Trial Manoeuvre
Vector length   6   min (T)
Past Positions 8.0   min (T)
Trails length 0.0   min (T)

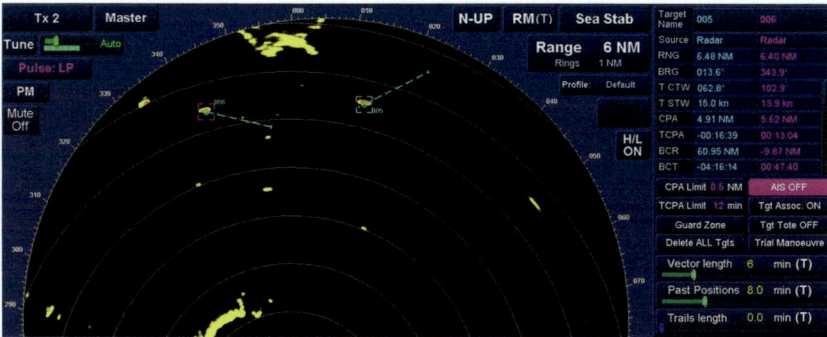

FIGURE 15.5a Radar screen with radar acquired targets. AIS off.

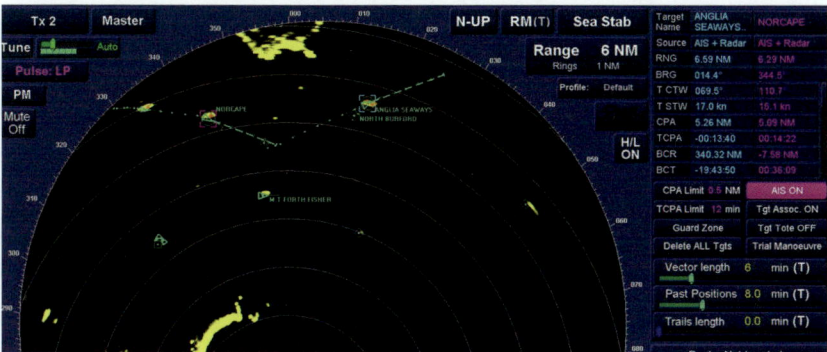

Tx 2 Master   N-UP RM(T) Sea Stab

Tune Auto
Pulse: LP
PM
Mute Off

Range 6 NM
Rings 1 NM
Profile: Default

H/L ON

Target Name	ANGLIA SEAWAYS	NORCAPE
Source	AIS + Radar	AIS + Radar
RNG	6.59 NM	6.29 NM
BRG	014.4°	344.5°
T CTW	069.5°	110.7°
T STW	17.0 kn	15.1 kn
CPA	5.26 NM	5.09 NM
TCPA	-00:13:40	00:14:22
BCR	340.32 NM	-7.58 NM
BCT	-19:43:50	00:36:09

CPA Limit 0.5 NM   AIS ON
TCPA Limit 12 min   Tgt Assoc. ON
Guard Zone   Tgt Tote OFF
Delete ALL Tgts   Trial Manoeuvre
Vector length   6   min (T)
Past Positions 8.0   min (T)
Trails length 0.0   min (T)

FIGURE 15.5b Radar screen with target association 'On'.

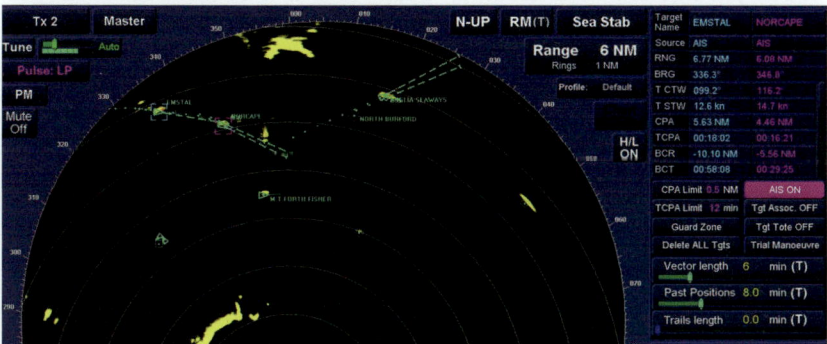

Tx 2 Master   N-UP RM(T) Sea Stab

Tune Auto
Pulse: LP
PM
Mute Off

Range 6 NM
Rings 1 NM
Profile: Default

H/L ON

Target Name	EMSTAL	NORCAPE
Source	AIS	AIS
RNG	6.77 NM	6.08 NM
BRG	336.3°	346.8°
T CTW	099.2°	116.2°
T STW	12.6 kn	14.7 kn
CPA	5.63 NM	4.46 NM
TCPA	00:18:02	00:16:21
BCR	-10.10 NM	-5.56 NM
BCT	00:58:08	00:29:25

CPA Limit 0.5 NM   AIS ON
TCPA Limit 12 min   Tgt Assoc. OFF
Guard Zone   Tgt Tote OFF
Delete ALL Tgts   Trial Manoeuvre
Vector length   6   min (T)
Past Positions 8.0   min (T)
Trails length 0.0   min (T)

FIGURE 15.5c Radar screen with target association 'Off'.

Target Association Target Fusion or Target Correlation

The data on which any target is presented on the display, when radar and AIS are on the same display is, as we have seen, derived from two completely different sources. The resulting calculation and data presented are different, the course, speed, etc. and vector displayed on the screen are derived for the radar target by a plotting calculation, those for the same target, but from AIS are from the information transmitted from the target ship.

This can be seen when an AIS target symbol and vector is separated from the radar image of that target. This may be because the AIS return is referenced to the target CCRP, whereas the radar image depends on the radar cross section (RCS) characteristics of the target. Also, as the AIS tracks the target vessel with reference to the ground (GPS), and the radar can be sea stabilised, the resulting vectors can display that difference (see Figures 15.5a, 15.5b and 15.5c). In Figure 15.5c, the two vectors for each target are shown. Both are 'true' vectors. The difference is caused by the tidal stream.

Target "association" (or fusion, or correlation) links a Tracked Targets to the AIS target of what appears to be the same ship. This is based on certain parameters or criteria for association which are either a default condition or those set by the operator. When an AIS target and a radar tracked target are within the limits set on those parameters one symbol represents the target, one vector is produced and the readout data is for that one target. When such association is set, the AIS symbol is aligned with the radar echo, provided it is within the parameters of association.

Setting Up the Target Association Parameters

The parameters or criteria by which AIS and radar tracked targets can be linked or associated are a default condition or can be set by the operator. The functions for this will be found in different places in different makes of radar; in one make, the KH set illustrated here, placing the cursor over the "Tgt. Assoc." Box on the screen, and pressing the middle button of the roller ball control opens the "Target Association Dialogue Box".

In this the parameters are:

Difference in bearing.
Difference in range.
Difference in COG (course over Ground).
Difference in SOG (Speed over Ground).

The dialogue box gives the option of changing the parameters, but not individually; in other words the parameters can be increased or decreased, but the difference mentioned above will change in proportion.

The default conditions are usually in the range of:

Diff. Range	150 m
Diff. Bearing	1°
Diff. COG	10°
Diff. SOG	1 to 2 knots.

Another criteria used in some makes of radar is target separation, usually less that 0.1 miles. These are upper limit conditions. If for example the radar tracked target's speed and that of the corresponding AIS target differ by more than 2 knots (presuming 2 knots is the setting), then the targets will be "disassociated" and two vectors will appear. Similarly if the target's course over the ground, calculated by the radar tracking, and that shown by AIS differ by more than 10°, the target will be "disassociated", and two vectors shown.

Associated targets may be designated as having radar or AIS priority. This means that the data presented on the readout for that target, and the vector will be that derived from the priority source, and the association function is there to ensure that the data from the other source is within the association limits. In the KH set the option is given to select AIS as "default" or radar as "default". The associated targets will then be shown on the radar screen as either a radar tracked target or an AIS target depending on the selection. The Target data dialogue box will show the target source as a RADAR+AIS or as an AIS+RADAR target, depending on the priority selected, whilst they are associated. Any Tracked Target and AIS target that meet the criteria will be associated as long as it continues within the parameters. If a target then fails to meet the association criteria it will be "disassociated" and shown as a separate Tracked Target and AIS target

While the functions of associating target and disassociating them and the criteria for such is usually set globally, association can be often overridden for individual targets if necessary.

This function only operates where a radar tracked target and the corresponding AIS target have been associated and allows the target to be shown as either a RADAR target or as an AIS target. It does not change the target information source if it is a RADAR target or AIS target only.

Accessing this function is usually a function associated with the particular target's data, and the source of such data. Going into the function there should be an option to designate the source of the data for that target, either radar or AIS. Selecting one or the other will give that data, and vector for that target, which then should appear on the readout for that target.

In tidal waters great care needs to be taken in the use of these functions, and particularly when the data sources are different for different targets. The dangers of misinterpreting "true" vectors in tidal waters is discussed elsewhere, but it cannot be emphasised enough particularly where so many different setting can be applied to the display.

Switching off the AIS

The ethos, and regulations, state that the AIS system should always be switched on and operating correctly at all times, this is regardless of whether the vessel is alongside, at anchor, underway, etc. It was however recognised immediately on its introduction that AIS could pose a security threat to ships (it could also pose a commercial/competitive threat to individual shipping companies operations). When there is a serious threat to the security and safety of the ship the master may decide that the transmissions of the AIS may be temporarily stopped by switching the unit off. A typical example would where there is a clear threat to the vessel in waters where pirates are known to be operating.

If this is the case then this action will need to be recorded in the ships logbook. When the threat subsides then the system should be restarted as soon as possible.

On some older models of AIS equipment,when switched off, the ship's static information will be lost and will default to pre-set up configuration. However the vast majority of AIS will now retain all static information when switched off and within two minutes of switching on own ship's data should be transmitting. It would be good practice to check the AIS unit to ensure that all data about ownship is correct at this stage.

Other examples of where AIS may be turned off are in oil terminals and other port operations. The ISGOTT (International Safety Guide for Oil Tankers and Terminals) Guide recommends that all radio transmissions be limited to a maximum of 1 watt power. When normal VHF transceivers are set at 'low' power then this is within these limits. For some AIS systems a special 'hazardous cargo' setting can be selected, allowing 1 watt transmission power. However the majority of AIS units at low power only reduce radiation to 2 watts power. This falls short of the ISGOTT recommendations and as a result ships may be requested to turn off their units by the shore authorities. This is becoming more commonplace in the tanker industry.

Other port authorities may require AIS to be switched off where it is envisaged that flammable gases may be present.

Some units when travelling less than a set speed, or selected as moored or anchored, will automatically reduce its VHF transmission power to 1 watt.

As a matter of interest it is sometimes disputed as to the real potential risk (or otherwise) of VHF transmissions in these hazardous areas, regardless of transmission power, but ships' officers need to be aware of the envisaged risk and obviously comply with the relevant port or terminal regulations.

The AIS data from ships is transmitted by several websites on the internet. This practice has been condemned by the IMO as it is considered to pose a security risk. However, it has not deterred the practice.

A description of the different AIS equipment

Most seagoing commercial and other ships will be equipped with a Class A unit. This provides the facilities described above. Class B units are to be found in leisure craft and such like and the main purpose is to make them 'visible' in an AIS context.

Class A units

Class A devices are designed to meet the current IMO Performance Standards. SOLAS Chapter V (Safety of Navigation) dictates their carriage requirement. Carriage of Class A units may be required for other vessels as domestic or regional carriage requirements dictate.

Class B units

Class B devices may not necessarily meet all the performance requirements specified by IMO MSC Resolution 74 (69) Annex 3. They are designed on operate harmoniously with Class A units. The Class B units may be used on craft not subject to SOLAS.

Other variants are under development specifically for base stations, aids to navigation and search and rescue but will all be derived from one of the existing standards and inter-operate with them.

Other developments

The development of Long Range Identification and Tracking (LRIT) is ongoing. This will extend the range at which ships can be identified and tracked, ultimately globally, whereas AIS is limited to 'line-of-sight' of the marine VHF bands. One area of development is satellite AIS where suitably equipped satellites can receive AIS signals from any station within its 'footprint' (the area of the earth's surface it can 'see').

While the purpose of this book is concerned principally with the shipboard functioning and impact of the system, it is important to realise the wider implications. Apart from the security threat already mentioned, some other uses and applications area as follows:

- legal evidence and accident investigation;
- sharing of data between VTSs and with national administrations;
- gathering information on the presence and pattern of traffic;
- planning of aids to navigation;
- fleet management;
- risk analysis; and
- generating statistics.

Conclusions

Summarising some of the benefits or otherwise of AIS for the for the OOW/bridge team:

- It can detect vessels that might otherwise be hidden in radar shadow.
- It gives real-time information about other ship's movements (course, speed and probably rate of turn).
- It automatically swaps information on destination, ETA, loading condition and other data with nearby ships.
- VHF voice traffic is reduced as much data is exchanged automatically.
- At present the system relies entirely on GPS. Possibly another GNSS satellite system such as Galileo may be used in the future and a ground based system such as e-Loran.
- The vulnerability of GPS to spoofing and jamming should always be borne in mind. Any failure in GPS will automatically mean failure of AIS.
- Data received cannot be interrogated to prove reliability.
- The question of sea-stabilised data, derived from radar as against the ground stabilised data from AIS should always be borne in mind.
- Not all ships/boats have AIS fitted.
- AIS can be turned off on a particular ships, intentionally or inadvertently.
- There is a need for regular checking to ensure that the data being transmitted is correct. If ownship were to transmit incorrect data then all receivers will receive incorrect data. It can be created by incorrect settings on bridge equipment. The BIIT (Built in integrity test) cannot check contents of data!

- It is an open system, the data on a ship can be readily accessed, even on the internet.
- In heavy traffic situations there is possible display overload for OOW.
- It is a VHF system and is thus subject to radio attenuation.
- Vessels transgressing regulations can be readily identified.

Chapter 16

Radar for Navigation

This chapter will:

- Examine radar as a navigation tool.
- Explain the techniques used in radar navigation.
- Look at the integration of radar with electronic navigation charts (ENCs).

As stated earlier, the radar is a 'blunt' instrument; its degree of positioning accuracy compares relatively poorly when compared with current global positioning systems and other electronic navigational systems currently being developed (e.g. E-LORAN, Galileo, etc.). However, it has one advantage over other navigational systems in that it is independent of transmissions from other sources. Provided the equipment is functioning and being operated correctly its positional information is a very valuable accurate navigational tool correct within its levels of accuracy.

Position Fixing

In using the radar for fixing the ship's position the tools are normally the EBL (electronic bearing line) and the VRM (variable range marker). In many modern sets these are the last few 'hard' controls and can be operated by rotating knobs. Obviously there is an 'on/off' switch for both; these are often (but not always) a 'soft' dedicated on-screen function. In many sets the EBL and VRM can be manipulated by a 'click and drag' using the screen cursor. Alternatively there is the screen cursor itself. This in default setting usually gives range and bearing from the origin. However, it may be possible to set this to give Latitude and Longitude position or sometimes range and bearing from the 'floating' or offset EBL as can the VRM and EBL (see earlier chapters on controls and their uses).

In fixing the ship's position using the EBL and the VRM or cursor, the operator needs to bear in mind the factors affecting the position fix. First of all there is the requirement to ensure that the correct object has been identified. Distortion of the appearance of the coastline due to beam-width and pulse length has been discussed. It is not unusual for the outline of the coast to appear quite different on the radar screen from that on the chart, and to possibly mistake a headland, or small island for the wrong object. Small objects such as isolated beacons or lightfloats can be confused with other targets such as fishing vessels and other traffic.

Another important factor, in coastal waters is the effect of tide and waves. At low water, in certain areas, much of the foreshore along a coast may be exposed, thus showing as the land echo on the radar screen. Even with examining the chart where there is a wide expanse of 'drying' foreshore (shown in green on charts) it is impossible to gauge where the actual water's edge may be for any state of tide. For this reason, 'point' objects such as isolated rocks and beacons, or steep rocky headlands where there is no 'drying' foreshore should be used whenever possible. Floating navigational aids should be treated with caution for fixing purposes as their moorings allow for a certain movement from the charted position due to tidal effects. However, the scope for such aids is quite small, and in offshore coastal navigation the error is usually negligible. As some of them are often easily identifiable, usually with a RACON beacon, they can be used, in the absence of anything more suitable, as a range and/or bearing target to fix the ship's position. However, floating aids can be out of position, often due to bad weather. This might be promulgated by the coastal state authorities in the form of navigation warnings and generally offshore and exposed buoys should be verified for their position before full reliance is placed on them.

Landfall

The echoes produced by coastlines can vary considerably, particularly at longer ranges. With modern navigational systems making a landfall is not usually the uncertain situation it sometimes was in days gone by. However, the fact that the coastline may be below the radar horizon when higher land is being shown on the screen should be borne in mind. Even in modern times, with GPS or other systems giving apparently precise and reliable positional information, the prudent navigator, on approaching land, will check the position by radar (and other methods such as the depth sounder!). Therefore he/she should be fully aware why the outline of the land appearing at long range may be difficult to reconcile with the chart, and that the positional information derived from the radar may differ by quite an amount from the position indicated by the other methods, including the depth sounder.

It is usually fairly futile to try and get positional information (range usually) from higher ground behind the coastline. Even examining the chart and noting the height of land, hills or mountains, the radar observer can rarely be sure of what high land the radar is actually

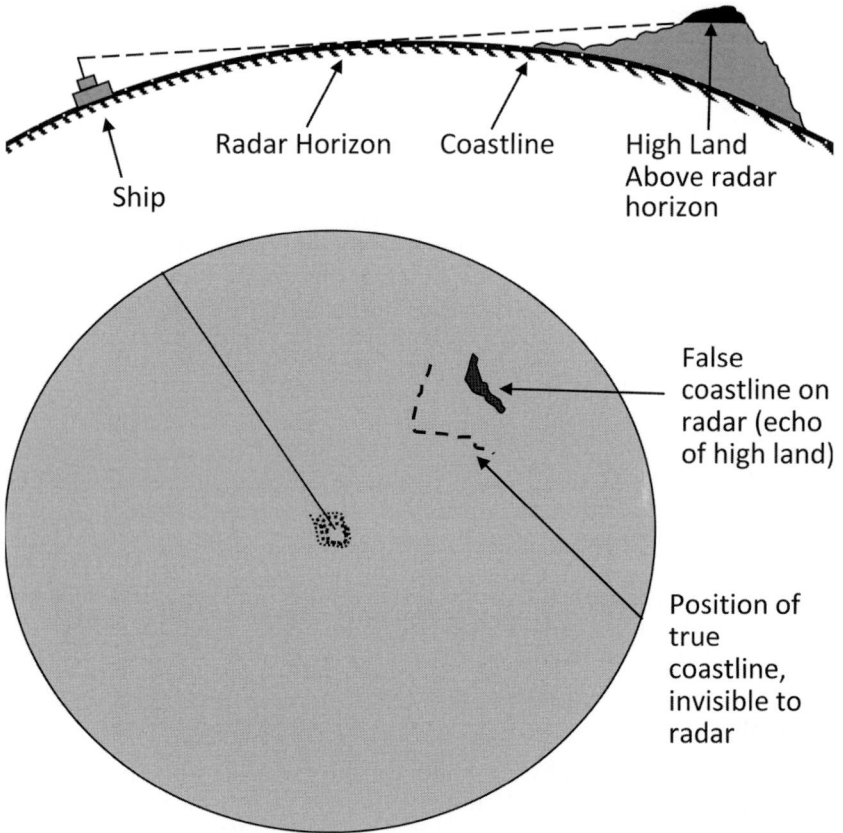

FIGURE 16.1   Showing land below the radar horizon.

getting echoes from. Sometimes the sailing directions can help where observations have been made by other navigators in the same situation. Examining the chart may make something fairly obvious, such as a low foreshore with a distinct and pronounced rise of land some distance behind it (see Chapter 6, 'The Radar Horizon').

Coastal Navigation

Even at shorter ranges, certain coastlines should be treated with a great deal of caution. A low foreshore may be below the radar horizon, but the land behind it may follow the line of the coast fairly precisely, such as an embankment, a road, a line of buildings. This can give the unsuspecting officer of the watch an outline that apparently matches the chart's coastline, and thus lead to erroneous and dangerous assumptions of the ship's position being further from the shore.

Tidal Effect

In coastal waters, the stage of the tide can alter the outline of the coast as it appears on the radar quite considerably. Obviously, where there are wide expanses of 'drying' areas, the OOW may be alerted to

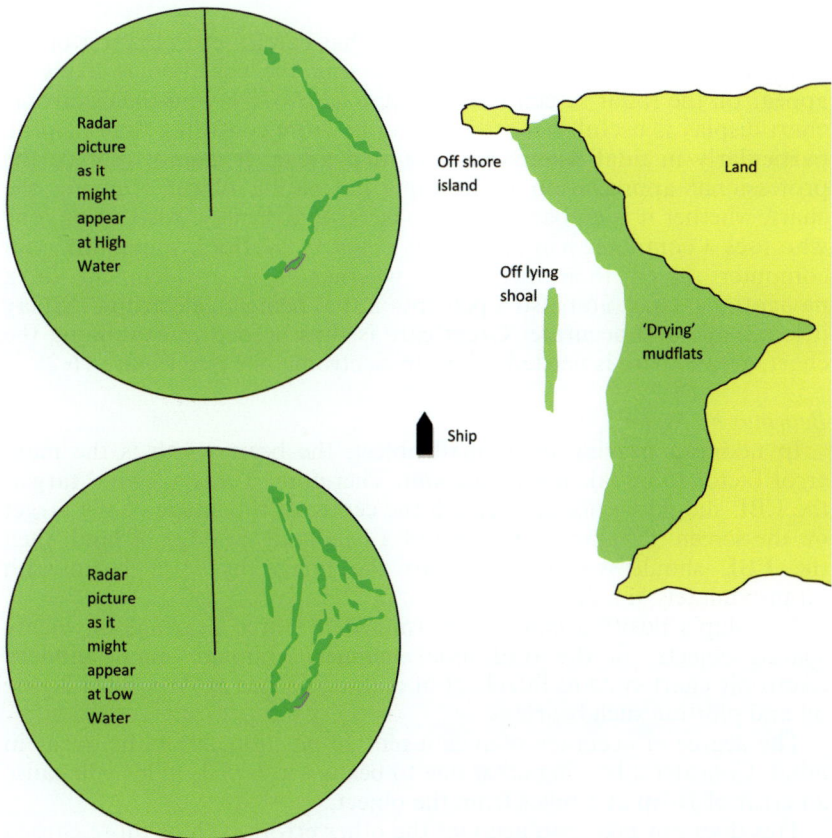

FIGURE 16.2 Showing differences of land echo on radar between high and low water.

the fact that the height of the tide is going to affect the radar outline of the coast, but it is in the situations where it may not be so obvious that the officer needs to be wary. For example a small off-shore island that appears quite separated from the mainland on the chart may be joined to it on the radar screen at low water, or when there is breaking waves in the strait between them. Inlets might disappear on the radar at low water for similar reasons, equally rocks and shoals can disappear at high water. Similarly breaking waves give a good radar response, so that the edge of drying areas can cause considerable distortion to the appearance of the coastline on the radar screen. Similarly shoals which in themselves do not project above the surface, even at low water, can cause tide rips or breaking waves and change the radar profile of the coastline.

In all of this, there can be no doubt that careful examination of the navigational chart with regard to the way the coastline is likely to appear on the radar screen is essential. Radar overlay on the electronic chart display is useful, but at times can be more confusing than helpful, particularly in tidal waters with 'drying' areas. It cannot replace the professional appreciation of a ship's navigating officer studying the chart, whether it is on paper or in an electronic display. Also, as anyone who uses a computer will agree, it is easier to overlook something on a computer screen than on paper, whether it is a document or a navigational chart and also a perception that being an electronic display it is reliable and accurate. Great care is thus needed in examining the chart, greater care is needed if it is an electronic one (see Figure 16.2).

Bearings by Radar

In taking a bearing of a small object, the beam-width is the main error factor to be taken into account. Therefore, if it is a 'point' target, the EBL should be placed through the centre of the image of the target on the screen. If taking a bearing of a headland, or edge of land, then the EBL should be brought into the image by ½ a beamwidth (approximately ½ a degree).

The ship's position can be determined by three bearings of widely spaced objects, in the traditional manner, although many modern electronic chart systems fall short of making proper provision for laying off and plotting such bearings.

The degree of accuracy of such a plotted position should be borne in mind. Consider a bearing error due to beam-width of 1°. This will cause an error of 161m at 5 miles from the object.

This does not take into account the other errors such as those caused by ship movement, beam-width and pulse-length distortion, correct location of the object (assuming correct identification), etc. As shown in

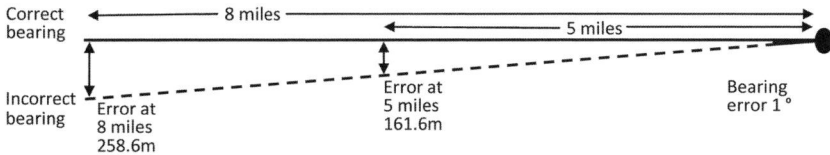

FIGURE 16.3   Bearing error.

Figure 16.3, the further away the object, the greater the possible error, therefore closer objects should be chosen where possible.

Generally a range measurement yields better accuracy, and the error is not so dependent on the distance from the object. It depends upon the pulse length in use. A pulse-length of 1 μsec. (m.sec.) will give an accuracy of 300 m as explained in the section on range discrimination (see Chapter 1, 'Radar Basic Principles'). This is a pulse length usually associated with the longer range selection on a radar such as 24 and 48 miles. At short ranges a pulse length of 0.05 μsec. is common and this gives an accuracy of 15 m. This also of course does not take into account the other errors mentioned above. It does mean that generally speaking, and taking into account other aspects of good navigational practice, such as getting a good "cut" between position lines, that three ranges should give a better "fix" than three bearings.

In measuring range, the VRM should touch the leading edge of the target fairly; not just skim it. Bear in mind that if the object is a small 'point' target, the position of the target is in the middle of the image appearing on the screen, but range measurement should still favour the leading (closest) edge. In fixing the ship's position using bearings and ranges, generally speaking the closer the objects the better the accuracy, as discussed above. However, in taking a series of such bearings and ranges the movement of the ship needs to be taken into account. The bearing or range of the object that is moving fastest relative to the ship needs to be taken last. For example if three bearings are being taken, those somewhat ahead and astern, which are changing less rapidly, should be taken first and the bearing of those near the beam last as these are changing most rapidly.

Once again, many modern electronic chart displays have difficulty in laying off three ranges and plotting a 'fix' (see Figures 16.4 and 16.5).

Chart Overlay

The ability to connect the radar display and electronic chart display is becoming increasingly common and in due course will be a

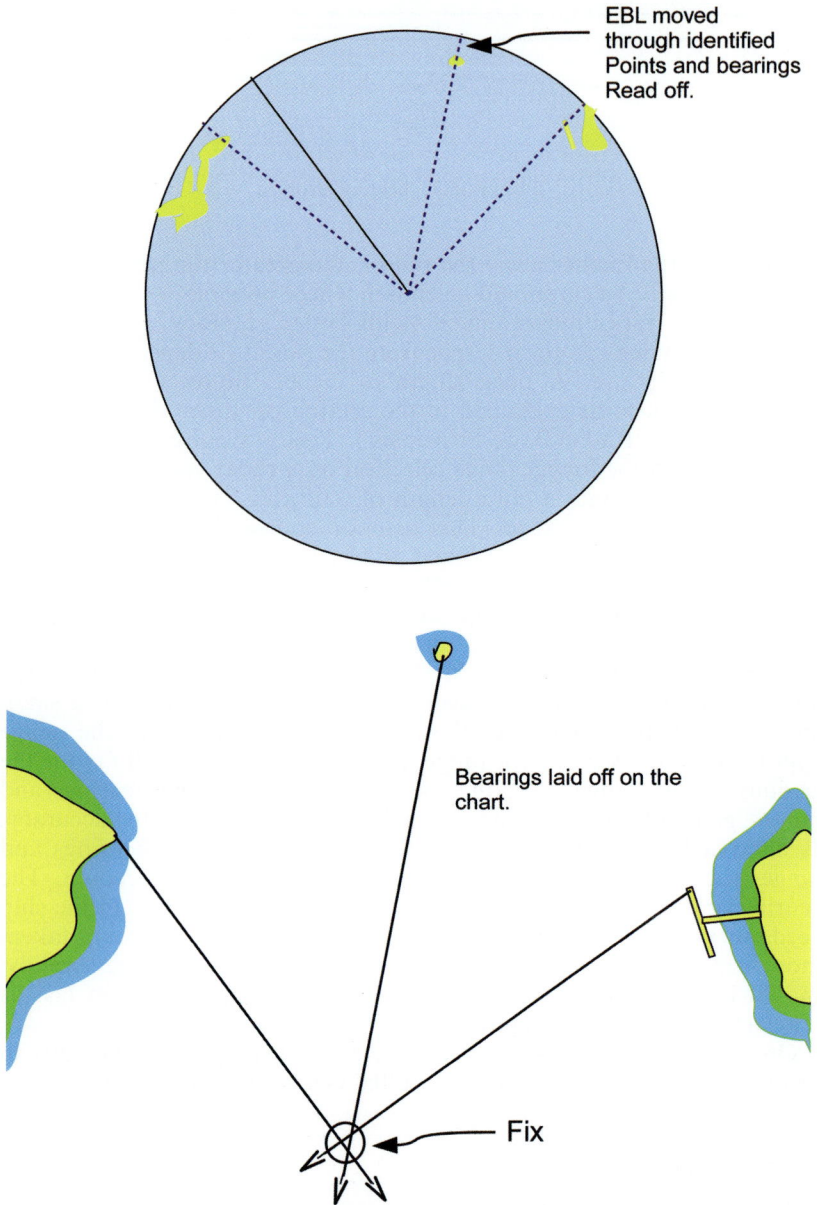

EBL moved through identified Points and bearings Read off.

Bearings laid off on the chart.

Fix

FIGURE 16.4   Showing three bearings for a fix.

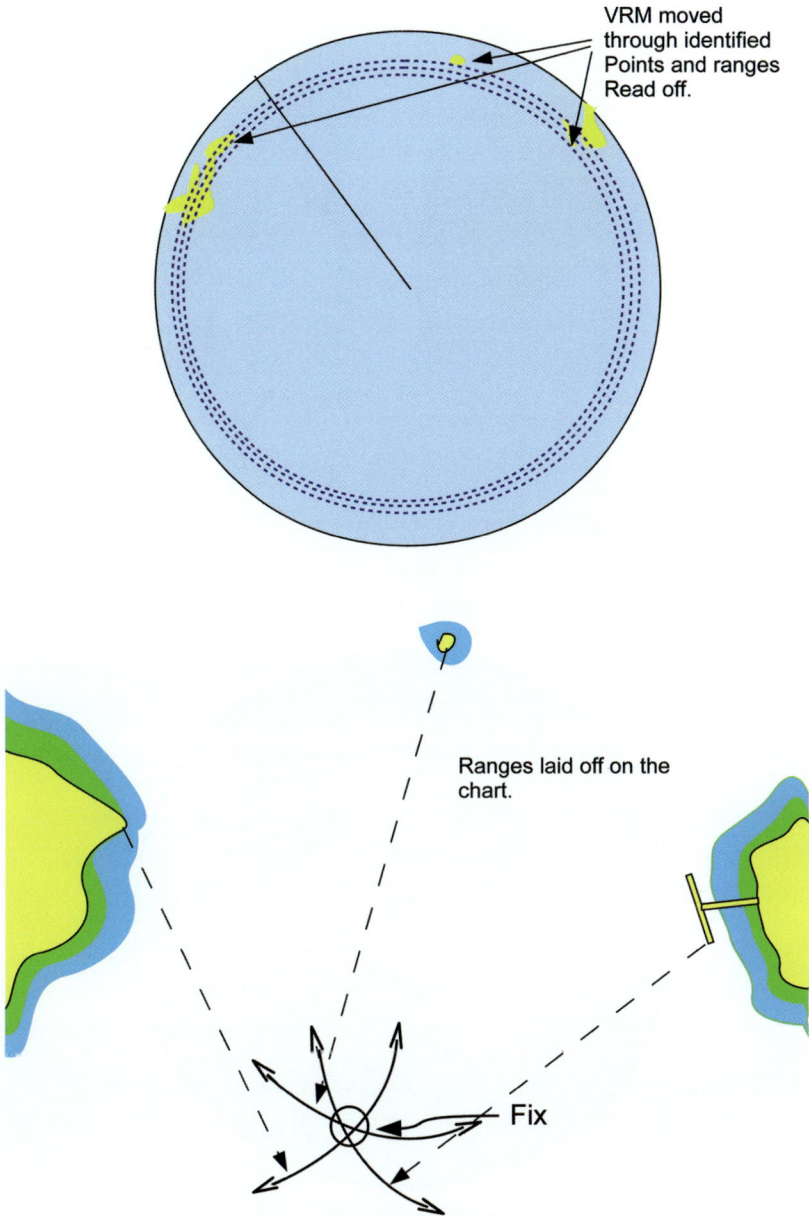

VRM moved
through identified
Points and ranges
Read off.

Ranges laid off on the
chart.

Fix

FIGURE 16.5   Showing three ranges for a fix.

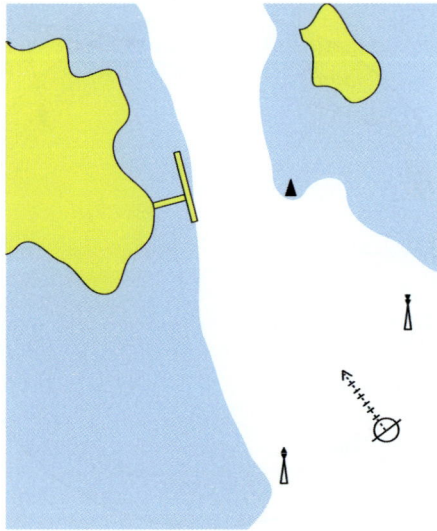

FIGURE 16.6a   Showing chart.

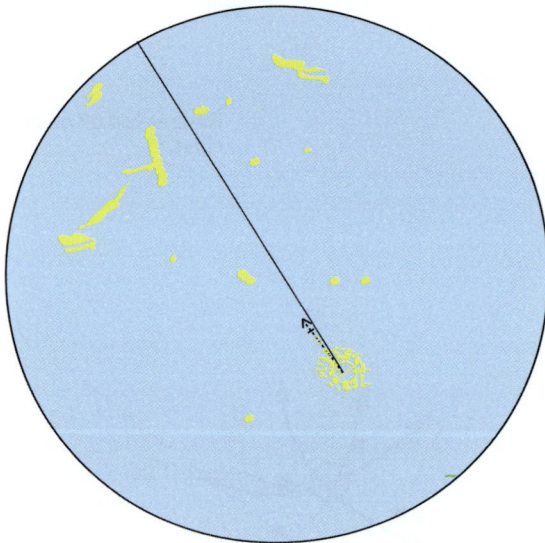

FIGURE 16.6b   Showing radar 'picture' of the same situation.

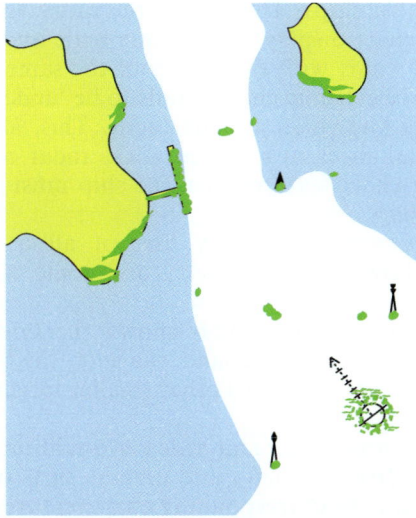

FIGURE 16.6c   Showing radar 'picture' overlaid on the chart.

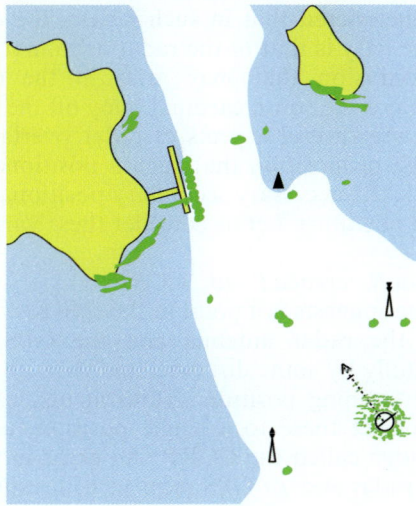

FIGURE 16.6d   Some error in positioning radar overlay.

requirement for most ships. This allows the radar video to be overlaid on the electronic chart display at the same scale and located to match the features of the chart with the echoes of the same features from the radar. Also certain electronic chart details to be 'underlaid' on the radar display, in what is known as a chart radar. This, allows the radar to provide a positional check at a glance; if the radar echoes line up with the corresponding chart features then the ship must be at the position indicated on the chart.

The series of figures here show first of all a representation of simplified chart showing land, a jetty and some navigational buoys (Figure 16.6a).

The second figure, Figure 16.6b shows the corresponding radar picture. It shows echoes from the land, the jetty, the buoys another ship on the port bow and some various other smaller targets (probably small craft).

The third figure shows the radar video overlaid on the chart (Figure 16.6c). The echoes from the jetty and certain parts of the land can be seen to coincide with the corresponding chart features as do the echoes of what are probably the buoys.

In the fourth figure there is obviously an error (Figure 16.6d). This could be due to one or more causes. A malfunction in the GPS receiver, incorrect adjustment to comply with WGS84, and incorrect adjustment in relation to the CCRP.

It should be remembered that in such a case, the correct range and bearing of, say, the jetty is that of the radar echo, not the chart feature. Also note the confusion that may arise in the case of correctly identifying the echo of the south cardinal buoy off the starboard bow.

However, there are several aspects of radar overlay that need to be considered, and the proposition that such a positional check might be sufficient to provide the necessary secondary positional reference to that of satellite derived positions. Let us consider these various aspects.

Consistent Common Reference Point (CCRP)

This, as the name suggests is a point in the ship to which all positional data is referred; the radar antenna and the GPS antenna can be separated horizontally by some distance, and these in turn can be some distance from the conning position. Adjustments, usually carried out on installation, correct these to a single position, usually the central position in the bridge called the CCRP. An error in the adjustment to the CCRP for the radar and/or GPS receiver will result in a discrepancy between the radar image and the corresponding chart features. This will normally only be apparent at very short ranges/large scales and where the horizontal distance between their antennae is large.

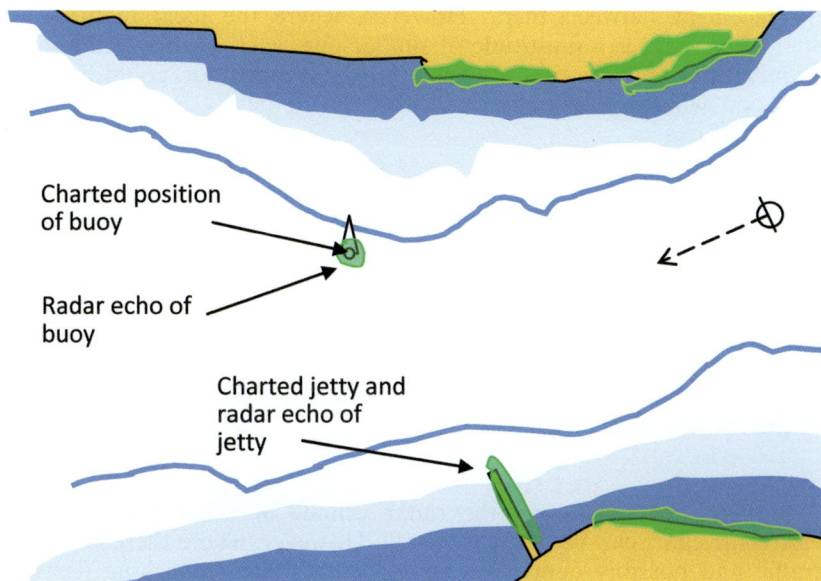

FIGURE 16.7   Showing example of radar echo size on electronic chart.

Echo size

The size of the radar echoes and distortion produced by pulse length and beam width can make the matching of radar image and chart features imprecise. This is particularly so with "S-band" (10 cm) radar, or with "X-band" on long pulse length (see Figure 16.7).

Range

The distance from the features to be matched can reduce the accuracy of the positional check to an unacceptable degree. This is related to the echo size. The ship might be in waters with shoals and other hazards nearby which are well offshore and the imprecise radar image of the distant land does not give adequate accuracy when overlaid on the electronic chart.

Correct identification of radar echoes

This is obviously crucial and in many situations is not a problem, the disposition of land, buoyage, jetties and other features are such that they can be readily identified and matched with the chart. When the radar overlay coincides with the chart features and it is easy to see any displacement between them. However, where the local features are sparse, or there is a multitude of similar ones such as in a well buoyed channel it is possible that an error in the GPS position could have the situation where some radar echoes match or nearly so to the wrong chart features.

False echoes and clutter

False and spurious echoes as well as sea and rain clutter can make the radar video so confused as to make it nearly impossible to match with the chart features. At the very least a profusion of such unwanted echoes can clutter up the chart display with the resulting danger to loss of situational awareness. Under such circumstances the radar video will probably be switched off and thus it gives no positional backup. (This shows the "honesty" of radar with the viewed image being a reflection of the quality of the information.)

Traffic

In many circumstances the radar echoes of other traffic on the electronic chart can be of great benefit. However, where there is a lot of traffic and possibly concentrations of echoes such as fishing fleets or pleasure craft, these like the false echoes and clutter can clutter up and confuse the chart display.

Other Errors

Any error due to the positioning of the scanner with reference to the CCRP will cause a corresponding error in the radar image *vis-a-vis* the chart. If the heading marker is not correctly aligned then there will be an azimuthal error in the radar image and any error in the time-base will proportionally put the radar image out of scale with that of the chart.

Chart Radar

The requirement, under the present IMO regulations is that chart features reproduced on the radar screen should not overwhelm the radar picture. In other words the colour and intensity of the chart features should be more subdued than the radar image. Also, the number of chart features which can be reproduced on the radar is limited. The purpose of the chart underlay on the radar is to improve situational awareness in dealing with traffic and collision avoidance. Monitoring the ship's position and progress should be done on the electronic chart display.

Even with the subdued colours and intensity, if there are too many chart features then the anti-collision plot can be confusing with possible dangerous consequences.

Bearing in mind these potential problems radar overlay on the chart and chart underlay on the radar are very useful features, the prudent navigator should be guided by a knowledge of these limitations.

Parallel Indexing (PiLs)

One of the most useful practices for coastal navigation and pilotage is the use of radar applying parallel indexing. This practice was very quickly adopted by the navies in the early days of radar, but was more cautiously adopted by merchant ships' officers in later years. In its use in early radar, in addition to the mechanical bearing cursor in older radar sets (with engraved parallel lines, centred on the centre of the screen), the main accessory to the radar display was a 'reflection plotter'. This allowed the officer to draw and plot on the face of the radar screen with a chinagraph pencil. It also required the radar to be gyro stabilised, something that was not always available in the early days of radar aboard merchant ships. The reflection plotter was a useful and versatile tool, but it had limitations. The main one was that whatever was plotted or laid off on it applied to only one range scale; if the operator changed scale, then the work on the screen had to be re-drawn.

Modern sets have electronic lines available to the operator. The number of such lines varies with the make of radar, but usually at least three are available. Each line can be rotated to any bearing, and placed at any distance from the origin marking the position of 'Own Ship'. The length of the lines is indefinite in most radars, and go from edge to edge of the screen at any range scale. Normally in the default mode, such lines, once placed on the screen by the operator remain fixed at the distance from the origin, and referred to 'true' and will remain fixed in azimuth but some radars refer to relative and the lines will rotate with ship's head. The point made earlier and repeatedly is that ships' watchkeeping officers must know their equipment.

The simplest application of parallel indexing is its use in maintaining the track passing along a coastline or off a headland. If, for example the track takes the ship 5 miles off a particular headland or other prominent radar target, then the index line is set at 5 miles on the appropriate side of the ship, and in the direction of the track. The ship is then steered to keep the echo of the object on the line. The ship may not be steering the actual direction of the track, but may be steering to counteract a tidal stream or current, and the OOW may have to adjust the course periodically if the echo of the object moves off the line. Similarly the OOW may have to alter course to allow for traffic. A glance at the radar, and the position of the PiL (parallel index line) in relation to the reference object will show by how much the ship is off track.

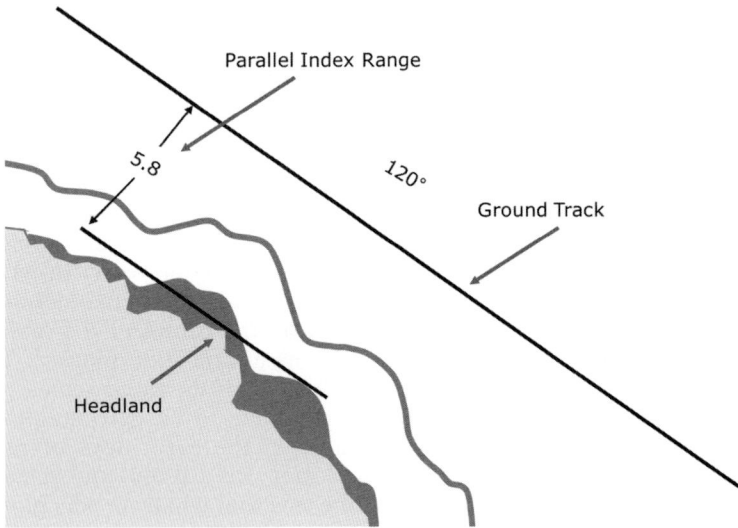

FIGURE 16.8a   Showing parallel index lines.

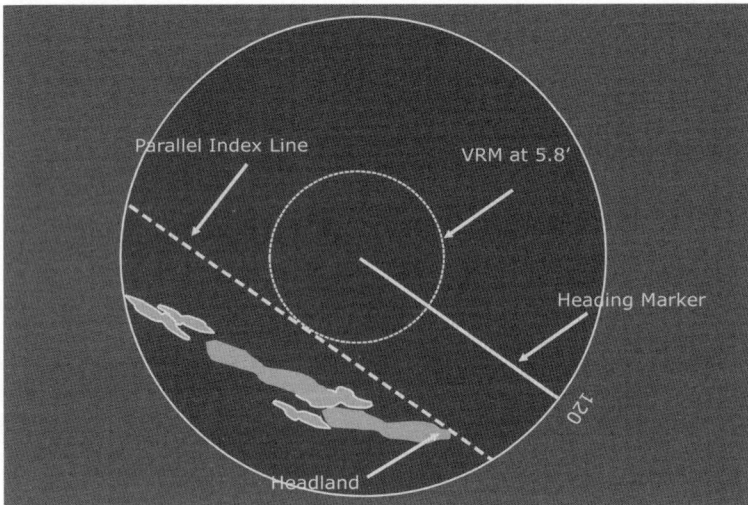

FIGURE 16.8b   PiL on radar screen.

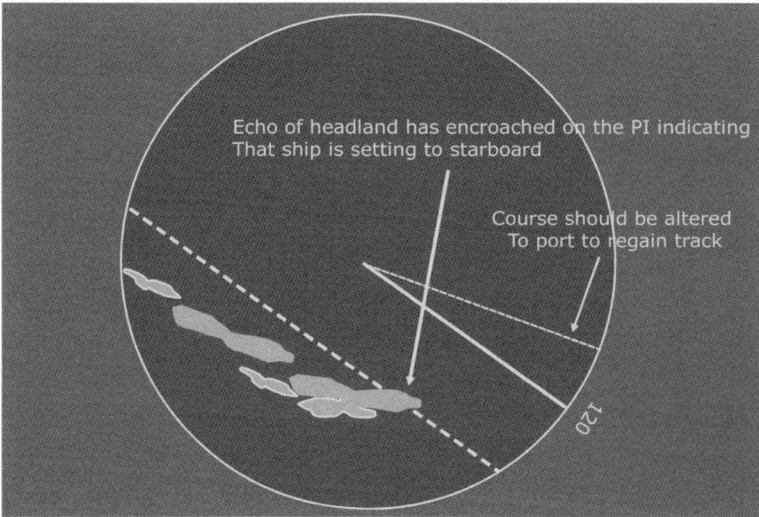

Echo of headland has encroached on the PI indicating
That ship is setting to starboard

Course should be altered
To port to regain track

120

FIGURE 16.8c   PiL indicating ship being set.

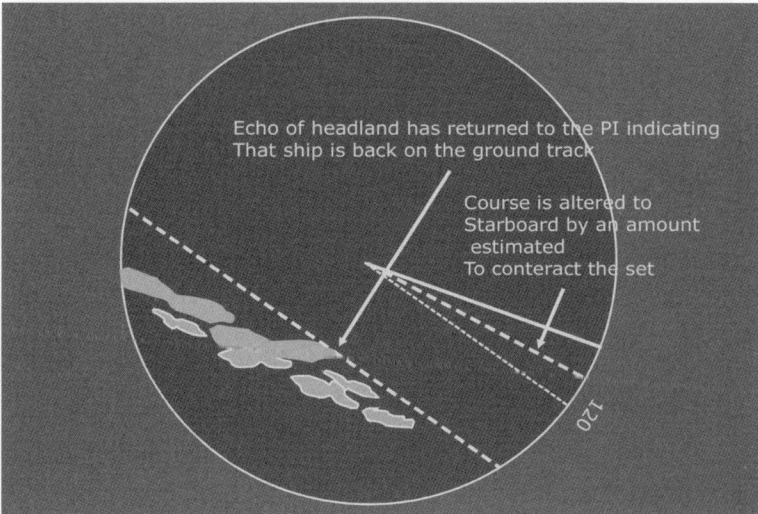

Echo of headland has returned to the PI indicating
That ship is back on the ground track

Course is altered to
Starboard by an amount
estimated
To conteract the set

120

FIGURE 16.8d   PiL indicating effect of corrective action.

The use of trails in parallel indexing can be used to advantage as, at a glance, one can also see if one is moving further away from the track, running parallel or returning to the track. It also gives an indication of the degree by which Own Ship is being set and thus the amount by which the course needs to be adjusted to counteract it.

Provided the correct setting up of the PiL line has been carried out, and the correct index target or object identified, this technique gives almost immediate and accurate positional information regarding the ship's offset from the planned track. It will also show fairly quickly any trend or divergence from the track. As it is in 'real time' it would certainly show any such trend more immediately than plotting a series of fixes over a period of time. However, it should be used in conjunction with fixes by other means, such as visual bearings when possible, or GPS or other navigational systems. Good navigational practice requires checking one system against another. Remember, parallel indexing, in its most basic application does not give the ships progress along the track (see Figures 16.8a, 16.8b, 16.8c and 16.8d).

Even in situations where the ship's position is being plotted continuously and automatically on the ECDIS, PiLs should be used. The fundamental reason for this is that as stated above, the positional information derived from radar is independent of external sources. The position on the electronic chart is derived from GPS or other external systems. Any discrepancy will become immediately apparent, although it is remotely possible that the error from the external source is directly along the line of the parallel and therefore would not be readily apparent.

In the older technology of reflection plotters and chinagraph pencils, the parallel index lines had to be prepared for a particular range scale, or several different lines for more than one range scale. In the latter case, the screen could become cluttered, and there was a consequent increased danger of error. However, one redeeming feature was the ability to put small notes or 'tick' on the index lines, particularly 'count down' distances to go to an alteration of course or to an anchorage. The modern system does not allow for this and once again the navigator must develop his/her skills in manipulating the VRMs and EBLs and other graphics features of the particular set.

A second 'clearing' index line is sometimes useful to ensure a safe clearance from some danger, or shoal waters. In this, it should be remembered that this is laid off in the reverse order from the distance taken from the chart. If, for example, the track is passing 5.8 miles off the headland, and a clearing PiL of 3.5 miles is required, in other words the ship is not to close the headland closer than 3.5 miles, the clearing PiL is set at 3.5 miles from the origin and appears to be on the 'wrong' side of the track PiL (see Figures 16.9a and 16.9b).

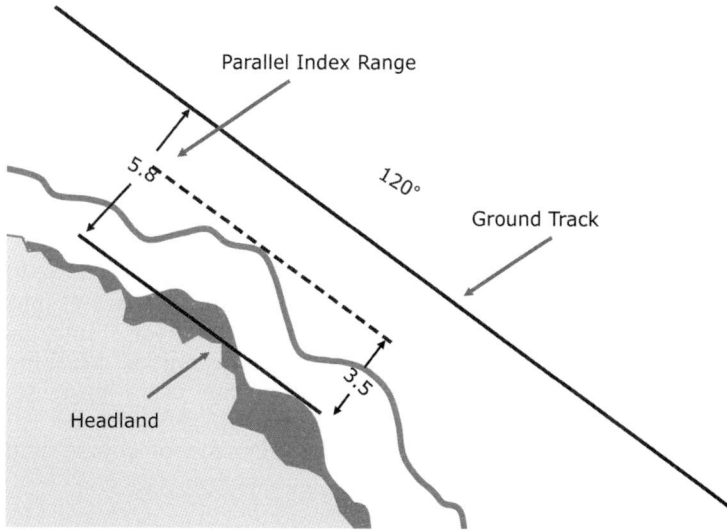

FIGURE 16.9a   Showing parallel index and clearing PiL.

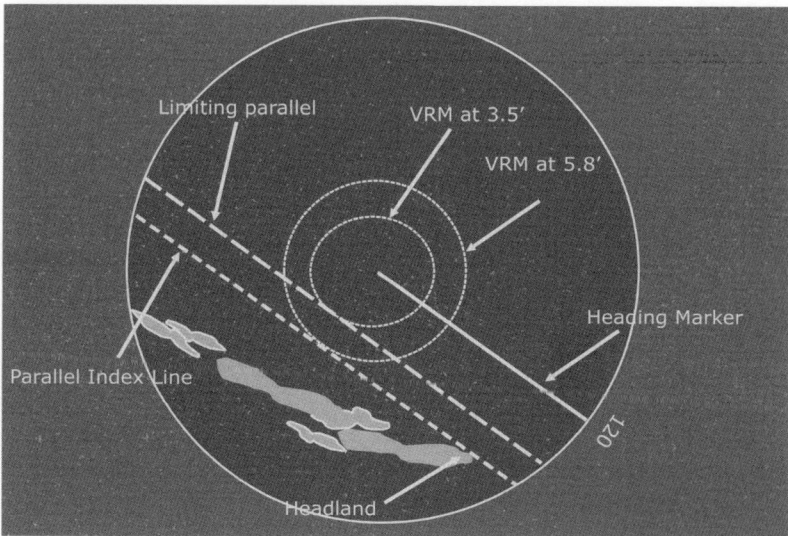

FIGURE 16.9b   Showing parallel index and clearing PiL.

Apart from maintaining track, parallel indexing is essential for other aspects of 'blind pilotage'. The marking of 'wheel-over' (WO) for an alteration of course, a succession of tracks entering confined waters and clearing parallels are all part of a previously prepared passage plan. It is dangerous in the extreme to attempt to do it 'on the fly'. As the number of lines available on a particular radar may be limited, there will be a need to move the lines for new PiLs as the ship progresses from one track to another. The details of the passage should be written into a navigator's notebook showing clearly the parallel index ranges, the clearing index ranges and the objects to which they apply. To make the task easier, a diagram or simple hand drawn map, or series of maps, is the best way of noting this. This will be part of the overall information in such a notebook giving details of the passage plan. Marking the WO is discussed below under 'alteration of course' (see Figures 16.10a, 16.10b, 16.10c, 16.10d and 16.10e) (see page 266).

The section on precise anchoring and alteration of course contains more detail on this topic.

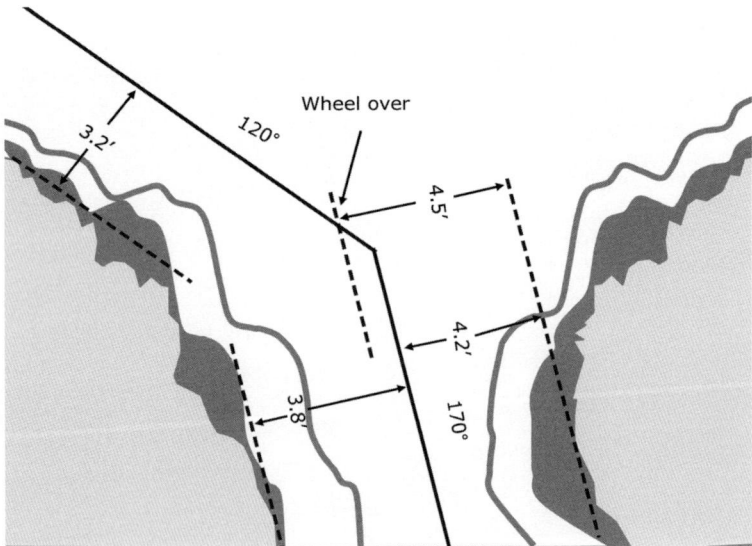

FIGURE 16.10a   Showing use of parallel indexing for alteration of course.

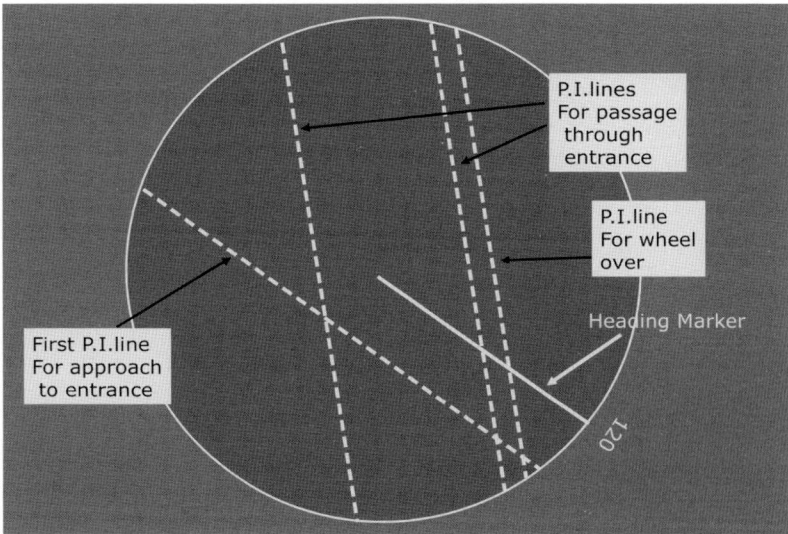

FIGURE 16.10b   PiLs prepared for alteration of course.

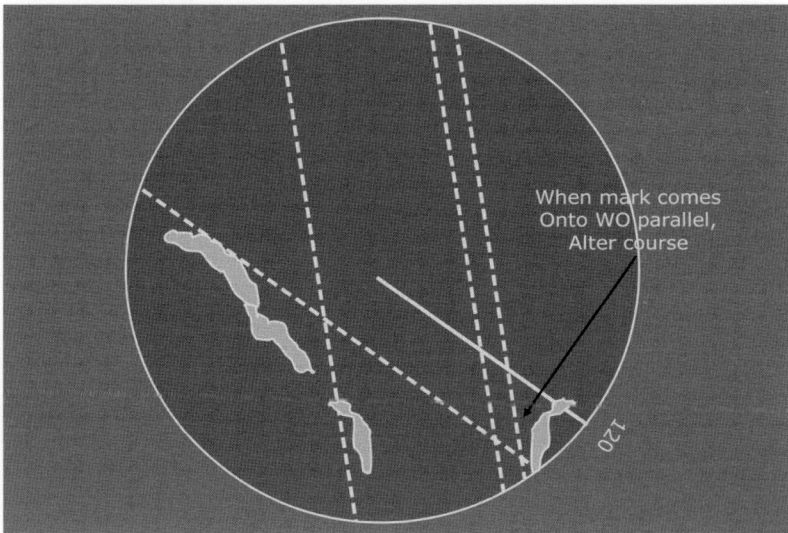

FIGURE 16.10c   Approaching 'wheel over'.

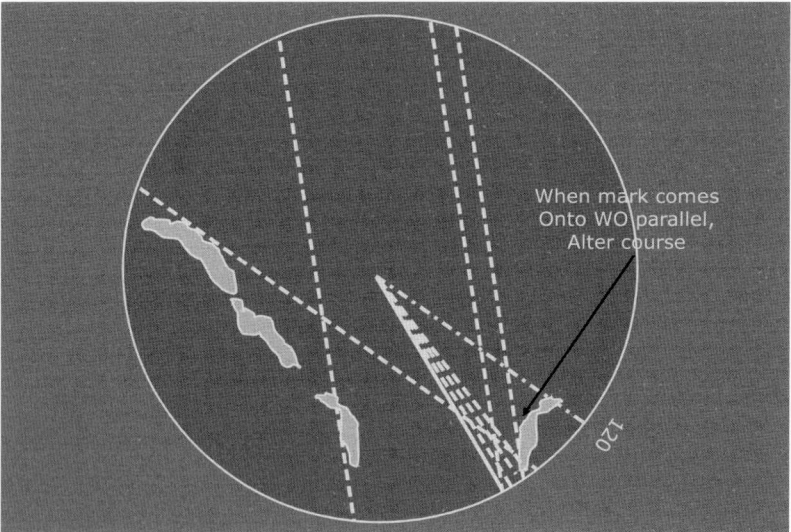

When mark comes
Onto WO parallel,
Alter course

120

FIGURE 16.10d   On 'wheel over'.

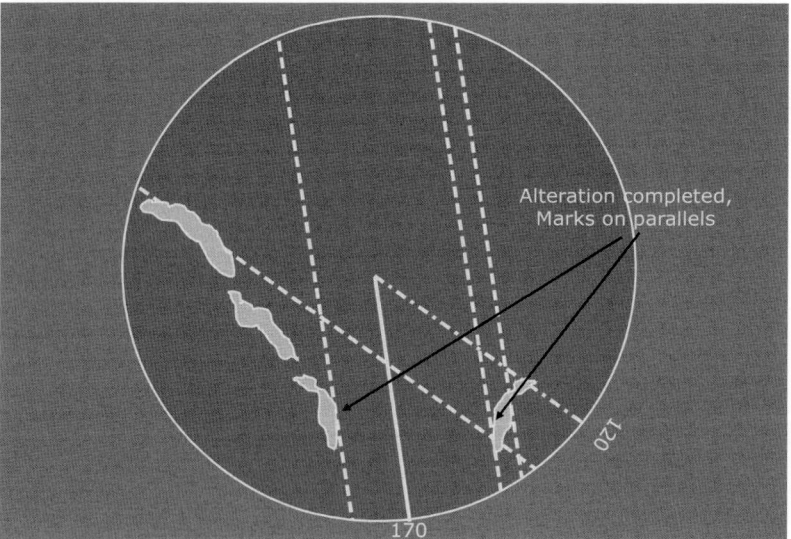

Alteration completed,
Marks on parallels

120

170

FIGURE 16.10e   On new course.

Navigation Lines and Maps

Depending on the make of the radar set, a number of navigation lines, or an application to draw simple 'maps' is often available. The use of these is different from parallel indexing in so far as the lines, or outlines drawn are usually related to geographical features, for example a traffic separation scheme. They must therefore be fixed geographically. This is either done by the GPS or other navigation system, or by automatically tracking a known stationary target and 'locking' the map to it. The map or series of lines and other symbols can of course be often be fixed to the origin and so a more complex series of parallel index lines can be stored. The operator should be fully satisfied that this is the case before using this function. Sometimes if the map or series of lines is not fixed geographically, or to a tracked stationary target, they may not 'lock' to the origin and drift. This of course is obviously dangerous; the navigator could keep 'dragging' his map back into position, but this is not very good practice and prone to error.

The tools provided for drawing these lines or maps vary from manufacture to manufacture of radar. Usually they allow two methods of inputting the necessary data; either alpha numerically entering Latitude and Longitude figures or 'drag and drop' using the roller-ball. In many sets the process is less than 'user friendly'; it is tedious and time consuming and sometimes difficult to save the work as it progresses. Often, after having spent some considerable time and care in tediously transferring a series of Latitudes and Longitudes, the pressing of a wrong button in error can lose the whole work necessitating starting all over again. However, once the map has been successfully drawn, it can usually be stored and retrieved when required. The use of such maps, even where there is electronic chart underlay, offers the advantage of plotting a navigational feature that is applicable to the particular ship or voyage. They can also simplify some navigational feature or task that might be more cluttered or complex than necessary if using electronic chart underlay.

In drawing and storing such maps great care and attention to detail must be taken. The process is often so tedious that it is easy to make a mistake. Preferably another person should be asked to check the work. It is also the duty of the OOW, who is responsible for the safe navigation during his/her watch to double check any such maps or lines in use, particularly those drawn by someone else, possibly in the distant past who is no longer serving in the ship.

In using maps, or otherwise ground referencing or locking the display to a stationary radar target, care needs to be taken in selecting such a target, and one may not always be available. Ideally it should be a small

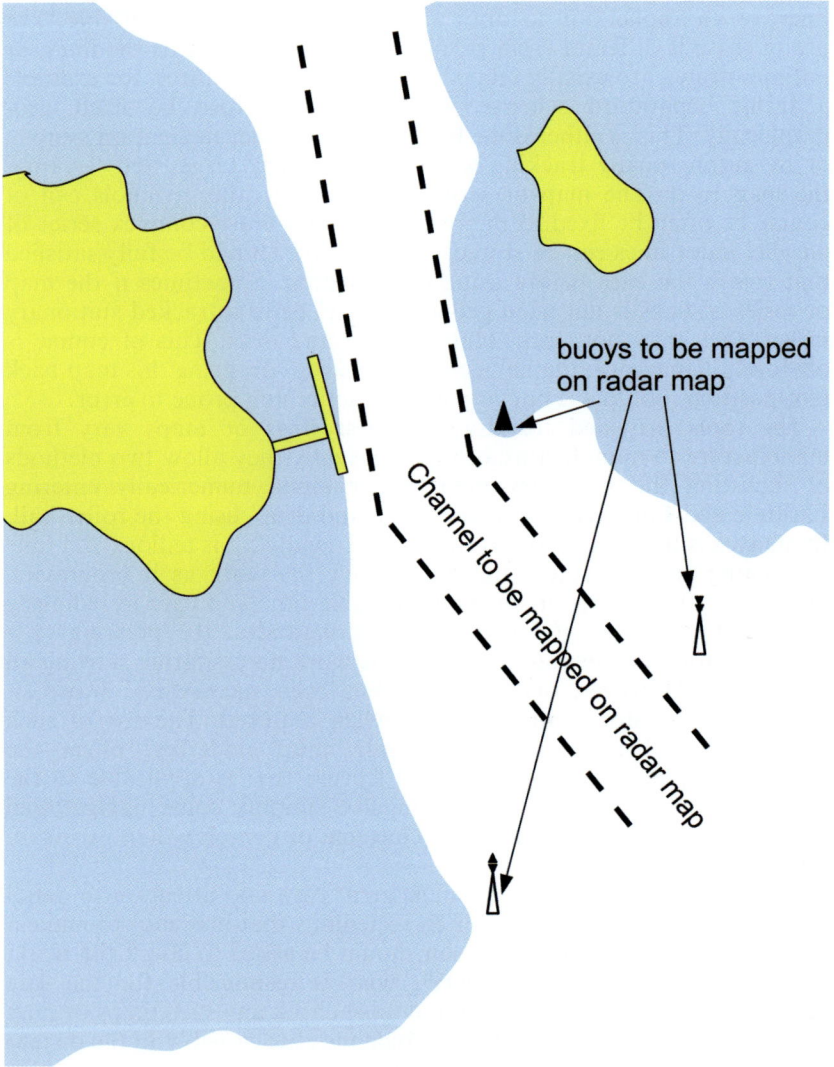

FIGURE 16.11a   Sequence showing simple radar map.

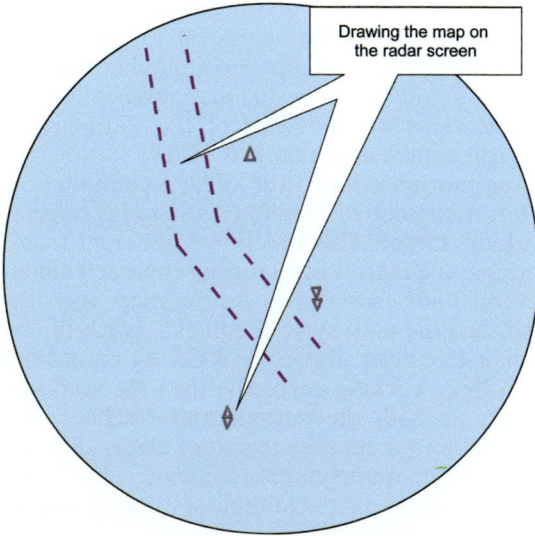

FIGURE 16.11b   Map drawn on radar screen.

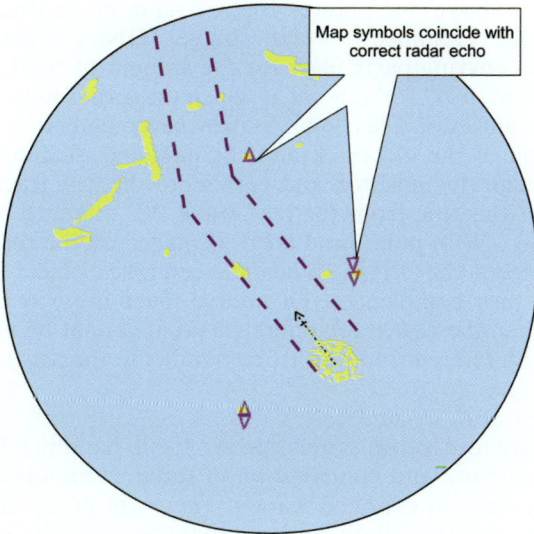

FIGURE 16.11c   Map in use on radar.

target with good radar response, and needs to remain outside the area of clutter. It should obviously be identified as ground fixed and thus stationary. Isolated rocks, beacons and similar are most suitable. Navigational buoys are suitable in the absence of anything else. Knowledge of the exact co-ordinated of the stationary target is not important once it is established that it is in fact stationary; its function is to provide true motion data to the radar's computer. The target so selected often has a distinctive symbol on the radar screen to identify it as the 'ground-lock' target. This function is different from locating the map. If for example the map is marking an approach channel to a port, and one or several buoys are drawn on the map, then the map needs to be located so that the map symbol coincides with the relevant radar target. If the map has been drawn in WGS 84 co-ordinates (lat. and long.) and the GPS is working correctly, then the map symbols should coincide or nearly so with the relevant radar echoes. If there is any discrepancy beyond an acceptable degree of error, then the map needs to be moved so that the map symbols coincide with their radar echoes and the radar ground locked on a suitable stationary target.

Alteration of Course

The use of a PiL in marking a 'wheel over' point is illustrated in the series of figures above (see Figures 16.9a, b, c, d, e). As discussed in the next section, the 'Distance to Next Course' (DNC) is extracted from the ship's manoeuvring particulars for the amount of rudder to be used in a constant rate turn, or the point at which the turn is to be commenced in a constant radius turn. This point, for whichever method, is laid off from the intersection of the old track and the new one. A line is drawn on the chart through this position and parallel to the new track. The cross index range of this line from the new track PiL reference point marks the 'wheel over' (WO) point, and a PiL is placed on the radar screen at this distance from the origin. When the reference object impinges on this line, the turn commences. Of course there is no reason why an object other than the new track reference point cannot be used for WO, but the PiL with reference to it must be parallel to the new track.

Altering Course in Restricted waters

The topic here touches on some aspects of ship handling. Our purpose is to examine the use and contribution of radar in monitoring turns as a ship alters course in confined waters. There are many fine works on the subject of ship handling and manoeuvring which is a complex subject in its own right particularly in dealing with ships in confined and shallow waters.

There are two requirements to getting a ship safely from one track to another. One is to alter course in such a way that the ship comes on to the new track accurately, and the other is that she follows a pre-determined path during the turn.

The two methods of allowing for, and controlling an alteration of course:

constant rate,
where the rudder is put over to a set amount, and kept there until
    steadying up on the new course, or
constant radius,
where the ship is kept to an arc of a circle using varying rudder
    amounts.

In smaller ships the second of these requirements may not be so critical and once the turn has been started at a suitable point, the process of getting her onto the next course can be adjusted during the turn often by watching the parallel index lines of the new track. In larger ships, and even in smaller ones in very confined waters, a more precise track needs to be planned and achieved during a turn or alteration of course.

Traditionally planning a turn was (and is) done by referring to the ship's turning circle and determining a 'wheel-over' position, and by using a standard amount of rudder, say 15°. The ship then was expected to track around a segment of a circle roughly corresponding to her turning circle. Allowances can be made for tidal stream, usually in determining the position of 'wheel-over'. The problem with this is that there are too many variables and unknowns. Often the only turning circle data produced for the ship is taken during her trials and is that for full rudder at a fairly high speed, with the ship in ballast condition, and in deep water. In any particular instance the tidal stream data may not be accurate, and the effect of wind on the ship during the turn is undeterminable. The effect of different loaded conditions of the ship will also affect her behaviour; the windage may be different, the under-keel clearance will be different, the mass(weight) of the ship will affect how she turns as will her trim and the pivot point about which she turns. Certain bearings can be checked to ensure the ship is not getting into danger during the turn, but otherwise its position during the turn can be somewhat unknown and the navigator is watching the next track data to ensure an accurate outcome of the manoeuvre rather than the precise position during the turn. This technique is known as 'constant rate' (see Figures 16.12, 16.13, 16.14).

What is far more determinable and controllable is a constant radius turn. This is a segment of a circle, whose radius is decided by the navigator and which can be plotted accurately on the chart. It should,

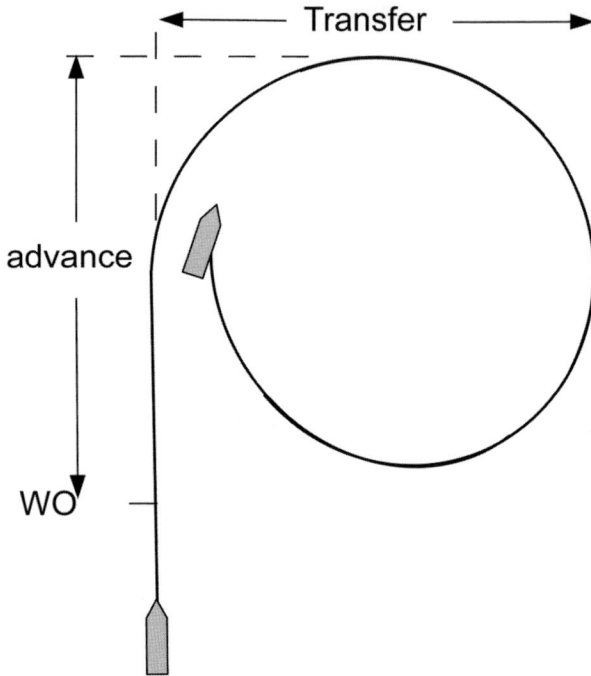

FIGURE 16.12   Turning circle.

of course be not of a lesser radius than the ship's smallest turning circle. It is centred either on a suitable navigational mark, an accurately positioned buoy for example, or an imaginary point plotted geometrically for the intersection between the two tracks. Most electronic chart systems will plot such a curved track with the facility to input the required radius. This then can often be displayed on the radar screen and the ship conned along the track usually with the radar in 'true-motion' mode. The 'rate-of-turn' (RoT) required should be determined and then carefully maintained by watching the RoT indicator and adjusting the rudder angle. The factors affecting the RoT are:

- the fall-off in speed which usually occurs during turns,
- the extent of the turn,
- initial speed,
- windage,
- amount of rudder,
- skill of the helmsman, and/or adjustment of the autopilot.

As the speed will drop during the turn, it follows that the RoT will also drop. It therefore requires an adjustment of the RoT to maintain the radius. If, for example, the radius is 1 mile, and the speed entering the turn is 14 knots, the RoT will be 14° per minute. If the speed during the turn drops to 13 knots, the RoT should be 13° per minute to maintain the radius. (if possible there is much merit in having turn radii of either ½ mile or 1 mile in passage plans as it is very easy to work out the required rate of turn. At ½ a mile the RoT will be twice the figure for speed in knots.)

With very large ships and ships which are limited in the rate of turn (RoT) which can be applied, such as large cruise ships which cannot be heeled more than a very few degrees, the subject of controlling the ship during a turn and ensuring that it follows a predefined track has made the subject increasingly topical. The use of electronic charts has made such monitoring possible, and allowed the following of a precise curved track. This should be complemented by check bearings, and often shore monitoring such as from VTS in certain ports.

However, as radar is a completely independent system and does not directly rely on GPS it offers a reliable method which can be applied in many cases to support and confirm the monitoring of the ship's track during a turn.

Constant Rate Turn

The constant rate turn is derived from the ship's manoeuvring data as mentioned above. Ideally this is shown for a number of different rudder angles. In such turns, the ship will describe a slight spiral track if the rudder angle is maintained. However, for many merchant ships, the available manoeuvring date is quite limited. As stated above, there will be a limited number of trial turns carried out and these will be usually carried out in ballast conditions and in deep water. The variations imposed by different loaded conditions, and the under-keel clearance (UKC) can only be approximated. In Figure 16.13, WO is 'Wheel Over', the position that the helm is applied depending on the amount of helm to be used. Ideally there should be a selection of such diagrams for rudder angles of 5°, 15°, 20°, 30°. 'Advance' is the distance the ship advances along the direction of the original course during the turn, and transfer is the lateral displacement of the ship from the original course during the turn.

One of the phenomena of this is that the actual speed of the ship has very little effect on the dimensions of the turning circle; in other words the diameters will be largely the same if the ship is 5 knots or 15 knots.

In commencing a turn, a factor in determining where the WO

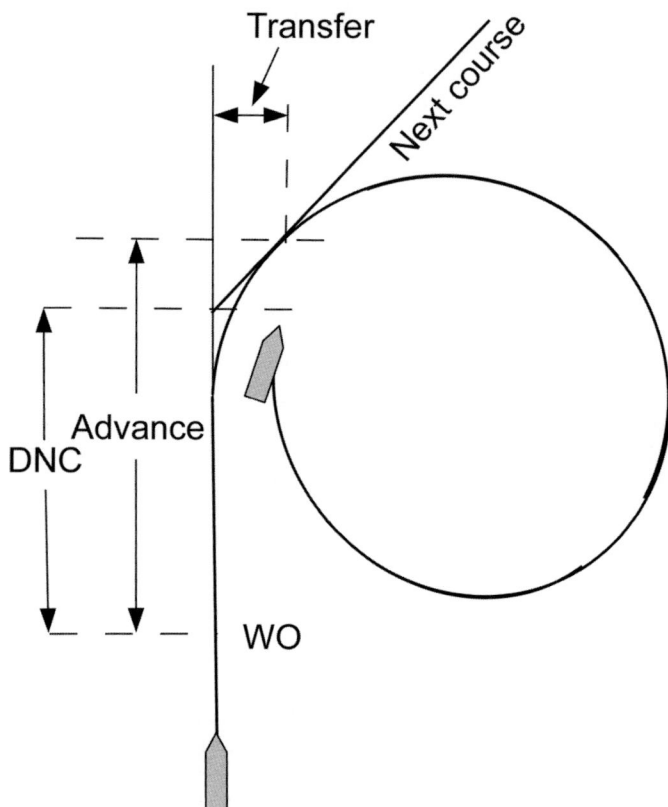

FIGURE 16.13   Showing advance and transfer.

position is, in relation to the amount of advance and transfer is the directional inertia of the ship. Generally speaking, long narrow ships have greater directional stability and therefore will delay for some time after rudder is applied before the turn starts. On the other hand, beamy ship are often described as directionally 'unstable' and require little amount of rudder to start a turn.

Factors that will affect the diameters of the turn are rudder angle, and UKC. Most manoeuvring data is derived from deep water trials. Where the depth under the keel is approaching or less than the draft, there will be a substantial increase in the diameters, increasing them by say 100%. The bridge team can be led into an erroneous situation

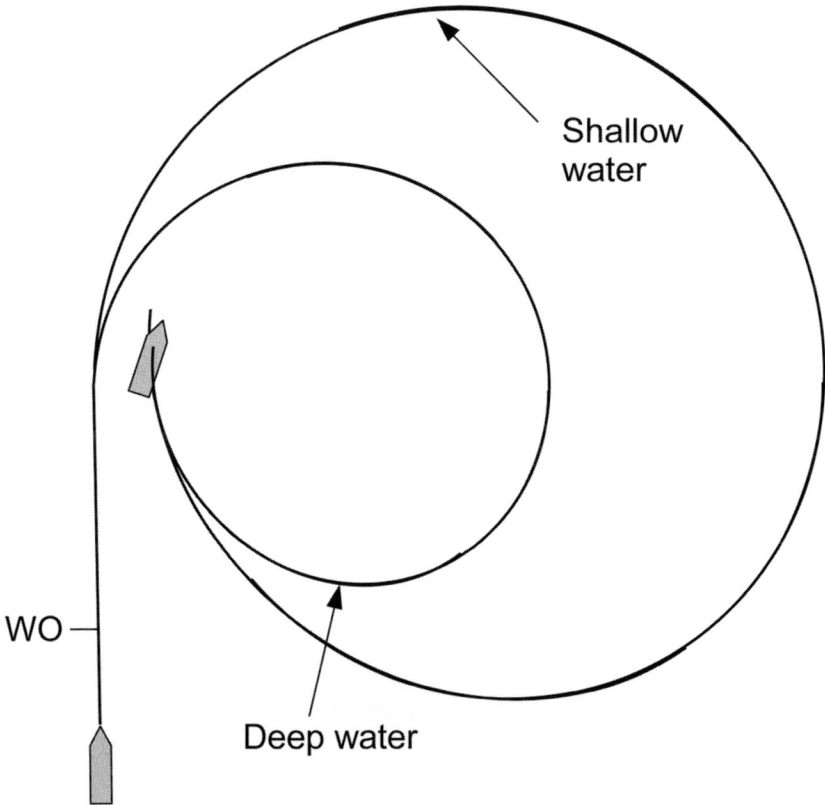

FIGURE 16.14   Showing increase in turn radius in shallow water.

because the rate of turn will not change; in other words the ship is turning just as fast in shallow water as in deep, but due to the smaller drift angle, is taking a much wider curve for a given rudder angle.

Another phenomenon is that the ship's speed will fall dramatically in large alteration of courses, for a given propeller RPM (revolutions per minute). Contrary to what one might expect such speed reduction will be more pronounced in deep water than in shallow due to the smaller drift angle.

The use of the turning circles as provided in the ship's manoeuvring data can be used to plan the alterations of course and establish 'Wheel Over' lines, as in parallel indexing. Let us say that an alteration of course of 40° is being planned, and to use 15° of rudder.

Probably the most useful bit of information out of this is DNC (Distance to New Course), to establish the Wheel Over line.

The wheel over line should be a line, parallel to the new course or as close to parallel as possible, as this will eliminate the cross track error, if any, on the original course going on to the new course (see Figure 16.15).

The wheel over line, or position will be affected by any tidal stream or current. This need to be allowed for where it is required to fall

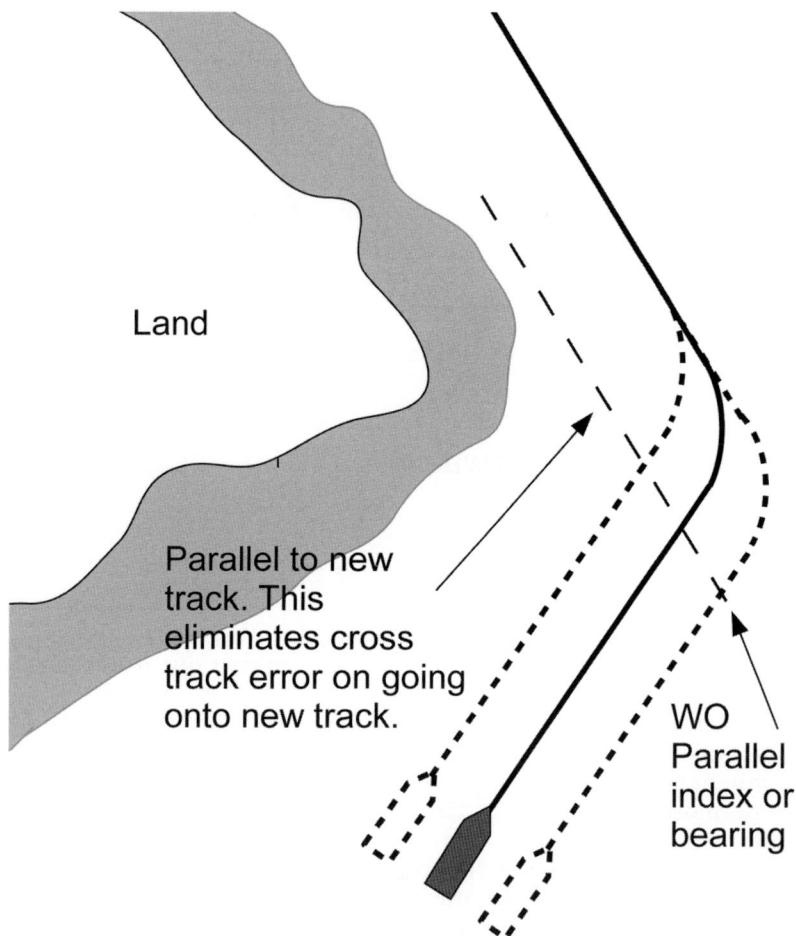

Land

Parallel to new track. This eliminates cross track error on going onto new track.

WO Parallel index or bearing

FIGURE 16.15   Showing alteration of course and WO bearing.

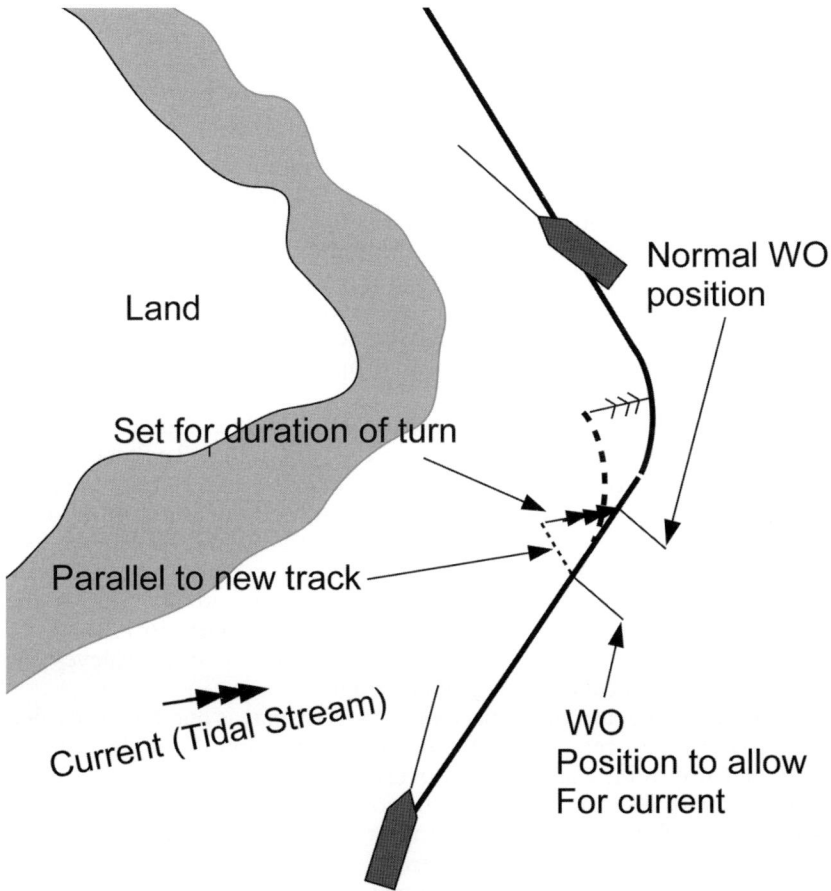

FIGURE 16.16   Showing WO adjusted for tidal stream.

accurately onto the new track, as the ship comes onto the new course.
In this, the alteration of course will be the difference between courses to
steer to counteract the current before and after the alteration (see
Figure 16.16).

Constant Radius Turn

A more accurate way of achieving a turn is to use rate-of-turn
(RoT) to achieve a constant radius. Once the radius is bigger than the
tactical diameter for the maximum rudder angle then the navigator can

choose the radius of the turn. This gives greater control and possibility of monitoring the track than the constant rate method.

RoT ($°$/min.) = V/r,
where V is speed in knots, and
and r is radius in nautical miles.

As the speed will drop during the turn, it follows that the RoT will also drop. It therefore requires an adjustment of the RoT to maintain the radius. As discussed earlier, if the radius is 1 mile, and the speed entering the turn is 14 knots, the required RoT will Initially be $14°$ per minute. If, as expected the speed decreased during the turn, the RoT will have to be decreased accordingly by adjusting the amount of rudder angle. If the speed during the turn drops to 13 knots, the required RoT should be $13°$ per minute.

If the turn radius is 0.5 miles, then the RoT will be double the speed. If the ship enters the turn at 14 knots then the initial RoT will be $28°$ per minute. If during the turn the speed falls off to 12 knots then the RoT will have to be adjusted to $24°$ per minute to maintain the 0.5 mile turn radius.

A Rate of turn indicator is required to carry out this manoeuvre. Also an auto-pilot with a rate of turn setting and/or track mode connected to and electronic chart can be used (see Figure 16.17a).

The track must be monitored to ensure that the ship achieves the planned geographical curved (ground) track.

Monitoring the turn on an electronic chart display is one way, provided the accuracy and reliability of the positioning input (usually GPS) is verified. Confirming that the radar overlay on the electronic chart display coincides with relevant and correctly identified geographical features is one method of ensuring the correct relative geographical location of the curved track, and the ship's progress along it. In ensuring that the overlay and chart coincide, well defined objects need to be selected (see the section on radar navigation and chart underlay).

Concentric Indexing

Radar comes into its own in 'Concentric Indexing', a procedure developed by Captain Paul Chapman, Senior Marine Pilot of Marine Safety, Queensland, Australia. There are several ways in which radar is used for this purpose which are discussed. It must be said however that familiarity with the features of the particular radar set is essential and the technique should be tried and practised in less critical situations before it is used in earnest. The wide variations in functions, such as VRM, EBL, navigation lines, trails, stabilisation, in modern radar set

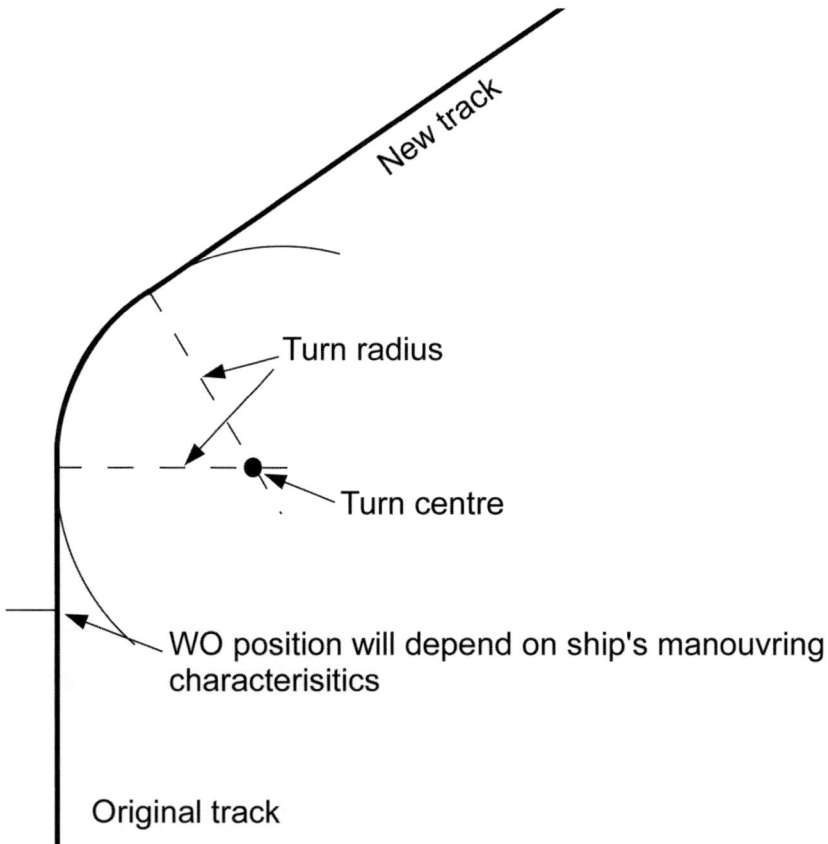

New track

Turn radius

Turn centre

WO position will depend on ship's manouvring characterisitics

Original track

FIGURE 16.17(a) Concentric indexing.

make practising this essential and even to get some modern sets to imitate old, basic head up displays with trails 'relative' to such a display can be quite difficult.

Methodology

The first method is to set up a radar display in head-up unstabilised, with the VRM offset to either port or starboard, that is, on the inside of the turn, by an amount equal to the radius of the turn. It is important to ensure that the offset VRM remains relative to the origin; some radar displays, even in head-up unstabilised mode will have the offset VRM

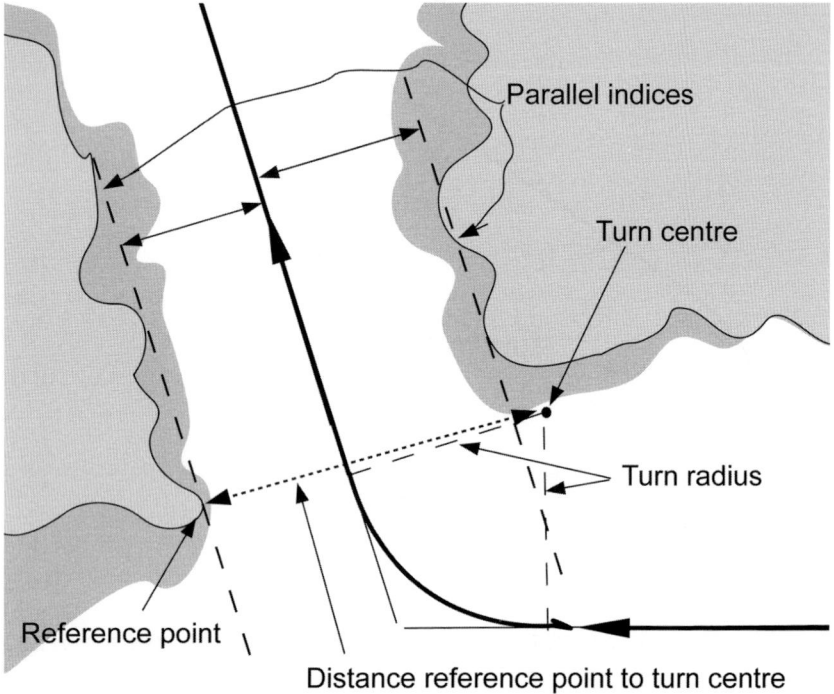

FIGURE 16.17(b)   Concentric indexing chart.

referenced to 'True' and it will not stay in the correct position relative to the origin. The VRM range is then set to the distance of some prominent and easily identified radar feature from the turn centre. Ideally this feature, called the 'turn reference point' should be outside the turn. Ideally it should be well placed to give parallel index range on the new track after the alteration. However this is not essential if other identifiable marks are available for PiLs on the new track (see Figure 16.17b).

Where an offset VRM is not available, an offset EBL with a range marker on it can be used. This is located at the turn centre, with the range marker at the distance of the reference point from the turn centre. This then will have to be rotated as the reference point rotates around the origin during the turn.

During the turn the reference point should track around the offset VRM, or should remain on the range marker of the offset EBL. If it encroaches on the VRM or range marker, the ship is turning too slowly, and conversely if it moves away from the VRM or range marker the ship is turning too quickly.

Starting and completing the turn can be a problem and depend largely on the ship's manoeuvring characteristics as mentioned earlier such as its directional inertia, the radius of the turn and amount of the alteration of course. Tidal stream and leeway will also affect the timing and positions of starting and completing the turn. The WO (wheel over) manoeuvre should be started at some point shortly before the reference point comes onto the offset VRM, or range marker on the offset EBL.

Exiting the turn should be monitored by parallel index lines for the new track. Shortly before the PiLs come onto their reference points for the new track the wheel should be eased so as not to over-shoot the new course. Again, this will depend on the ship's directional stability and the avoidance of excessive rudder movements. Visual bearings and references, soundings and of course GPS position on the ECDIS should all be used; the topic here is a radar technique but it should not be used to the exclusion of other techniques (see Figures 16.18a, 16.18b, 16.18c and 16.18d).

The technique can also be used in 'North Up' mode, but is less intuitive and not as easy to follow, and hence more prone to error. The plan of the turn is the same as for 'Head Up' in plotting the turn centre and reference point. The turn centre is then located abeam of the origin, at the turn radius, with the offset VRM centred there. It is essential that the offset VRM (or EBL) remain in that position relative to the origin, that is abeam, as the heading marker changes direction during the turn. In other words, the centre of the VRM (or EBL) will move in an arc around the origin as the ship alters course. As in the 'Head Up' mode, the reference point should stay on the VRM (or EBL range mark); if it encroaches on the VRM the ship is turning too slowly and if it moves off it the ship is turning too quickly. As can be seen, there are more 'moving part' in this mode and while it is usable, it loses the simplicity and ease which 'Head Up' gives (see Figures 16.19a, 16.19b, 16.19c, 16.19d and 16.19e).

It is important to observe that in using this technique, whether 'Head Up' or 'North Up' that the ship must be precisely on the approach track to the turn, any cross track error in the ship's approach will displace the turn centre accordingly, and its relationship to the reference point. If the turn can be centred on a buoy or other suitable point target, then any such cross track error will be apparent and can be corrected before the turn is started.

On page 283 et seq. is a series of radar images taken during an approach to south Dublin Bay, on a simulator, approaching from the

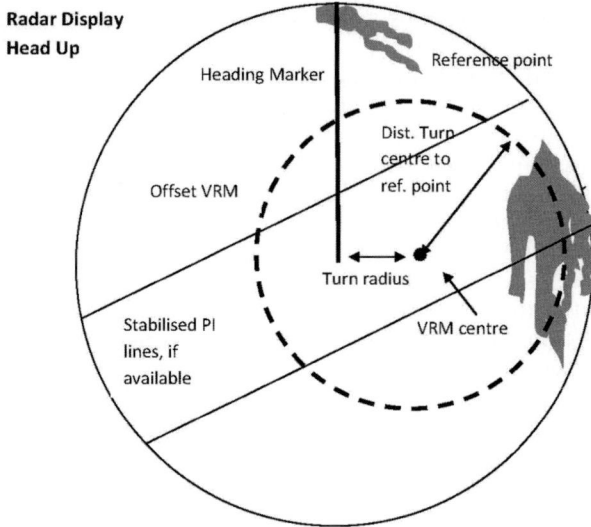

FIGURE 16.18a   Showing stages of concentric indexing turn (stage 1).

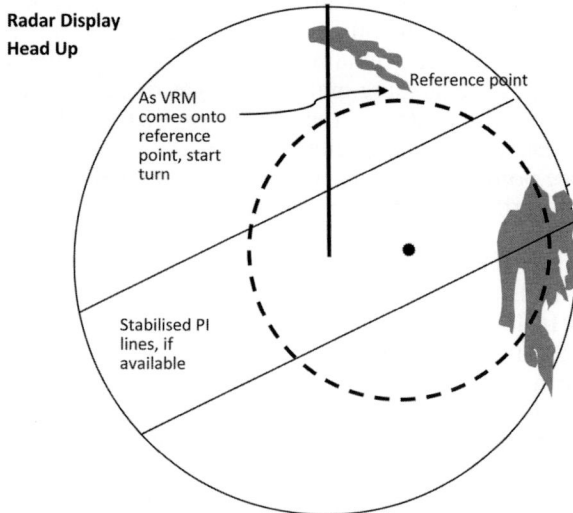

FIGURE 16.18b   Showing stages of concentric indexing turn (stage 2).

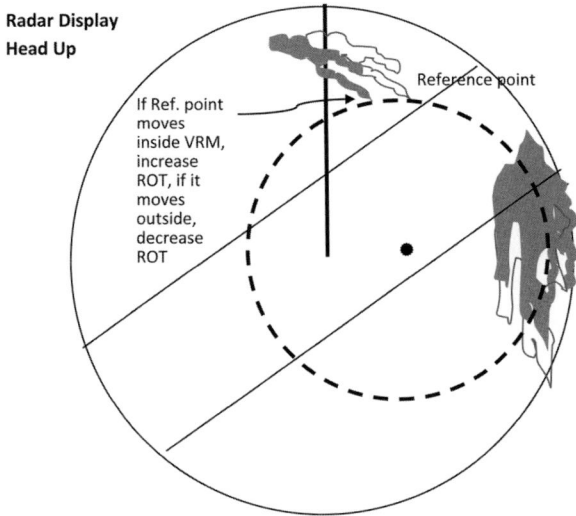

FIGURE 16.18c   Showing stages of concentric indexing turn (stage 3).

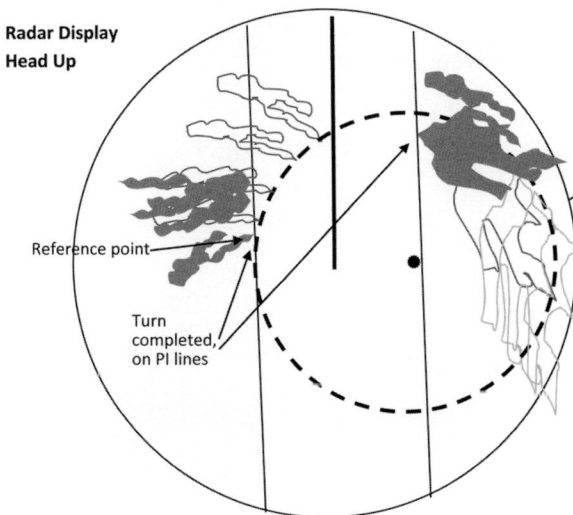

FIGURE 16.18d   Showing stages of concentric indexing turn (stage 4).

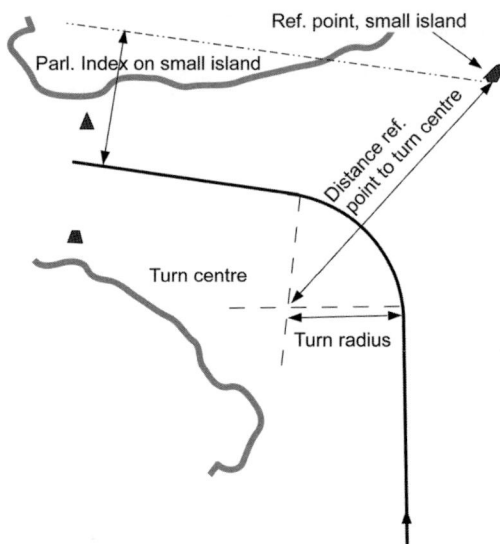

FIGURE 16.19a   Showing sequence of North Up concentric index turn.

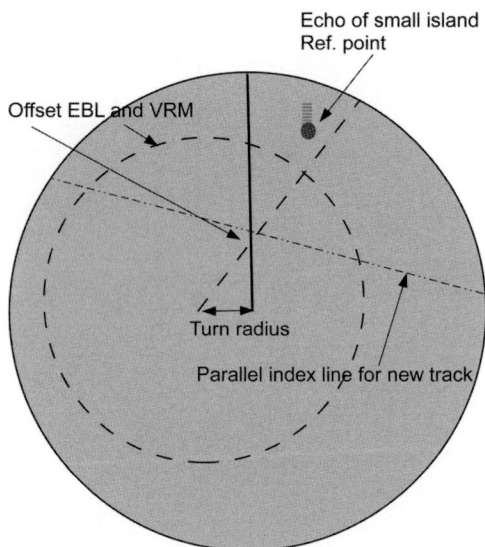

FIGURE 16.19b   Radar screen, North Up concentric index turn, stage 1.

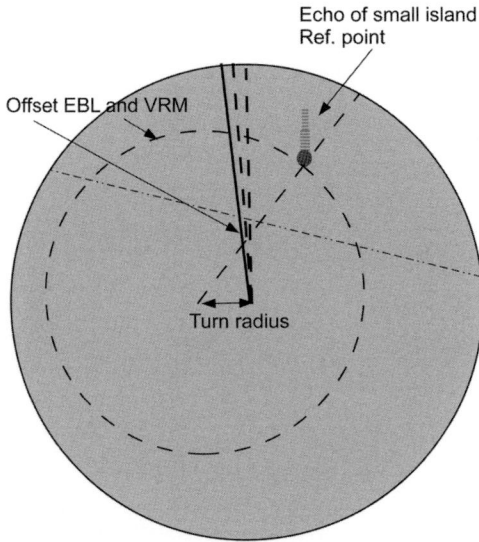

FIGURE 16.19c   Radar screen, North Up concentric index turn, stage 2.

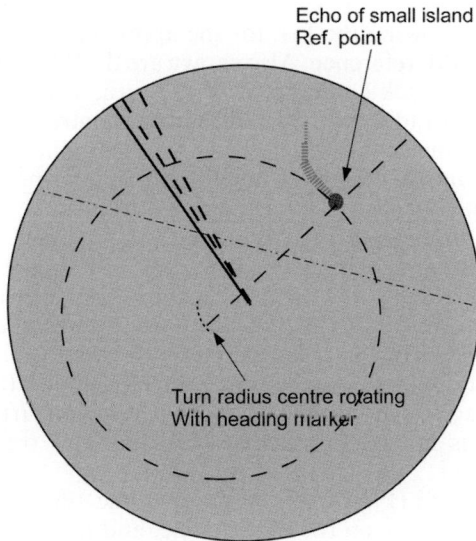

FIGURE 16.19d   Radar screen, North Up concentric index turn, stage 3.

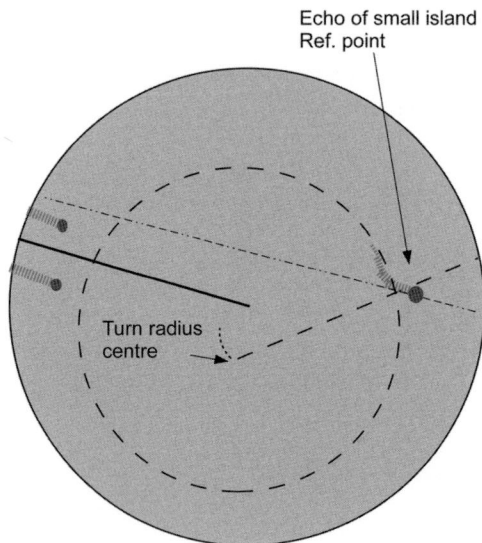

FIGURE 16.19e   Radar screen, North Up concentric index turn, stage 4.

north. The first diagram shows the planned track on the electronic chart (Figure 16.20a), showing the PiL for the approach to the turn, with the Muglins Rock as PiL reference. Also shown are the PiLs for the wheel over (WO) and for the track coming out of the turn, also referenced on the Muglins. The turn radius in 1.0' and the turn centre is shown, with the distance (2.9') to the turn reference point, the Muglins again. The radar is in 'Head Up' mode. It should be noted that in some radars the PiLs may not remain in 'true' orientation, as shown here. This is a major disadvantage in using 'Head Up' mode if this is the case.

The first radar screen image (Figure 16.20b) shows the ship some miles before the turn, showing the PiLs and the turn reference index circle, centred on the offset EBL to starboard, by 1.0', the turn radius.

The second radar image (Figure 16.20c) shows the situation shortly before the WO PiL comes onto the WO reference point (the Muglins).

The third image (Figure 16.20d) shows the turn started, as the turn reference point is not on the index circle, the RoT (rate of turn) will have to be reduced.

The fourth image (Figure 16.20e) shows the turn proceeding, with the reference point now on the index circle, and therefore the ship is on the planned curved track. The PiLs which are of no further use have been deleted.

The fifth image (Figure 16.20f) shows the ship on the PiL for the next track.

The next series of diagrams (Figures 16.21a to f) shows the same turn carried out with the radar in 'North Up' mode.

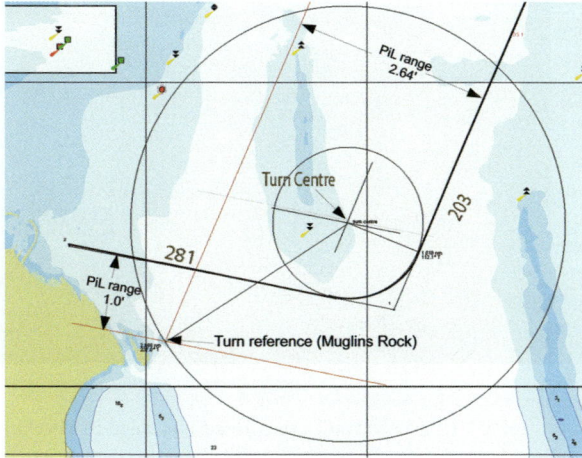

FIGURE 16.20a   Sequence of images from simulator showing concentric index turn 'Head Up'.

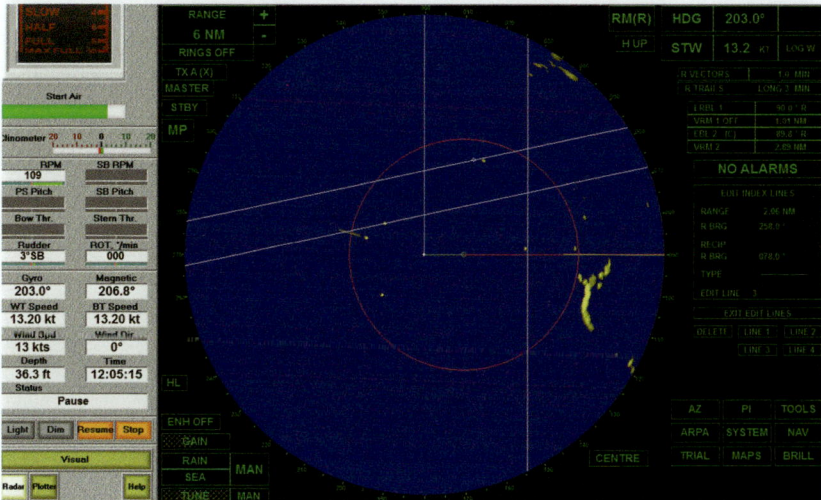

FIGURE 16.20b   Concentric index turn Head Up, stage 1.

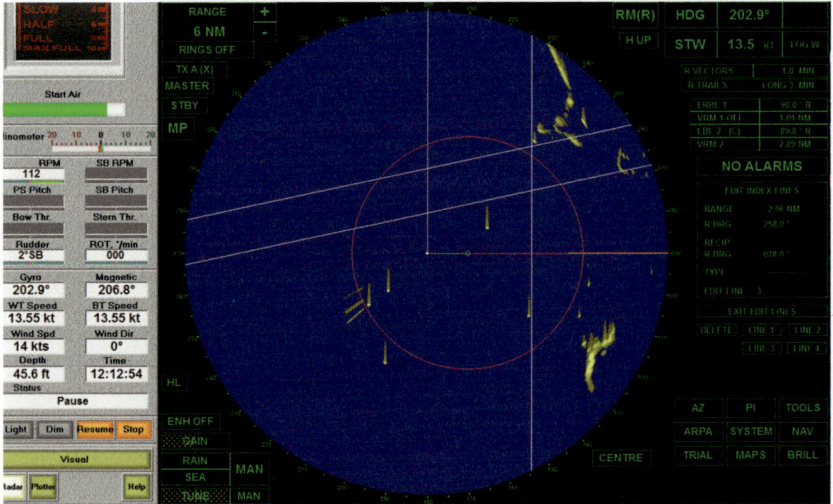

FIGURE 16.20c   Concentric index turn Head Up, stage 2 (wheel over).

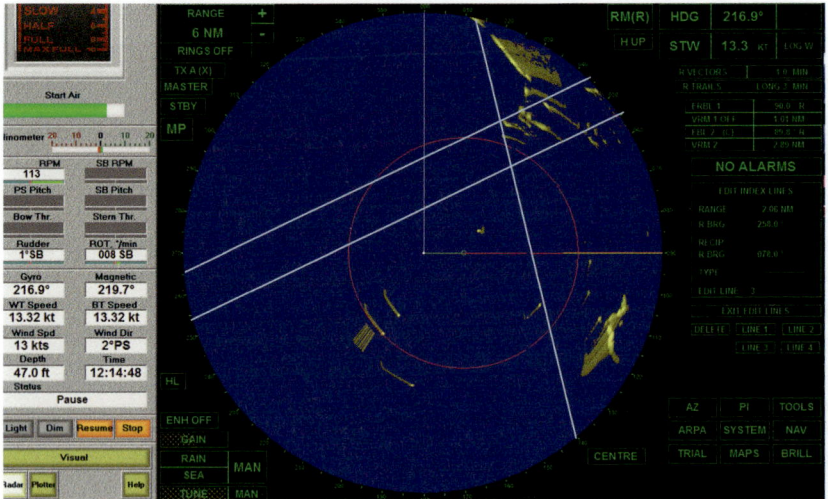

FIGURE 16.20d   Concentric index turn Head Up, stage 3 (turning too fast).

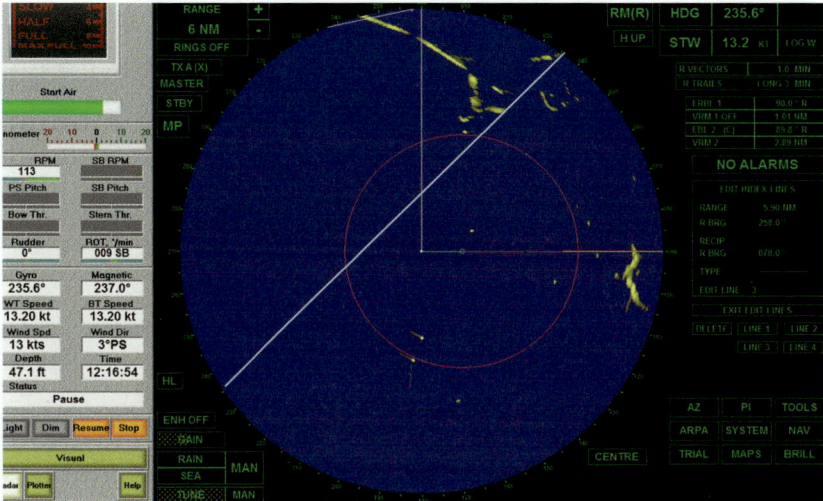

FIGURE 16.20e   Concentric index turn Head Up, stage 4 (turning correct).

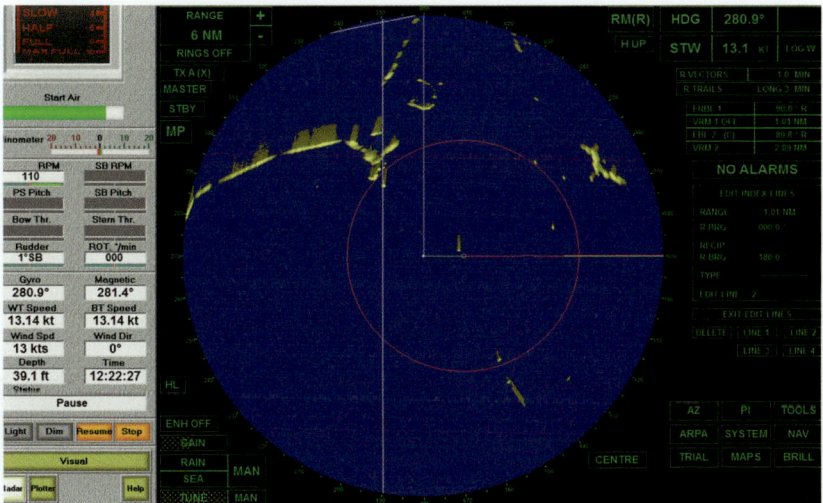

FIGURE 16.20f   Concentric index turn Head Up, stage 5 (on new course).

FIGURE 16.21a   Sequence of images from simulator showing concentric index turn 'North Up', stage 1.

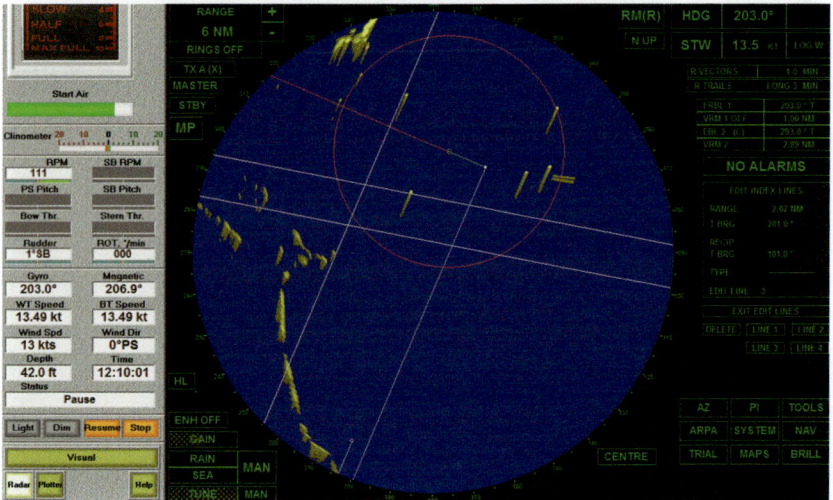

FIGURE 16.21b   Concentric index turn North Up, stage 2.

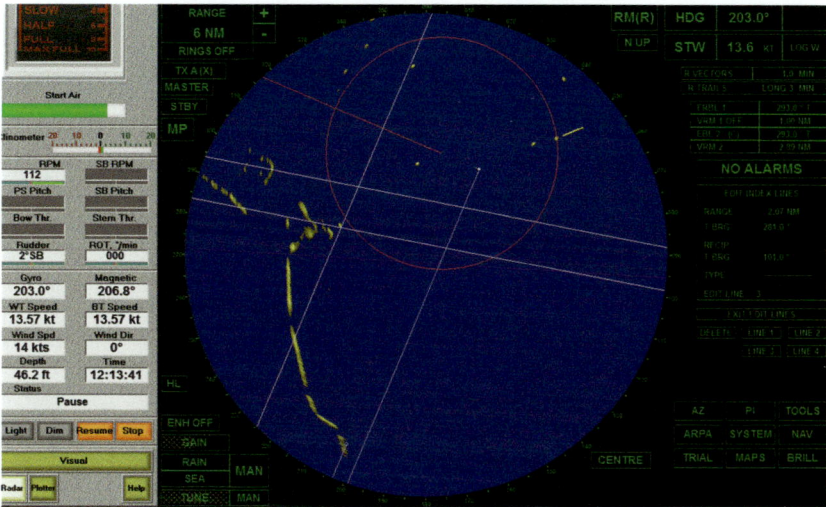

FIGURE 16.21c   Concentric index turn North Up, stage 3 (wheel over).

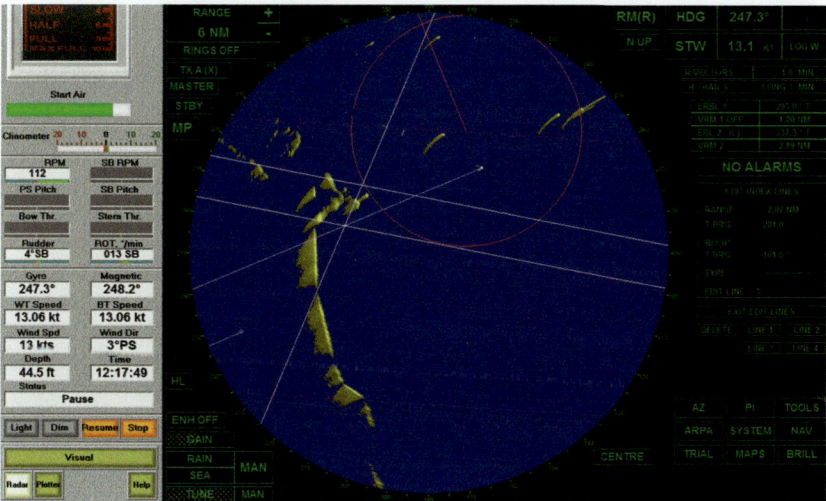

FIGURE 16.21d   Concentric index turn North Up, stage 4 (turn correct).

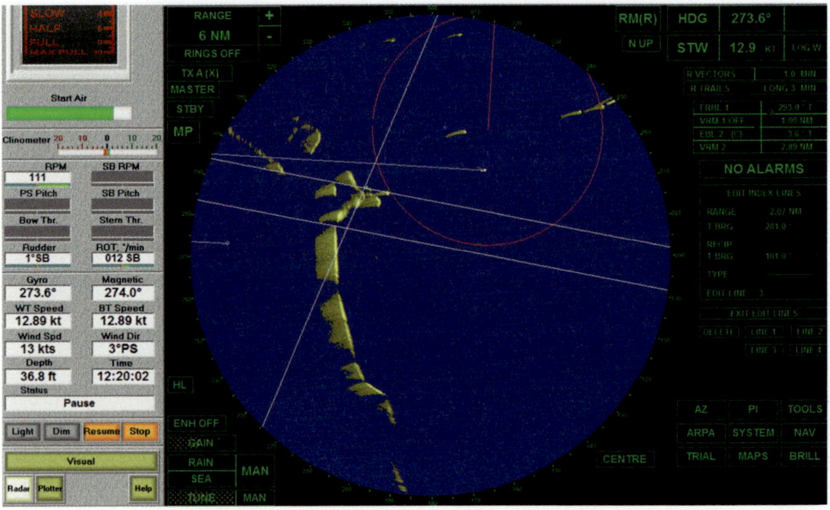

FIGURE 16.21e   Concentric index turn North Up, stage 5 (turn correct).

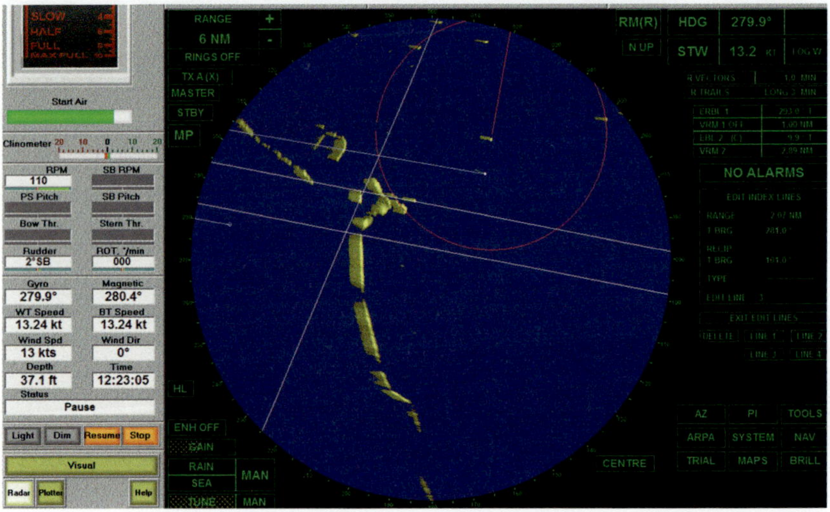

FIGURE 16.21f   Concentric index turn North Up, stage 6 (on new course).

Curved Heading Marker

Some modern sets include what is described as a curved heading line in their applications. This allows the origin to be tracked along a curved track which can be located geographically. The inputs to such an application are:

alteration of course,
turn radius or Rate of Turn,
time delay, and
speed.

On activating this function, a curve will appear at some default position on the heading marker ahead of the ship. The required settings, under the headings mentioned above, can be set in a dialogue box, or in some applications, for example the alteration of course, by 'dragging and dropping' using the roller-ball and buttons.

Such an application can be used in relative motion, either 'North Up' or 'Course Up' or in true motion. However, it must be ground locked or referenced. In this it is preferable to ground lock to a stationary radar target, provided a suitable one is available, rather than GPS. The reason for this is to provide an independent back-up navigational system with local reference.

Depending on the application or the manufacture of radar, the curve can be moved by clicking and dragging to some other part of the screen, off the heading marker.

Some radars and ECS (electronic chart systems) have curved EBLs that stem from the heading marker curve along the user defined radius to whatever is the user defined bearing, typically the next course. This as another cross reference for commencing a turn.

Precise anchoring and alteration of course (Dead Ranging or 'count down')

The use of parallel index lines, EBL and VRM allow for precise and controlled approach to an anchoring position, or to an alteration of course (track). Taking the case of anchoring first. Figure 16.22 below show, first of all the approach track to the anchor position. The length of the ship from the stem to the conning position ('stem to standard (compass)') is laid off back to give a 'Let Go' (LG) position. When the conning position, the Consistent Common Reference Point (CCRP), is here the anchor is at the precise anchoring position and the order to drop the anchor is given. The distance to go is laid off in cables back from the 'Let Go' position.

On the radar screen (see Figure 16.23) the parallel line (PiL) for the

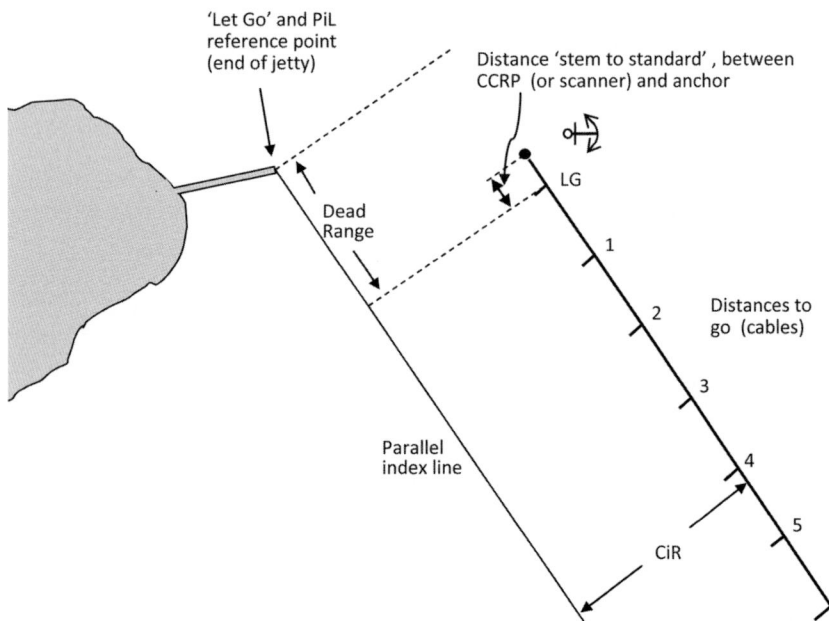

FIGURE 16.22   Planned precise anchorage.

approach track is laid off, as shown. The EBL is placed at the true bearing at which the PiL and dead range reference point, in this case the end of the jetty, should be when the CCRP is at the 'Let Go' position. The VRM is offset along the EBL at the range at which the reference point should be when the CCRP is at the 'Let Go' position.

When these preparations are made the ship is then conned along the approach track keeping the PiL reference point (the end of the jetty) on the PiL. The VRM is continuously adjusted the keep it on the dead range reference point (the end of the jetty); the VRM read out gives the distance to go.

Wherever possible good seamanship suggests that a ship approaching an anchorage should, where possible stem the current (tidal stream), or the wind if windage is the greater cause of drift. However, in some circumstances this may not be possible and the approach may have to be made across the current. In this case the anchor position may not be precisely where planned (shown by the dotted outline of the ship in Figure 16.24). Calculating or estimating the drift angle can be difficult

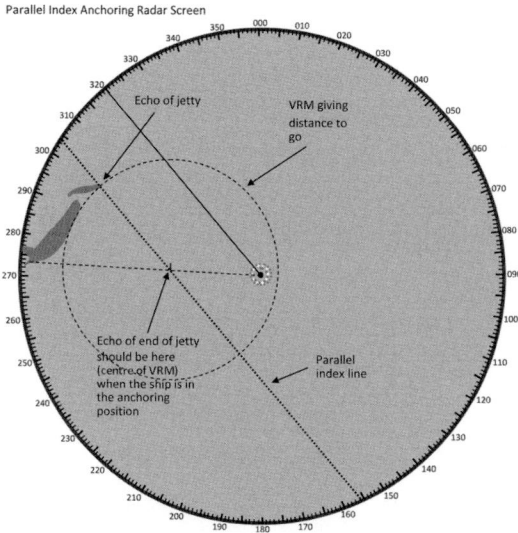

FIGURE 16.23   Radar screen for precise anchorage.

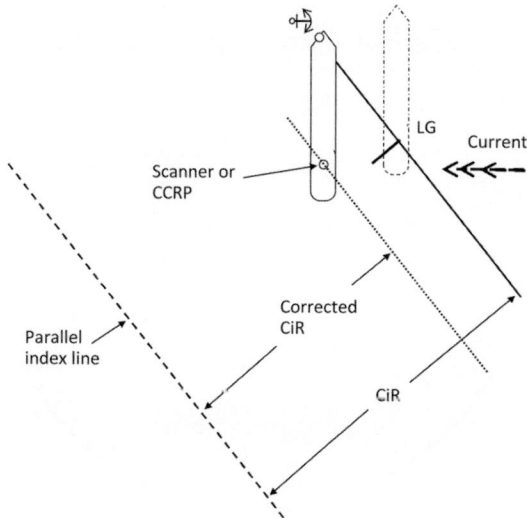

FIGURE 16.24   Allowing for drift in precise anchoring.

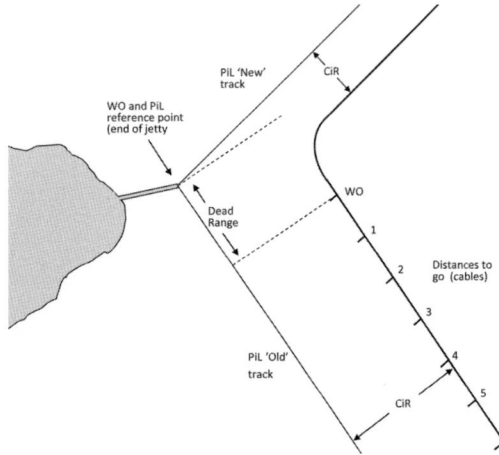

FIGURE 16.25a   'Dead Ranging' for wheel over.

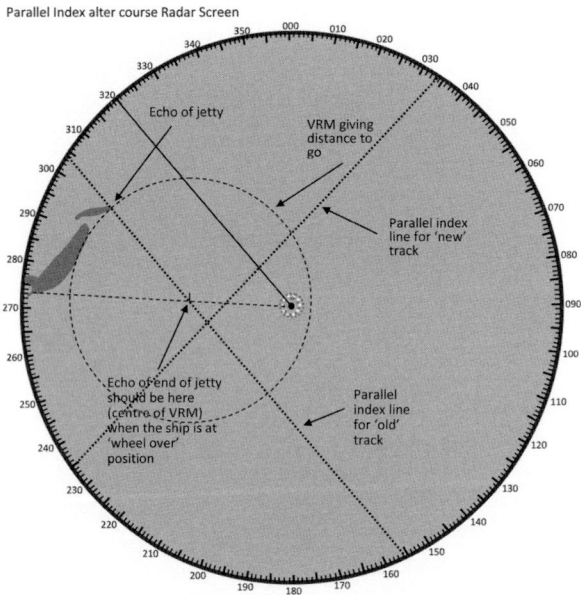

FIGURE 16.25b   'Dead Ranging' for wheel over.

as it will increase as the ship's speed is reduced on approaching the anchoring position. Some attempt can be made to get the anchor into the precise position by reducing the PiL cross index range (CiR) by a small amount, depending on the estimated drift angle for the last few cables of the approach. In very manoeuvrable ships it may be acceptable to overshoot the anchor position while letting go the anchor to achieve a precise anchoring position. In very large ships this is not the case and the final approach is very slow and the ship must be stopped with reference to the ground when the anchor is dropped or 'walked out'. This however is the subject of ship handling and outside the scope of this book.

An approach to an alteration of course can be similarly planned and precisely conducted by parallel index lines and EBL/VRM as shown in the following Figures 16.25a and 16.25b.

Track Planning

In many configurations, particularly in integrated systems, the passage plan/route created on the electronic chart can be displayed on the radar, or in some cases actually laid out or planned on the radar (chart radar) itself. Planning a passage using the tools of electronic chart displays is outside the scope of this work but should be familiar to the navigator. Different manufacturer of radars and electronic chart systems approach the task in different but similar ways, usually by a form of computer spread-sheet in which a series of way-points is entered in Latitude and Longitude co-ordinates with other details such as speed, times, etc. This is usually supported by a 'drag and drop' facility by which more local adjustments can be made to the track. The radius of turns at alterations of track can be applied either generally or individually to different track changes. The result is a ground referenced track which appears on the radar screen if required. This track is usually GPS referenced to WGS 84, and there may be difficulty in locking it to another ground reference such as a stationary radar target. In this case care needs to be taken with regard to the reliability of the chart with reference to WGS 84, and the ships position locally.

With such tracks, the ship can be either steered or conned along the track, including the turns, or it can be connected to the autopilot in 'track' mode and the ship will follow the track.

Conclusions

In this chapter we have looked at the use of radar as a navigational tool. Summarising its use for navigation:

- The traditional use of the radar is by fixing the ship's position on a chart by a series of ranges and bearings.
- The radar lends itself to more dynamic methods such as parallel indexing and concentric indexing.
- It provides an independent fixing system to that provided by other systems or methods.

The impact of modern technology, particularly developments in electronic navigational chart systems has had a major impact on radar for navigation. New applications and techniques are continuing to develop as we see in concentric indexing. It is in these dynamic modes that radar is best, for example in parallel indexing. While it is still used in the traditional mode of fixing the ship by series of ranges and bearings it lends itself better to dealing with moving situations and the use of graphics and chart over and underlay. In 'blind pilotage' it provides the main tool allowing safe navigation in conditions of poor visibility or at night in circumstances where there are few if any navigational marks. It is however only one tool in the range of aids to navigation such as visual aids and depth sounder. These aids, no matter how technically sophisticated cannot replace the skills and experience of the navigator.

Chapter 17

Radar and Collision Avoidance

Introduction

The correct use of the radar equipment is the most important part of the OOW and bridge team's relationship to it and most critically in its use for collision avoidance. As stated elsewhere a sufficient understanding of the working of the equipment, its capabilities and limitations is essential to the user to get the best out of it. But more importantly this is necessary to avoid errors and misunderstandings of the presented information. Also, in this is the correct setting up and use of the controls, and understanding what they do. Taking all of this for granted, the way the equipment is used, and what information it provides for the purpose of collision avoidance will now be examined.

In this chapter we will look at the following details of radar and its use in collision avoidance:

- A brief look at the history of the use of radar for in collision avoidance.
- The tools available on a modern radar set.
- The application of target data.
- Appreciation of the situation.
- Methods for defining safety margins.
- Aspect.
- Avoiding close quarters situations.
- Traffic Separation Schemes.

In studying the history of radar aboard merchant ships it became obvious that radar had a huge impact on navigation and ship handling in the sense of avoiding collision. In conditions of reduced visibility things could be 'seen' that otherwise would be invisible, and it was now possible for ships to proceed where otherwise they would be compelled to stop or at best proceed at minimum speed. At the time it was assumed that in such conditions ships would still proceed cautiously and that radar would give an extra degree of safety. But this assumption ignored human characteristics. It was soon obvious that whatever the collision regulations said, ships were proceeding in thick weather, in congested waters at much their normal speed, or possibly only making a token reduction of speed.

It also became apparent in those early days of radar, in analysing the collisions which occurred, that the limitations of the equipment was not the major fault. It was the limitation of the training of the bridge officers in its use. Collisions occurred in situations where both ships were fully aware of each other's presence in ample time. Often in these cases if neither had taken any action, or ironically had been unaware of each others' presence they would have safely passed each other. It was in misinterpreting the presented information, and taking the wrong avoiding action based on it that caused many of these collisions.

The Figures (Figure 17.1) show a situation which occurred all too often. Both ships have 'Head Up' unstabilised radar displays. Both are in restricted visibility. Ship A, seeing the echo of Ship B coming onto the edge of the display, fine to port, alters course to starboard to put the target further out to port. No plot is carried out, and the assumption is that the other vessel is on a reciprocal course. Ship B meanwhile sees the echo of Ship A broad to starboard and assumes it is going well clear. No plot is carried out here either.

A while later, Ship A seeing the echo of Ship B close to the heading marker again alters course more to starboard with the intention of increasing the passing distance. Ship B sees the bearing of the echo of Ship A drawing right and assumes that as the bearing is opening that it is going clear.

In the final stage Ship A again attempts to increase the passing distance by altering course to starboard, probably quite alarmed by the loudness of the fog signals from Ship B. Ship B finally becomes aware that Ship A is closing and shortly before the collision is very surprised to see the port side of Ship A rather than its starboard side.

The second figure (Figure 17.2) shows the positions of the ships related to the three pairs of radar screen pictures.

These observations are made on the basis that although radar and watchkeeping have changed considerably since those early days, the fundamental points on interpreting the information and taking action based on it have not. On the bridge of a modern ship, the radar provides the main 'situation display' to the master/OOW. Whatever might be said about the primacy of lookout it cannot be denied that the information gleaned from other sources is usually related to the radar display. This is particularly so in complex traffic and navigational situations. For example, a target appearing on the edge of the radar screen will then be sought by eye for further identification. Something which has been reported by a sighting (the lookout) and not picked up by the radar observer, will result in careful examination of the radar screen until an echo appears and is matched with the visual observation.

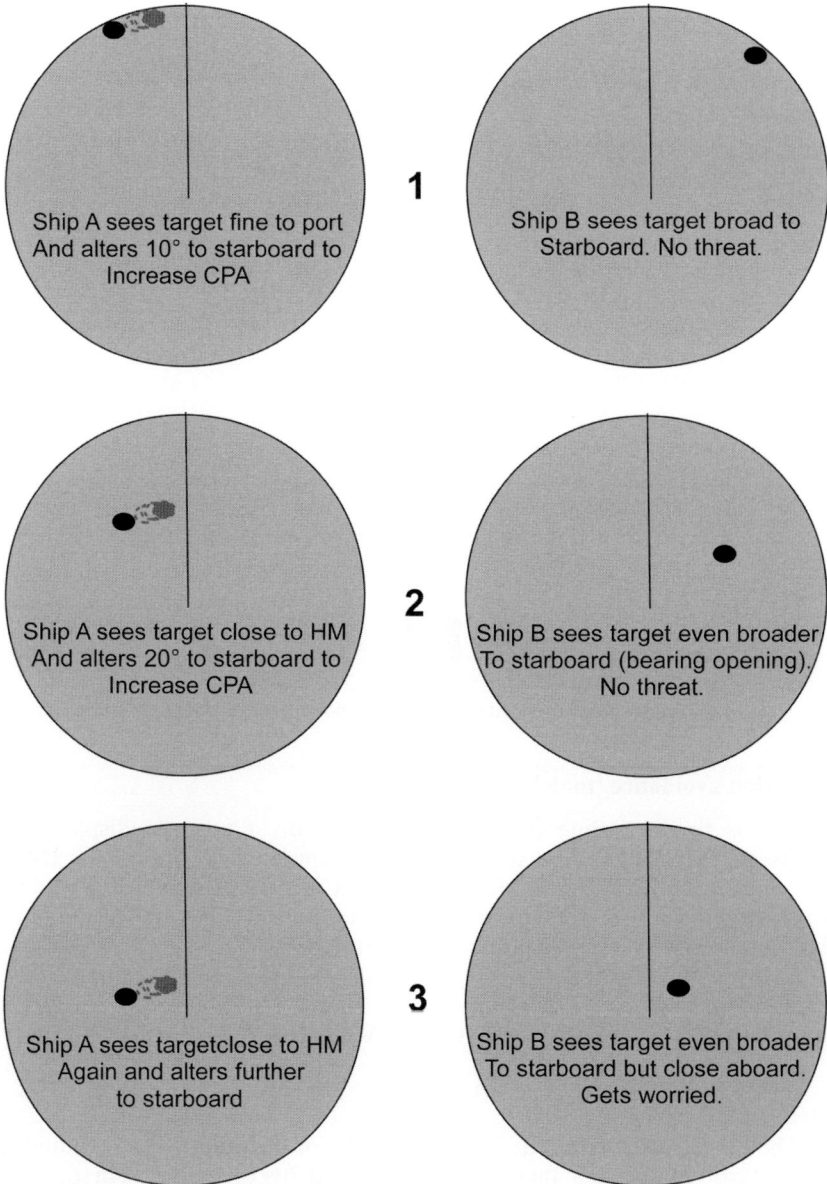

Ship A sees target fine to port
And alters 10° to starboard to
Increase CPA

Ship B sees target broad to
Starboard. No threat.

Ship A sees target close to HM
And alters 20° to starboard to
Increase CPA

Ship B sees target even broader
To starboard (bearing opening).
No threat.

Ship A sees targetclose to HM
Again and alters further
to starboard

Ship B sees target even broader
To starboard but close aboard.
Gets worried.

FIGURE 17.1   Radar Assisted collision radar screens.

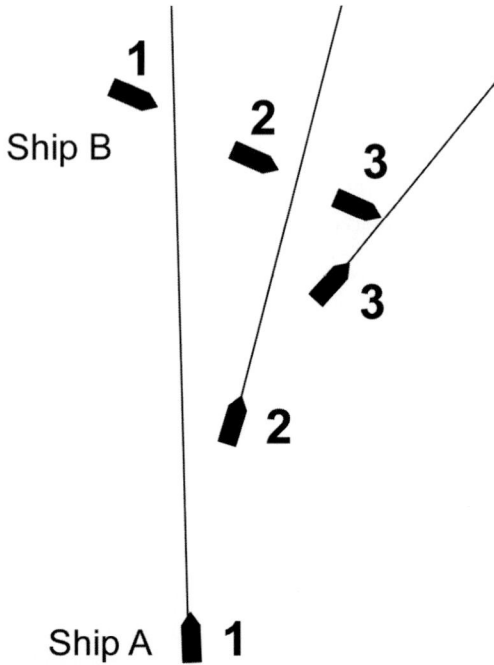

FIGURE 17.2   Radar assisted collision tracks of ships.

Collision avoidance 'tools'

Let us first examine the tools presented on a modern display for collision avoidance. The minimum requirements are quoted in IMO Resolution MSC.192(79). Most equipment fitted in ships of 500 gt and over exceed the requirements of this circular and the only limitation may be the absence of 'Trial Manoeuvre' in radars for ships under 10,000 gt.

In discussing the features in each point below, it must be remembered that there will be differences from radar manufacturer to radar manufacturer, within the scope of the IMO circular.

Acquisition

The targets are acquired, either manually or automatically. Manually consists of marking a target to be acquired by the screen cursor and pressing an 'acquire' button. Automatically is by guard zone set up by the operator or by certain parameters set by the operator (falling within

CPA/TCPA limits) (see Chapter 12 on ARPA and Target Tracking). The IMO requirements are for the system to be able to track 30 (ships under 10,000 gt) or 40 targets. Most makes exceed this figure and allow 100 to 200 targets to be tracked.

Tracking

Within 3 minutes the target must show a steady vector, have an identification (e.g. a sequence number) and display the read-out of data. The readout may be available for one, or multiple targets simultaneously. The operator is able to select the target/s whose data is displayed. Some systems will automatically give the readout when a target causes an alarm, such as entering a guard-zone, or exceeding the CPA/TCPA limits set by the operator.

Vectors

Vectors are set by the operator; the length is set by time (e.g. 6 minutes) and either true or relative. As mentioned in the chapter on ARPA, care needs to be taken where vectors may revert to a default mode different from that set by the operator. Some makes of radar have the arrangement that if the display mode is 'true motion', the vectors' default is 'true'. If the operator changes the vectors from 'true' to 'relative' they will revert to 'true' after a few minutes. In such sets the reverse also happens; if the display mode is 'relative motion' the default for the vectors is 'relative' and they will return automatically to this when changed to 'true'. Bridge watchkeeping officers should be fully aware of this as there have been serious situations where 'true' vectors have been mistaken for 'relative'.

Past movement of the target

Trails and history dots (or small circles in some older sets) show the past positions and hence trends of the target's movement. Modern sets allow for 'True' or 'Relative' trails to be selected, no matter what the mode of the display (True Motion or Relative Motion). The length of trails may be set by the operator. In some equipment trails will start to be generated once the 'trails' function is activated but will start anew if the range scale is changed, or the mode, 'True Motion' to 'Relative Motion', is changed. Many OOWs like to have the settings displaying relative vectors and true trails as this give much data at a glance. It is better than the reverse, true vectors with relative trails, as the relative vectors give a better appreciation of CPA and TCPA.

History dots also give information on the past motion of the target. True history dots will give the best appreciation of recent manoeuvres by the target. In some equipment history dots will only start to generate when the function has been activated by the operators, in others the last five or six dots will appear immediately.

AIS targets

When AIS is switched on by the operator, targets transmitting an AIS signal will appear on the screen, navigation marks (AtoNs) as small faint diamonds, ships as small green triangles (see Chapter 15 on AIS and symbols). It requires an acquisition, or activation function by the operator to get further data such as a vector and readout data on any particular AIS target. This activation will often cause the name of the target vessel to appear beside the AIS symbol for the target. In some equipment an AIS target entering a guard zone will activate it. Trails will not apply to AIS targets, but history dots may.

AIS and radar acquired targets can be combined by a process called 'association', 'correlation', 'fusion' or similar terms. This function avoids a target which has been activated by AIS and also acquired by the tracker(ARPA) having two vectors. Care needs to be taken that the true vector is sea stabilised for anti-collision purposes. (see the chapter on AIS and note particularly the warning from the UK MCA about collision avoidance based on AIS data).

Guard Zones

Many modern sets allow flexible guard-zone applications. Some allow for additional 'exclusion' zones to be defined. These options are discussed in the chapters on ARPA functions.

CPA/TCPA limits

Setting these causes an alarm to activate if both limits are violated. There will be an audible alarm, a 'Collision Warning' appearing on the readout, and probably the relevant target vector changing to a red colour and possibly a flashing symbol over the target.

Trial manoeuvre

Inputs consist of change of course, change of speed and time delay. Usually the change of course is the most intuitive and easily applied setting. Care needs to be taken that the operator is fully conversant with the trail manoeuvre function of his/her particular equipment as there are variations. Usually it is used in the relative vector mode as this gives the clearest indication on the effect the proposed manoeuvre will have on the plot particularly for CPA.

The Application of data

Whether the information on a possible close-quarter situation is derived by manual plotting, by ARPA, ATA or other plotting aid or by AIS, several fundamental factors need to be considered:

1. The reliability of the information (e.g. the error inherent in the plot).
2. Target data such as CPA, TCPA, etc.
3. The possible courses of action to be taken to avoid a close-quarter situation.
4. The effect of such action assuming the other target/s continue as they are.
5. Possible action on the part of the other target/s.

Examining each of these in turn.

The reliability of the information

Accuracy and reliability. This is an assessment of factors such as whether the information was derived from manual plotting or from ARPA or other plotting aids. For example is the information being provided by a cadet plotting manually or is it from the latest model of target tracking radar?

Then there are the inherent errors in radar and ARPA which have been discussed elsewhere. Such inherent errors are accommodated by making enough allowances for them usually by taking early and substantial action where required.

Target data

Whatever the method of plotting, the basic 'report' is the first step in informing the OOW, master or pilot. Firstly, how close is a target likely to come to own ship and whether this distance is acceptable or not. This must take all factors into account such as the state of visibility, traffic, draught restrictions and room to manoeuvre, manoeuvring capabilities of own ship including restrictions imposed by type of propulsion (including sails) and the time in which this situation is developing.

The actual direction in which the target is moving, its range and bearing from own ship and its aspect are the next pieces of information required by the OOW to assess the situation. This assessment can be examined at two levels:

1. Dealing with the target in isolation. In looking at the CPA and TCPA it can be seen if there is a threat or not. Even if there is not, good practice would require, if possible, to have an appreciation

of the target's course and speed if it is likely to pass anywhere close to own ship. 'Close' of course is a subjective term and will depend on circumstances such as traffic density, closeness of other dangers and so on.

2. Assessing the target's movement in relation to other traffic and navigational features. This is where ARPA or ATA is almost essential. While action taken to avoid a close quarter situation can only be made on the basis of the latest information, the possibility that the target may change course or even speed because of other traffic or coming to a major navigational mark must be borne in mind. A 'chart radar' with chart underlay is a major advantage here as such things as traffic lanes, buoys and so on can be seen and assist in assessing traffic movement. The same applies to ECDIS with radar overlay.

The possible courses of action to be taken to avoid a close quarter situation

This is where the professional judgement, experience and knowledge of the Rules by the OOW are important. The Rules (IRPCS) list various conditions and circumstances governing their application. It should go without saying that the officer on watch, in fact all the bridge team should be fully conversant with these Rules. Those particularly relevant to this topic are listed in Appendix A, but all the Rules are equally critical including those referring to the sound signals, lights and shapes.

Some of the most important statements in the Rules in this regard are in Rule 8 'Action to Avoid Collision'. This emphasises positive action, made in ample time and substantial so that it can be detected by other vessels observing by radar or visually; a series of small alterations is to be avoided. It also suggests that an alteration of course alone may be the best option to avoid close quarter situations. In this regard, early substantial alterations of course, where and when there is room to manoeuvre and where the full benefit of the ship's speed and manoeuvrability are available are obviously better than allowing a situation to develop where room to manoeuvre is restricted, or where it is necessary to reduce speed or stop, thus putting own ship at the mercy of the actions of any other vessel.

Rule 19, for vessels not in sight of one another is particularly applicable to the use of radar and AIS. This Rule often causes confusion in the options which it seems to give, or to deny, in avoiding close quarter situations. Quoting para. (d)

'A vessel which detects by radar alone the presence of another vessel shall determine if a close quarter situation is developing and/or risk of collision exists. If so, she shall take avoiding action in ample time, provided that when such action consists of an alteration of course, so far as possible the following shall be avoided:

(i) An alteration of course to port for a vessel forward of the beam, other than for a vessel being overtaken; and
(ii) An alteration of course towards a vessel abeam or abaft the beam.'

The first comment to be made regarding the confusion which often exists is that in the circumstances referred to in this Rule (vessels not in sight of one another when navigating in or near an area of restricted visibility) there is no 'stand-on' vessel; every vessel, which has determined that a close quarter situation is developing, or risk of collision exists shall take avoiding action. This applies even to the case of a ship proceeding a 10 knots and with a radar target approaching from astern at 22 knots which is apparently on a collision course or going to pass very close. Doing nothing and assuming that the overtaking vessel is going to keep out of the way is not an option; if a close quarter situation is developing the Rule states that 'she shall take avoiding action in ample time ...'. Much will depend on circumstances, of course. What may be considered a 'close-quarter situation' in open waters may be acceptable in a traffic separation scheme. The range of visibility is another variable factor. In the scenario mentioned above, while the slow ship is required to take avoiding action while the fast ship is out of sight, once they come into sight of one another, the fast ship now becomes an 'overtaking vessel' within the meaning of the Rules for vessels in sight of one another and with the duty to keep clear, and the slow ship is required to maintain course and speed.

The other part of this Rule that tends to cause confusion is in situations where there is more than one target vessel involved. An example of this is where there is an approaching target forward of the beam with a small CPA. Another target is slightly abaft the beam to starboard, but posing no immediate threat, possibly on a similar course to own ship. It has often been said that in this circumstance, an alteration of course to starboard is not an option as it would involve 'an alteration of course towards a vessel abeam or abaft the beam'. This is not so, if the vessel in question, the one slightly abaft the beam is not posing a threat, a close quarter situation is not developing with this vessel. However, obviously the action taken to avoid the first target, must not now cause a close quarter situation to develop with the one slightly abaft the beam. This is where using the 'trial manoeuvre' function is useful, when it is available.

The Effect of actions taken to avoid close quarters situation

When it is available, trial manoeuvre should always be used before a proposed manoeuvre to avoid a close quarter situation is carried out. In the absence of trial manoeuvre, the OOW should have a clear mental picture of the effect of the proposed action on the plot. Some situations are easily determined. Take a target which is on a collision course, but is on a broad aspect and on the starboard bow, moving from starboard to port. It is clear that a broad alteration by own ship to starboard, putting the target broad on the port bow, will effect a new relative track going well clear. What may be less easy to mentally determine is the effect of a proposed manoeuvre on a target which is on the quarter and on a slow relative rate of approach. This more so if the proposed manoeuvre is being carried out to avoid another vessel. This is one of the reasons why developing plotting skills is so important, not so much to actually plot, although this might be necessary, but to develop the mental image of such situations.

There can be a brief period of uncertainty while a manoeuvre is being carried out. If the plot is in relative motion, the changes can have a confusing effect until own ship settles on the new course or speed. A true motion display gives greater oversight and continuity particularly for detecting changes by the target vessels during the course of own ship's manoeuvre. Also, during large alterations of course, the filtering process in an ARPA or ATA computer will return questionable target data, including vectors that seem to change direction, until such time as own ship has settled on the new course. The reduction of speed by own ship consequent to a large alteration in course will also be an unknown factor in the computation of the plot, whether automatic or manual.

Once the ship has settled on the new course/speed, the plot should continue to establish that the situation is developing as anticipated and to detect if any of the targets have taken any action that may cancel the manoeuvre of own ship. This is important as it is natural to assume that the action taken has resolved the situation and to move on to the next task, for example, getting back on course or dealing with other targets which have appeared. Targets need to be carefully tracked until well passed and clear.

Possible action by other ships

This of course is the great imponderable. In most cases other vessels will do as anticipated, but there are no guarantees that they will. The person in charge of the other vessel may not be keeping a proper watch and may be unaware of own ship's approach; the other vessel may not have a radar set or one that is properly functioning and in poor visibility be unaware of own ship's presence. The other vessel may be constrained by draught, a tow or the nature of her work. This may be promulgated by VHF radio, AIS and where applicable by VTS, but there is no guarantee that this will be so. The OOW in another vessel may have poor knowledge of the regulations and do something unexpected, or do nothing when something is expected. There may be some situation which is not apparent to own ship that causes a target vessel to take some action that is inexplicable to own ship's OOW. These are some of the many situations which may cause a target vessel which is developing a close quarters situation with own ship to do something unexpected. The only defence against this is taking early and substantial action which allows for further substantial action in the event of the target vessel manoeuvring which cancels own ship's manoeuvre.

Observing carefully developing traffic situations will warn the vigilant OOW of possible actions by target vessels. For example, when in a Traffic Separation Scheme (TSS) with some targets ahead proceeding in the same way as own ship, and a crossing vessel which initially plots as passing clear ahead of own ship alters course for the targets ahead. This will change the situation regarding own ship. Approaching some prominent landmark or navigation mark other traffic may alter course. A target vessel approaching another target vessel may alter course which again changes the plot. A vessel ahead slowing down or stopping can be a particularly dangerous situation as it can be a manoeuvre which is difficult to detect initially and own ship can find herself at very close quarters before it becomes apparent. Fishing vessels can be notoriously unpredictable, and of course not always identifiable as such, particularly on radar. Similarly pleasure craft should be given a wide berth, not so much that they may do unpredictable things but if under sail then they cannot manoeuvre very easily. Sometimes such craft have the best intentions of not impeding the passage of large ships and manoeuvre to keep out of the way, but consequently do something that confuses the plot on own ship. Also, the concept of distance and room to manoeuvre is different from the level of a small yacht to that of the bridge of a VLCC. The yacht can be quite close in the VLCC's estimation when it may take some unexpected action designed to keep out of the way of the larger ship.

While the above discussion has concentrated on the use of radar in restricted visibility, it is of course an essential aid in clear weather. The bridge team's eyes, looking out the bridge windows, are the primary source of information advising the OOW, radar and AIS complement the information so gleaned. The con (the person who currently is in charge of the navigation of the ship), whether it is the OOW alone, or a bridge team, should make use of all the available information. Some of it may be redundant (in the sense of more than is required), but the OOW can decide what is redundant when at least it is known. A crossing ship in clear daylight, with the bearing changing rapidly may obviously be going well clear. The ARPA plot tells the OOW how far away the other ship is, her course and speed, the actual distance she will pass clear (CPA) and the time this will occur (TCPA). The AIS also gives similar information as well as the name of the crossing ship, call sign, and a host of other data. Most of this information is more than the OOW normally requires to deal with the situation, but it serves to confirm what is so apparent. Under different circumstances the other

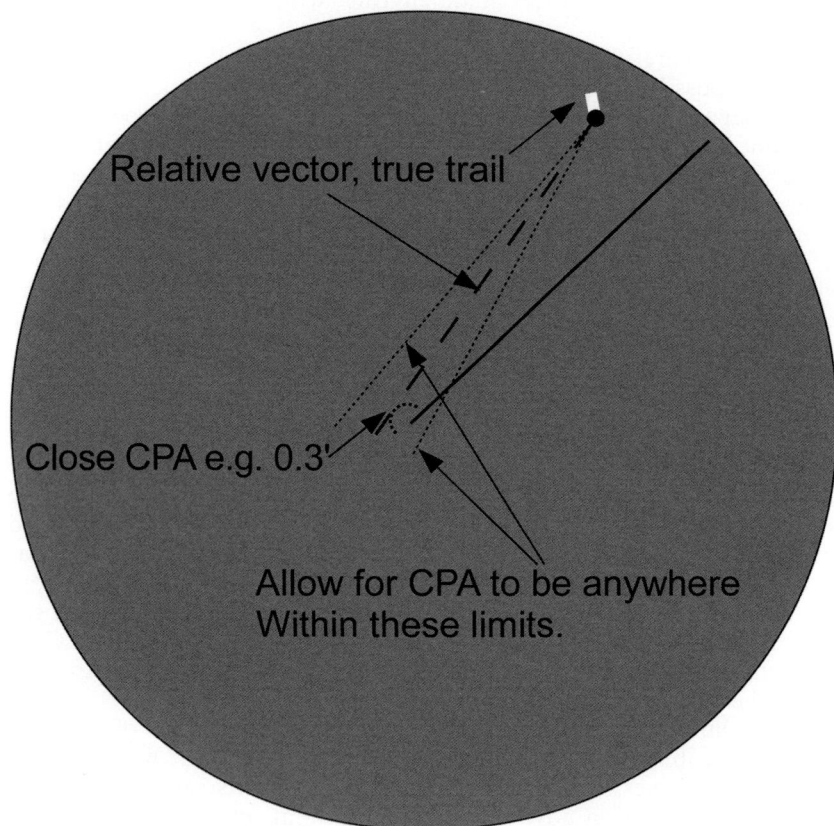

FIGURE 17.3   Allowance to CPA.

ship may not be so obviously going clear, and the data provided by radar/AIS may be needed to further assess the situation. An added bonus when the OOW uses his/her visual estimate of the situation and compares it with radar and AIS is that it helps to build up mental images for situations where the other vessel is not visible.

One of the dangers which the advent of such complete information has brought about is the tendency to pass too close to other ships. There is no hard and fast rule as to what is 'too close'. There are too many variables and different circumstances. While the information from ARPA radar and AIS gives apparently accurate and complimentary information, the OOW should be aware of the limitations of such information. Errors in radar data and ARPA are discussed elsewhere and such margins of error should always be allowed for. The ARPA may give a CPA of 1 mile, but this, under certain circumstances could be anything from 5 cables to 1.5 miles without any malfunction on the part of the radar/ARPA. Similarly, with small CPAs the CPA could be on the other side of own ship to that shown by the ARPA (see Figure 17.3).

As well as taking the points mentioned above into account, the prudent mariner should always allow for the unexpected. A classic case occurred some years ago in the Dover Strait TSS. In moderate to poor visibility a naval frigate was overtaking a VLCC on its starboard quarter. There was another large vessel on the frigates starboard bow and about the same distance off as the VLCC. For some reason the VLCC was unaware of the frigate. Unknown to the frigate's bridge team, a rogue crossing vessel had appeared ahead of the VLCC, which suddenly altered course across the frigate's bows resulting in a collision. Fortunately there was no loss of life or serious pollution.

Appreciation of the Situation

In radar and AIS plotting (in this the term plotting means the assessment of the plot, not the physical carrying out of plotting functions), there are two levels of assessment. The first is the taking in or reporting of the situation, assessing the threat, and carrying through a proposed course of action, usually an alteration of course. This might be considered the 'Operational' level function. The other is making a more thorough analysis of the plot, relating it to the current situation, applying the regulations and discussing all the possible options. This might be considered a 'management' or 'command' level function. This 'discussion' could be among the bridge team, but is more likely a mental exercise, and it should be practised by all bridge watchkeeping officers with some watchkeeping experience behind them. It encourages a more complete appreciation of the situation. In nautical colleges students for Chief Mate and Masters should be given exercises where they must write down their thoughts and assessment of the plot, the possible courses of action with recommendations for and against each possible course of action, and a decision on the course of action proposed with reasons as to why this is considered the best. Taking a simple case as example, a crossing situation from starboard to port, in restricted visibility, the target is about 8 miles away, aspect about red 40 degrees, and CPA 0.5 miles to port in about 20 minutes; own ship making 12 knots. There are no other targets nearby and no sea-room restrictions.

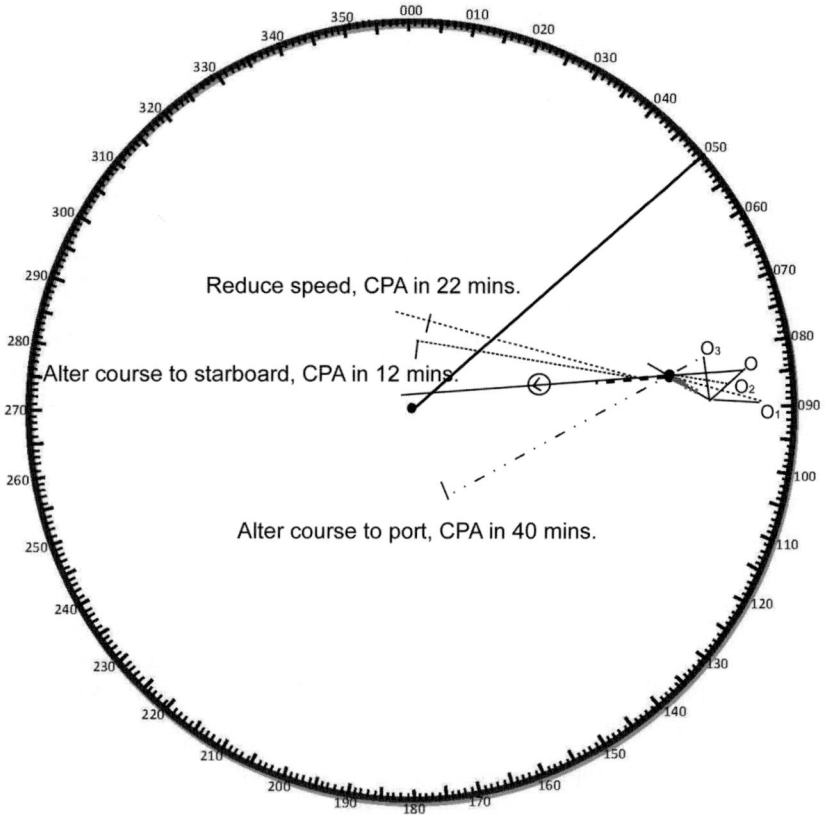

FIGURE 17.4   Showing a crossing situation for appreciation of the situation.

Having established all the data on the situation, and decided that action must be taken to keep the target at least 1.5 miles away, the immediate inclination is to make a bold alteration to starboard. Without any question this is quite acceptable and the course of action which would be expected without any discussion. But what other options are there? Examining the plot, it is found that an alteration of 15 degrees to starboard will get the target to pass almost 2 miles off. This will take the ship less off her track than the bold alteration. Why not do this? A reduction of speed or stopping could get the target to pass clear by 2 miles. Why not do this? What about an alteration to port? It is in formulating the answers to these questions that a more complete appreciation can be achieved.

The small alteration to starboard may be OK, but it is against the spirit and recommendation of the Rules in so far as it may not be so readily apparent to the other ship. As there is no sea-room restriction the argument about taking the ship off its track is a weak one, against the bold alteration that would show up clearly on the other ship's radar (presuming it has an operational radar). Similarly the reduction is speed would not be so readily apparent. This also has the disadvantage that it reduces own ship's manoeuvrability and ability to get out of the way if the target should do something unexpected. The Rules also comment on this. The alteration to port, presuming it can be achieved to keep the target outside 2 miles, is first of all against the recommendation of Rule 19. It also places own ship and target heading in somewhat similar directions and hence extends the time when both ships are near each other and delays getting clear and thus takes own ship further off her intended track.

Having examined all these options it is clear that the bold alteration to starboard, while maintaining speed is the best; it gives clear indication that avoiding action has been taken; it complies with the Rules in their several parts; it clears the situation the most rapidly and effectively and allows base course to be resumed as early as possible.

This is quite a simple case as an example and demonstrates the thought process that should be used and, if need be, articulated. It should be used in all cases as it helps to develop an analytical thought process that examines all the options in any situation and the application of the Rules, other regulations, advice, recommendations and constraints to the situation.

Methods of Defining the Safety Margins for Collision Avoidance

In assessing the threat posed by other traffic, traditionally CPA and

TCPA have been used to define the extent of the threat. These figures are good, and generally well understood by watchkeeping officers. However, they do not define a safety zone or envelope around the target in the most effective way that ensures a sufficient margin of safety. Such a margin can be quite large, taking the size and speed of modern ships into account. A large passenger ship moving at over 20 knots will take up quite a distance to take way off, but with her speed can manoeuvre to avoid situations quickly. Similarly a large tanker or bulk carrier proceeding at about 14 knots will equally take a large distance to stop but will take longer to achieve avoiding action for developing close quarter situations. In the chapter on radar plotting we have examined collision points and 'Potential Areas of Danger' (PADs). These give a better appreciation of the situation and possibilities for taking avoiding action, where necessary. Only some of these may be displayed on certain modern radars, and some may be available on some older ARPA models. Other sets show graphically the position/s of CPA/s in relation to the origin.

A study of these topics will help watchkeeping officers to develop mental pictures of the situation even if they are not displayed graphically. However, the value called 'Distance to Collision' (DTC) is quite easy to apply (TCPA × Own Speed) when the target is on a collision course with own ship. This gives a better idea of what room to manoeuvre is available in a possible developing close quarter situation. In the Nautical Institute's 'Managing Collision Avoidance' it suggests the minimum distance for taking avoiding action is seven ship (own ship) lengths to the collision point (see Figure 17.5).

Aspect

This is a critical piece of data in the assessment of a developing close quarter situation, and often glossed over in text-books and training. The fact that most modern radars do not provide a read-out giving this information, where they do give CPA, etc., including 'Bow Crossing' distance, points to the neglect of this data. In practice its use is often more of an intuitive thing, brought about by experience, than any quantifiable figure. An experienced OOW, looking at the true vector on his display, or looking at the disposition of the navigation lights of an approaching ship will take avoiding action where necessary. What alteration of course, and when taken will vary widely from one individual to another. The resulting passing distance will often be unknown precisely by the OOW other than it's a substantial and 'safe' passing distance.

This is where simulator training comes into its own. A series of

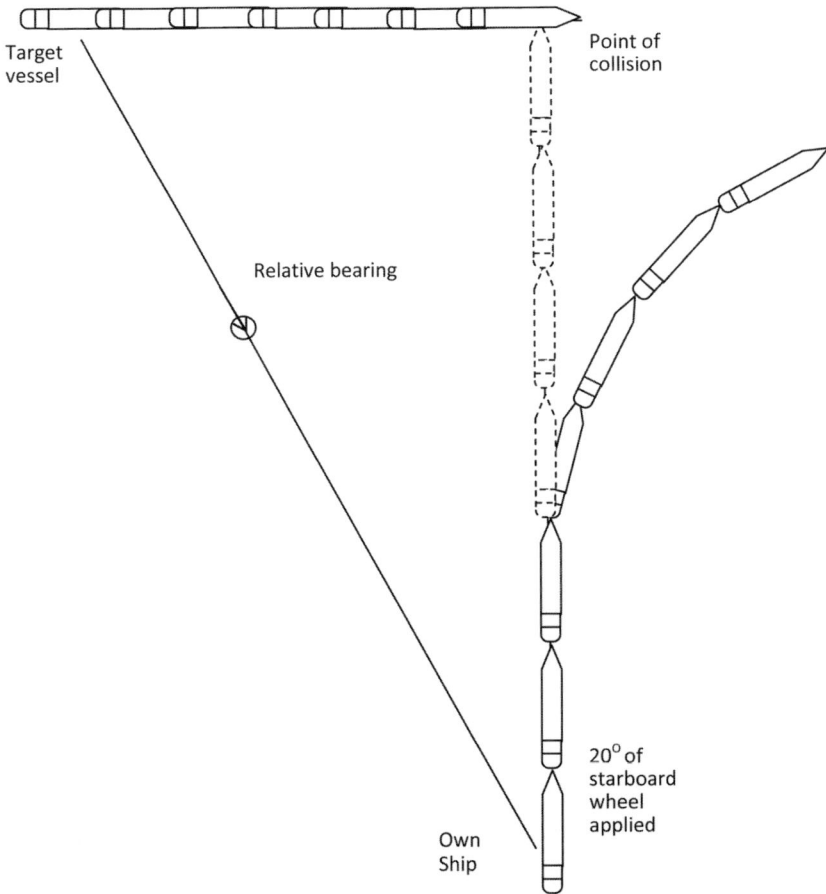

FIGURE 17.5   Distance to collision (DTC).

similar situations can be set up, with the variable being the aspect of the target vessel, where the 'Own Ship' is obliged to give way. An example of situations where the target vessel, on various relative positions around Own Ship, with an aspect angle of Red 10° is shown in Fig. 17.6. In each case the target's true speed is similar to that of own ship. The effect of an increase in target speed or an increase in aspect angle can be easily visualised in each of the four situations.

The purpose of such exercises is to illustrate, or train the OOW to

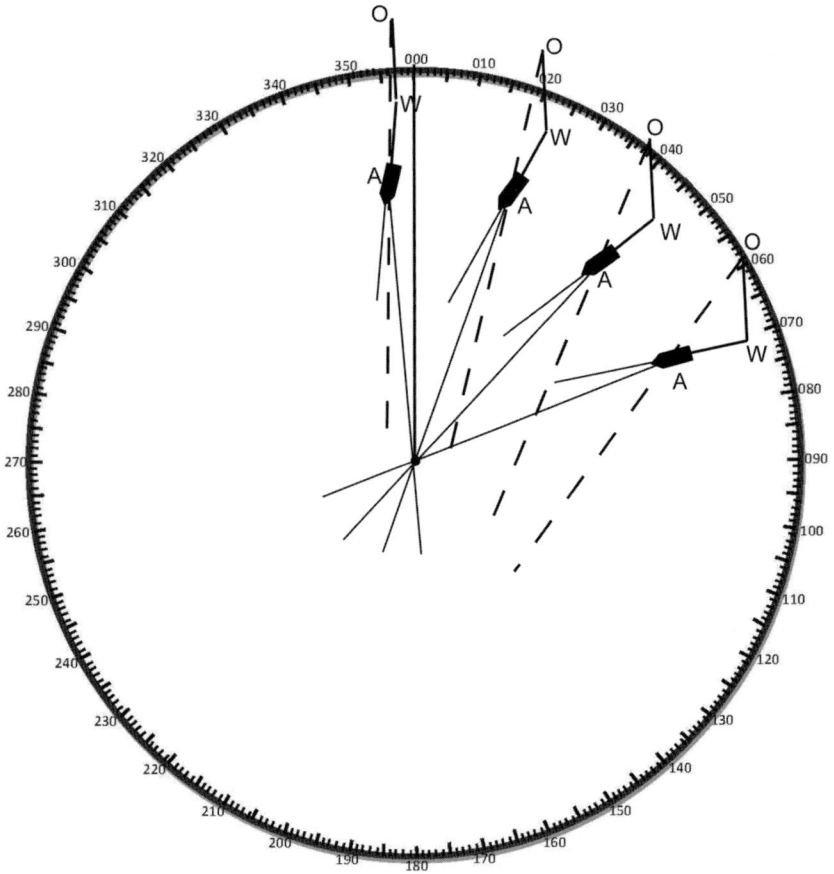

FIGURE 17.6   Showing aspects at different positions.

recognise the effect aspect can have on the CPA when all other things are equal. If the simulator allows, in many of the scenarios set up for this purpose should be supported by visuals, at 'night' preferably, but if time allows, by 'day', or a mixture, to relate the radar display to what can be seen visually. The one drawback in simulation is that visuals to not allow for the developing the skill of judging distances by eye, this is one of several shortcomings of simulation which cannot replace experience and seatime.

Avoiding Close Quarter Situations

A close quarter situation may be described as one where drastic action has to be taken to avoid collision. It is the situation where excessive amount of rudder has to be used, or an exceptionally large alteration of course has to be made, or engines have to be stopped or put astern. Other than discussing the action to be taken when a 'give way' vessel fails to do so, the actions to be taken in close quarters are beyond the scope of this book. Such discussion is in the subject of ship-handling, and while it is essential that OOWs should be fully aware and trained to cope with such situations (and simulation is essential for this), our emphasis is to avoid getting into close quarters (see below). It is worth mentioning here that several accident reports have remarked critically on some OOWs' lack of knowledge on the engine controls of their ships. Obviously OOWs should be fully conversant with such controls and the procedures to reduce speed or stop engines.

Early action to avoid a close quarter situation

The Rules do not define what is 'Early Action'; there are too many variables such as size and manoeuvrability of ships. Generally speaking it is the action taken when the OOW feels 'comfortable' with the action being taken and the effect it will have on the situation. It is where a moderate alteration (not necessarily a small alteration) of course is made, without undue amount of helm. The effect is an immediate opening of the CPA to an acceptable value, in open waters usually about 2 miles or more.

If the situation has gone beyond 'Early Action', this does not necessarily mean 'close quarters', but can be defined as 'Late Action' It does mean that more severe action has to be taken, such as a greater degree of helm and/or probably a bigger alteration or course. It may also mean a change of speed, but reducing speed should always be done with great caution. At this stage the OOW has to be taking action. There is no time for adjusting vectors or trying to make a new assessment of the plot and certainly not for trying to make contact (i.e. radio contact) with the other vessel.

Any action taken after 'Late Action' can be described as 'close quarters'.

The Admiralty Manual of Navigation, volume 4 gives two interesting tables examining the amount of an alteration of course needed to achieve a passing distance of 1 mile considering different speeds of own ship and target ship. These tables are reproduced in the Nautical Institute's 'Managing Collision Avoidance at Sea'. They show that a slower own ship in relation to the speed of the target vessel needs to

Minimum amount to Alter Course in a Give Way situation (target on a collision course)

Alteration at 5 miles to pass 1 mile clear

Give way ship	10 kts	15 kts	15 kts	20 kts	25 kts
Stand-on ship	20 kts	20 kts	15 kts	15 kts	10 kts
Relative bearing(degrees)	Amount to alter to starboard (degrees)				
015	35	27	22	12	18
030	35	27	23	17	13
045	39	30	25	24	
060	43	30	25		
075	50	37	26		
090	65	50			
100	75	60			
110	95	76			

FIGURE 17.7   Table amount to alter course in a 'Give Way' situation.

make a larger alteration of course than a faster own ship under the same circumstances.

Notwithstanding this fact, the need to take early and substantial action should mean that in all cases, and the more so when in restricted visibility, the initial alteration in course to avoid a close quarter situation should normally be more than may be suggested by such a table or by plotting or trial manoeuvre. The more it is obvious to the other vessel that action has been taken, the better. Once it is established that the target vessel is going clear, then own ship can start altering course back to the her original one.

A normal practice by give way vessels, particularly in clear weather is the technique of 'following her stern'. Own ship alters course to put the stern of the crossing vessel ahead or fine on the port bow and as the crossing vessel passes on her way, owns ship progressively alters course back to port, keeping the stern of the crossing vessel ahead or preferably fine on the port bow until own ship is back on her original course. This method is satisfactory under many circumstances, but it is 'rule of thumb' and does not give any indication of the passing distances. This technique also fails in circumstances where the crossing vessel is approaching at a small aspect angle, particularly if the initial relative bearing is small (fine on the bow). As discussed later, it is useful in complex traffic situations and it eliminates at least one target vessel from consideration as it is safely on the port bow and going away from own ship.

Practising 'following her stern' in restricted visibility should be

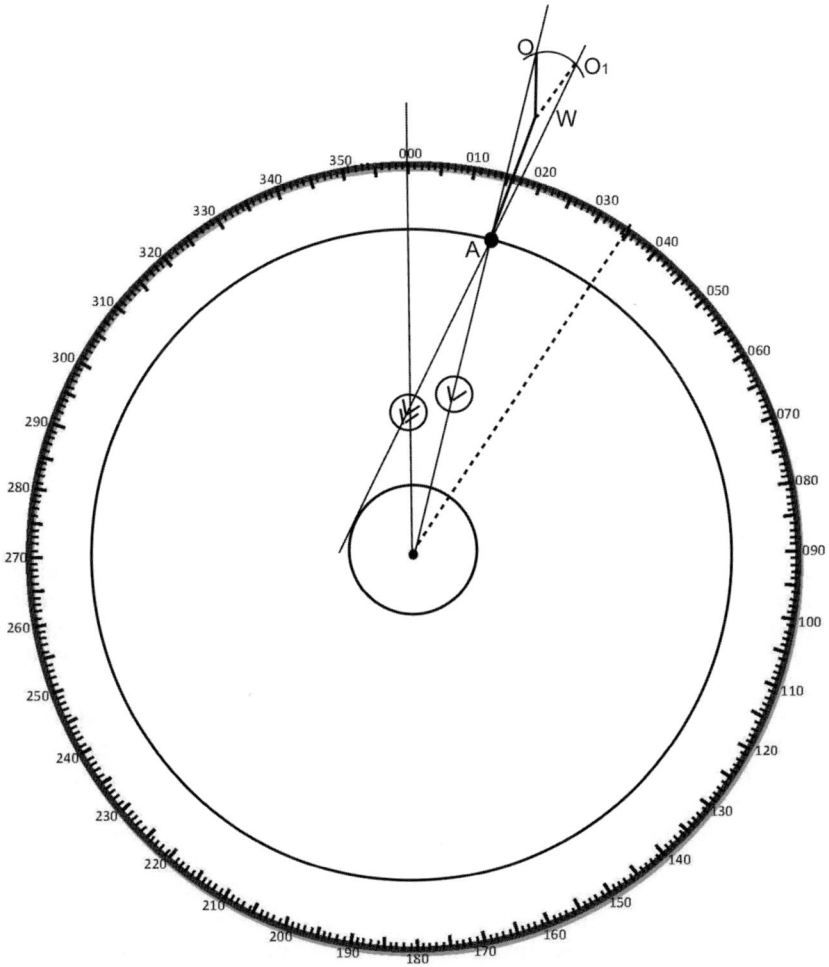

FIGURE 17.8 Showing minimum amount to alter course, first scenario from the table above.

treated with great caution and is generally not recommended except in certain busy situations as discussed later.

When Give Way Vessel fails to give way
This is probably one of the most worrying situations for an OOW.

It can develop quite fast and often there is no time to call the master, or even if called that he arrives on the bridge in time to be of any help. In any simulator course there should be several exercises involving this kind of situation. When should the stand-on vessel depart from the requirement to maintain course and speed? Under Rule 17a(ii) the stand-on vessel 'may however take action to avoid collision by her manoeuvre alone, as soon as it becomes apparent to her that the vessel required to keep out of the way is not taking appropriate action ...' and this is the Rule that allows the OOW to take action before a close quarter situation has arisen or Rule 17b comes into effect. How, or when does it become apparent that the 'vessel required to keep out of the way' is not doing so? The only basis on which the OOW can act is at a certain point to assume that the give-way vessel is not going to take action, but, in taking his/her own action making allowance for the give-way vessel making a belated alteration of course (or speed possibly, but unlikely). What need to be avoided is the situation where Rule 17b comes into effect; 'when from any cause, the vessel required to keep her course and speed finds herself so close that collision cannot be avoided by the action of the give-way vessel alone, she should take such action as will best avoid collision'. How does the OOW know when he/she is 'so close that collision cannot be avoided by the action of the give-way vessel'? This is emergency action, what in air navigation terms is a 'near miss' situation and much will depend on the OOW's instinctive reaction, rather than a carefully considered manoeuvre.

A good rule is to allow 7 ship lengths as the minimum distance from the point of collision at which action must be taken in a situation where the other vessel is not altering course, whether in restricted or clear visibility. For a VLCC or similar, this distance can be nearly 2 miles. This then needs to be converted into a time, in other words the latest time by which the OOW needs to take action. In a developing close quarter situation TCPA gives the time to collision, or near collision. Again, 'near collision' is very subjective, but in open waters with large ships, anything less than 1 mile can be considered as such.

If our VLCC is 400 m in length, and travelling at 15 knots, what is the minimum TCPA if 7 ship lengths is the minimum distance to the point of collision at which action must be taken.

The distance to the point of collision is $400 \times 7 = 2800$ m.
This equals $2800/1852 = 1.5$ miles.
Own ship will make this distance in $1.5/15 = 0.1$ hours, which is 6 minutes.

Therefore, in any situation, with this particular ship at this speed, 6 minutes represents the minimum TCPA at which some action must be taken.

Obviously, if own ship is obliged to take action under any rule, for example under Rule 19 in restricted visibility, action should be taken well in advance of this. However, under Rule 17a(i) where the stand on vessel is obliged to maintain its course and speed, then such a minimum TCPA represents a time when she must take action under Rule 17a(ii). It probably is reasonable to take such action some minutes before this actual time has arisen.

The obligation regarding the use of sound and light signals should not be neglected in any of these situations.

Multiple Targets and Complex Traffic Situations

The Rules (IRPCS) only deal with the responsibility to avoid collision between two ships. They do not cover situations where more than one ship may pose 'risk of collision'. Dealing with such situations involves applying the regulations where possible to each of several ships, the practice of good seamanship, experience and basic common sense.

Seamanship, experience and common sense suggest that avoiding, as much as possible, situations which may become critical is the best practice. Keeping a good look ahead and assessing traffic and traffic patterns and avoiding developing congestion where possible seems obvious. For example, concern has been expressed by several administrations where ships which plan their tracks on electronic charts increasingly seem to be following the same or very close tracks in certain coastal passages and traffic separation schemes. Good seamanship would suggest that where such tracks have been highlighted by the relevant authorities in coastal states, that they would be avoided, as much as possible in the planning stages.

Also, in congested waters positioning one's own ship well in advance of developing congestions is good seamanship. This, however seems to be causing difficulties where OOWs have been instructed to keep the ship within a certain distance off the track, say 0.5 or 1 mile. Masters need great care in framing their instructions to OOWs, either written or verbal, in such a way to allow them to exercise discretion and judgement to plan strategically. For example, a ship in an inshore traffic zone, passing close to the separation zone or line of the bordering lane, but going in the opposite direction to the traffic flow in that lane, needs to ensure that meeting any opposing traffic does not cause her to be forced into the lane.

Learning to assess the situation and decide on a course of action where there is a lot of traffic is something which must be developed with experience. Simulator training is of major assistance in this, but realistic as it may be, it needs to be complemented by the real thing, where the adrenalin flows and cold sweat breaks out. Junior officers or officers new to the ship or route, say an officer who has come from a tanker company to work in cross-channel ferries, need to be given the opportunity, initially under supervision, to develop their skills in this. Nonetheless much skill can be learned by well constructed simulator courses, either generic or designed for particular ships or trades.

Looking well ahead and keeping out of difficult situations is the key to avoiding stress. Apart from any health implications in avoiding stress, one is more likely to make a mistake or wrong decision when stressed. The term 'comfort zone' is sometimes used to describe the situation where there is no other target closer or approaching than the OOW can feel fully relaxed and capable of dealing with.

Carrying out the 'appreciation of the situation' exercise, as outlined above, can assist greatly in coping with the situations and also reducing the level of stress.

Traffic Separation Schemes

Under most circumstances a ship will enter the TSS at the start, proceed along the lane in the correct direction with the several other vessels going the same way and exit at the end. On the radar screen the true vectors will all point the same general direction and the relative vectors will be small, pointing in various directions. The OOW's task will be to avoid getting too close to other targets proceeding in the same general direction and to watch for rogue targets, crossing vessels, fishing vessels and such.

Targets proceeding in the same general direction
The relative vectors are the most useful here. They will show immediately targets overtaking or being overtaken. Those being overtaken are probably of more concern, but overtaking vessels need to be watched particularly as they may restrict the options to manoeuvre somewhere ahead. If an overtaking vessel is getting close, it may be an idea if possible to reduce speed slightly to let it get ahead, of even an alteration of a few degrees away from it will reduce own ships speed along the direction of advance and allow the passing earlier than otherwise (see Figure 17.9). This shows the effect of an alteration of course to port of 20°, on the relative vectors (red).

Where own ship may need to gradually move across a lane, say for example where it is necessary to leave the lane at a small angle a similar process can be carried out to ensure a reasonable clearance, particularly

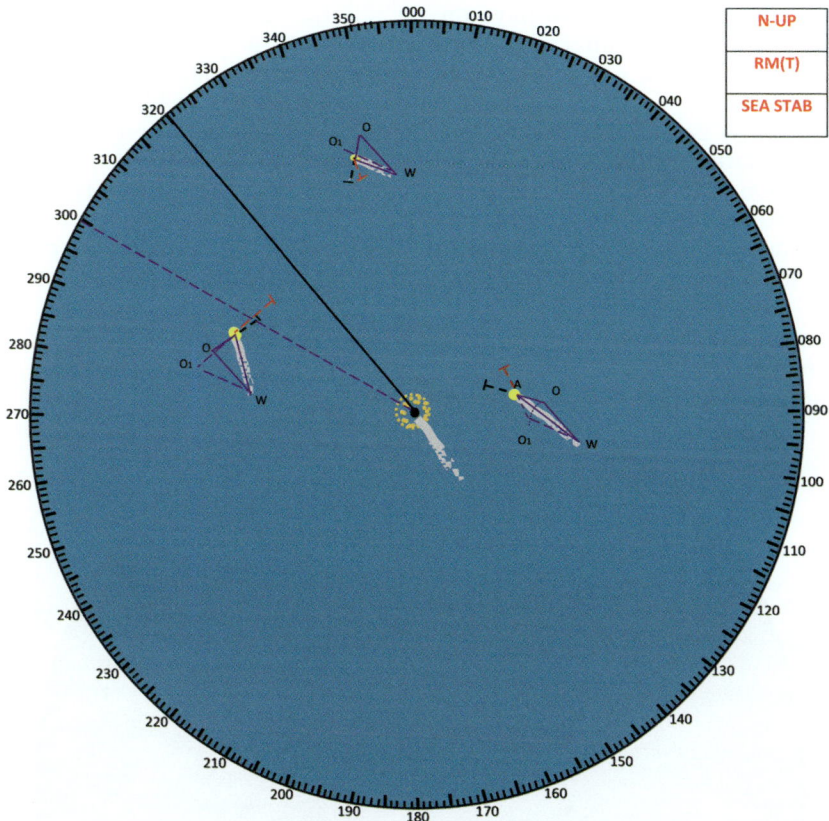

FIGURE 17.9   Traffic in lane, overtaking Sit.

with ships coming up on the quarter. Such a manoeuvre needs to be carried out well in advance of the position where own ship needs to leave the lane. By the time this point is reached, own ship should be positioned close to the border of the lane and with no other ship coming up on the quarter on that side.

Crossing vessels

The main restrictions on Own Ship in dealing with crossing vessels, when in a traffic lane, are those posed by the other vessels proceeding in the same general direction, particularly those close by on the quarters. In Figure 17.10, own ship must alter for the crossing vessel on her starboard bow. The proposed alteration of course is 52° (from 320 to

012). The effect this will have on the relative vectors is shown in red. Obviously if that alteration of course is made at this point, a close quarter situation will develop with the target vessel abaft the starboard beam (which is overtaking own ship as indicated by the black relative vector). A greater alteration to starboard or a lesser one will not resolve the situation with regard to the target vessel abaft the beam. A possible solution is to delay the alteration but by the time the overtaking ship has passed by a distance large enough to allow the alteration of course to starboard the crossing vessel will be very close. A reduction in speed seems the most obvious solution, or even a round turn to port to come

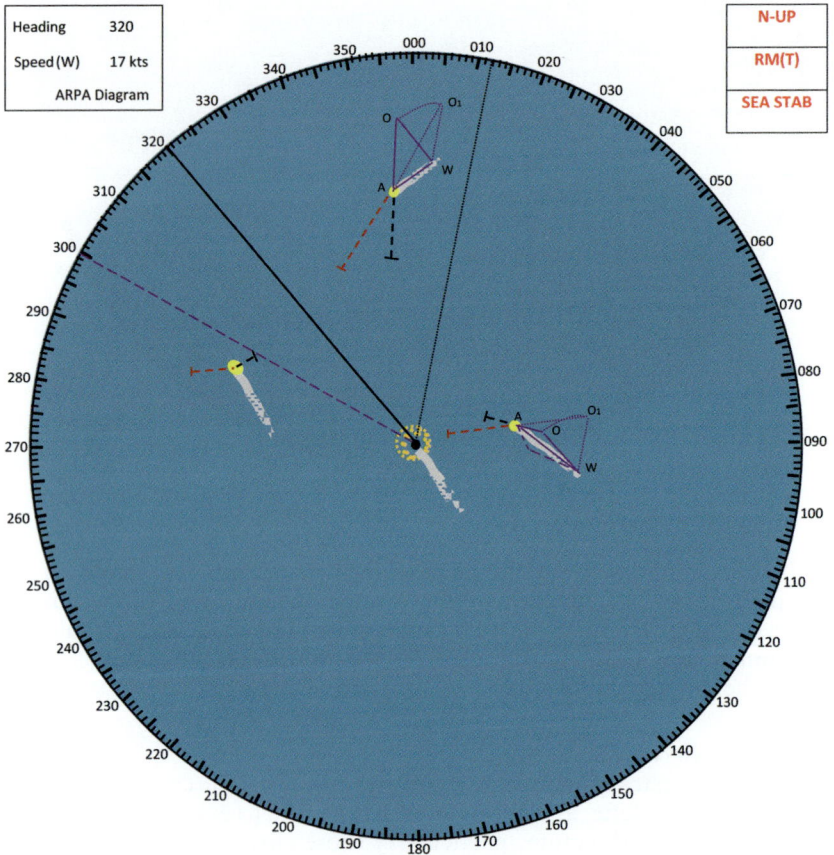

FIGURE 17.10   In a traffic lane dealing with crossing Sit.

up with the overtaking vessel ahead for the time being until course can be resumed.

Rogue targets

This is less common than it used to be, particularly in TSSs which have VTS or other supervision, and since the advent of AIS. However, it can and still does occur where a target is detected going the wrong way in a traffic lane. Where the inclination might be to contact the other vessel, or the coastal authority, this must be resisted until it is established that the target is going clear. The actions of other traffic in dealing with this situation can cause particular difficulties.

Other targets in the lane or approaching it

Similar to rogue targets, other traffic needs to be carefully monitored. As stated above, apart from dealing with it as a one-to-one situation, the other adjacent traffic and its possible actions need to be considered and watched carefully.

Crossing a Lane or Lanes

This can be one of the most fraught situations particularly for watchkeeping officers who may not have much experience of such situations. It is also more difficult in slower ships than in fast ones. Without the benefit of radar tracking it can be extremely difficult if not impossible in very busy traffic situations.

Crossing a lane with traffic from the port side

In clear visibility traffic from the port side should consist of 'give way' vessels. However, often the tension is more acute as one is depending on the other ship to alter, and wondering when it is or if it is going to alter at all. Occasionally one hears that some obviously incompetent people in charge of ships have the impression that being in the lane gives them right of way! Therefore the OOW or bridge team in the crossing vessel must always bear in mind that a 'give way' vessel may not do so, and consequently he/she must be ready to take evasive action. It is dangerous to be prescriptive in describing such situations, as each will depend on the particular circumstances, but probably the safest evasive action is to turn along the direction of traffic until the opportunity offers to resume the crossing.

In reduced or restricted visibility of course there is no 'give way' or 'stand on' vessel until they come in sight of each other.

In both cases it is best to study the vectors as early as possible. If possible, depending on sea room start the approach on the crossing course well before crossing the outer edge of the lane so as to establish the vectors. Then one can judge early which, if any, are likely to pose a problem. Quite a small alteration of course at this stage (very early!) may cause all the relative vectors to go clear with a reasonable CPA and this then can be the course to continue until a more direct (right-angle to the direction of traffic flow) course can be adopted without causing confusion to the other traffic. Trial manoeuvre comes into its own in this kind of situation.

In the following series of Figures, own ship is crossing the lane with several ships on her port bow.

From the extended relative vectors it can be seen that the targets which may develop close quarters are Nos. 3, 6 and 7. A trial manoeuvre proposing an alteration of course of 45° to starboard to 150° will cause these targets to go clear but may develop a close quarter situation with target 4. The next radar screen shows the situation after an alteration of course to starboard to 139°. In this it can be seen that all targets are going clear except for target 4. The last screen shows that own ship should pass ahead of target 4 by about four cables and it shows the trial manoeuvre to resume the course of 105° in about 20 minutes. It is clear from this, that a further alteration of course to

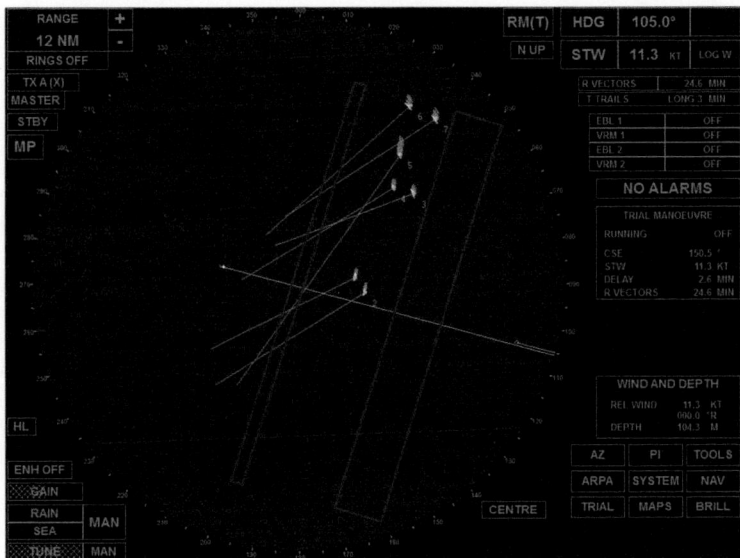

FIGURE 17.11.1   Traffic from port.

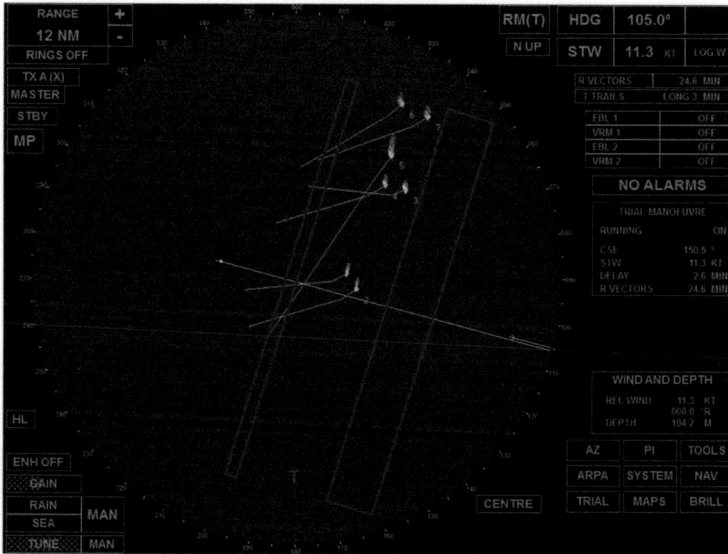

FIGURE 17.11.2   Trial manoeuvre.

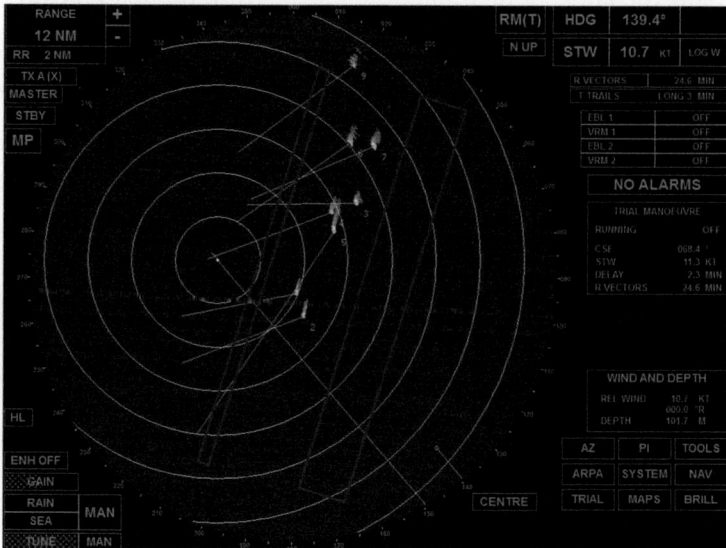

FIGURE 17.11.3   After alteration of course.

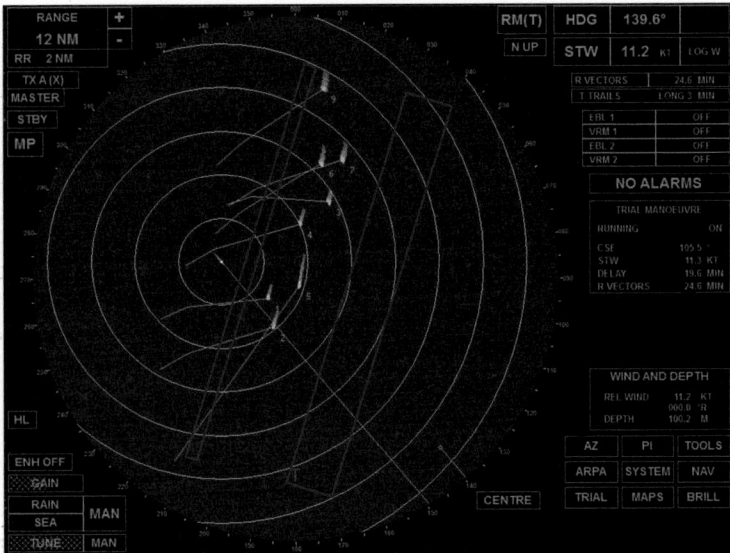

FIGURE 17.11.4   Trial manoeuvre to resume course.

starboard for about 5 or 6 minutes will increase the CPA of target 4 and then the course of 105° can be resumed. However, at this stage it may considered that a close quarter situation is developing with target 4 and that if it is in sight that course and speed must be maintained to allow target 4 to take action, or if not yet in sight not to resume the 105° course until target 4 is clear (Rule 19 not to alter course to port for a vessel forward of the beam, or towards a vessel abeam or abaft the beam).

It might be argued that in clear weather there is a requirement for Own Ship to maintain course and speed for the traffic approaching from her port side, or in restricted visibility to avoid altering course to port for a vessel forward of the beam. This, of course only applies where risk of collision exists; our proposed earlier manoeuvre is designed to eliminate, as much as possible such risk at an early stage; but, if at some stage the altered course brings Own Ship into a developing close quarters situation with a ship on her port side, as our example does, then in clear visibility she must maintain her course and speed and depend on the other vessel to take action, or turn to starboard or reduce speed if in reduced visibility.

This, of course is only an example of any number of possible situations and an example of possible actions in dealing with the situation.

Crossing the exit from a lane with traffic from starboard

In the next series of radar screens is (Figure 17.12.1), the traffic is approaching from starboard. From examining the plot it is clear that target 2 is the one most likely to come into close quarters. It is also obvious that any alteration of course to starboard to increase the CPA of target 2 is going to decrease the CPA of target 4. The trial manoeuvre shown (Figure 17.12.2) is for an alteration of course to starboard of 41° in 11 minutes. This will increase target 2's CPA to over 2 miles, but by the time target 2 is ahead (at 3 miles) target 4 is getting close (about 4 miles) with a CPA of 1 mile and target 5 is close behind. A delay in the trial alteration time would bring target 2's CPA closer but would reduce target 4's CPA (Figure 17.12.3). The next screen shot shows an alteration of 20° to starboard being made which increase the CPA of target 2 to almost 2 miles. This has made the CPA of target 4 very small (Figure 17.12.4). However, once target 2 has passed ahead, in about 15 minutes, own ship can start to alter back to port. At this stage target 4 should be about 5 miles away and any alteration of course will increase its CPA. By the time target 2 is clear target 4 should be going clear by over 2 miles.

There may be a reluctance to pass ahead of target 4, as own ship is the 'give way' ship in clear weather, or altering course to port for a

FIGURE 17.12.1   Radar traffic from starboard.

FIGURE 17.12.2   Trial manoeuvre.

FIGURE 17.12.3   Alteration in course delayed.

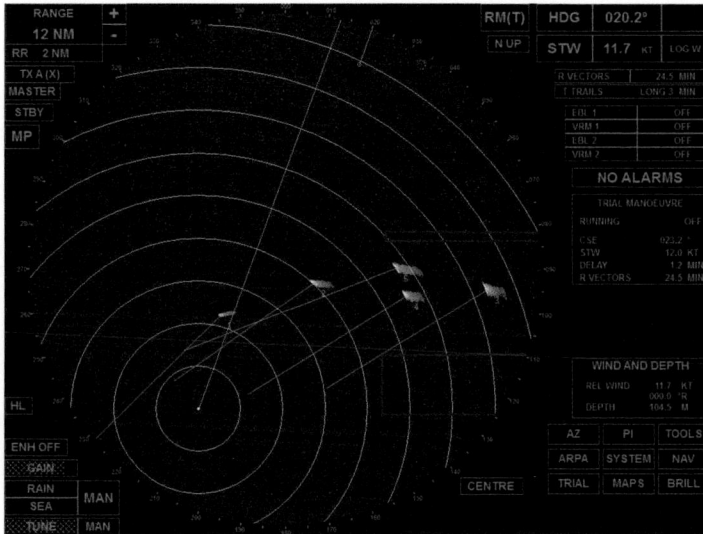

FIGURE 17.12.4   Altered course 20° to starboard.

vessel forward of the beam if in restricted visibility, but there may be little alternative. To alter course further to starboard for target 4 will mean remaining on such a course for target 3 as well and possibly other following traffic. While the scene illustrated in the series of radar screen shots depicts traffic coming out of a traffic separation scheme lane, it also apples for crossing a lane and own ship could end up going for a long period in the opposite direction to the traffic flow in the lane.

Precautionary Areas

While there is some kind of order in the lanes of a traffic preparation scheme, where several such schemes or traffic streams come together, such as in the approaches to a major port, then there can be a certain amount of 'free for all'. Sometimes the establishment of a 'round-about' can put some kind of order into the flow of traffic, but there are areas where there is no such arrangement for various reasons such as at a pilot station, or where the joining of different traffic lanes does not lend itself to such a rotation of traffic.

As before, studying the vectors gives the risk relative to own ship, but of course this can change very rapidly as the other vessels change course for navigational reasons, or to avoid other vessels. Careful judgement of the developing situation and appreciation of the likely movement of target vessels can allow early action to avoid close quarter situations to develop. For example, if the target vessel is approaching the pilot station, there is the likelihood that it will slow, and manoeuvre to embark or disembark a pilot.

A chart radar, or tracked target on the ENC or pre-prepared radar maps can help in determining the possible actions of target vessels. This coupled with AIS gives a great deal of information and assistance in avoiding difficult situations that the absence of such aids denies. Being able to see, at a glance where a target vessel is in relation to a lane, channel or major navigational mark gives much assistance in assessing the situation. Knowing the destination of a target vessel from AIS will indicate its likely route where several possibilities exist. Knowing, from AIS, that a target vessel is a deeply laden VLCC will suggest that it will take the Deep Water route where such exists as an alternative to a non-deep water one.

AIS and VHF Radio communication

The use of VHF radio, in conjunction with AIS which identifies a target vessel is another possibility in assessing the situation and deciding on a course of action. Using this has to be done with a great deal of caution and there are various publications warning mariners about the use of VHF for collision avoidance. It can be used to establish the broad intentions of a target vessel. For example it can establish if the target vessel is picking up a pilot or not. It can be used to warn another vessel that Own Ship is going to reduce speed or alter course for navigational reasons where such manoeuvres might be unexpected. It should generally not be used to make manoeuvring arrangements except in exceptional circumstances and certainly not when a close quarter situation is developing; under such circumstances it is a distraction and leads to confusion and the possibility of serious error. Likewise communication by AIS safety related messages should not be used, attempting to do so has contributed to 'AIS assisted collisions'. In this regard, it must be stated that it is still early days for AIS and what it implies for collision avoidance and its use and incorporation in the anti-collision plot is still evolving.

Conclusions

In this chapter we have looked at the practical use of radar for collision avoidance. The following points may be summarised:

- Training and knowledge in the use of radar for collision avoidance is essential,
- Full use must be made of all the 'tools' to make a full assessment of traffic situations,
- Skill should be developed in carrying out an 'appreciation of the situation' for traffic, particularly complex situations,
- Dealing with complex traffic situations in or outside traffic separation schemes requires training, practice and experience. Modern radar plotting aids are essential as is simulator training.

Chapter 18

New Technology Radar

This chapter takes a brief and very general view of New Technology (NT) radar. The following points are discussed:

- The radio regime requirements which are mainly responsible for prompting developments in this area.
- The comparison of the new technology and the old.
- Pulse compression and precise range measurement.
- Improved clutter rejection and small target detection.
- Different operator settings.

The term 'New Technology Radar' is used to describe the development of solid state radar and advances in-signal processing for merchant ship radars which comply with the specifications laid down by the IMO. The provisions of Resolution MSC 192 (79) 'Adoption of the Revised Performance Standards for Radar Equipment' is written in more general terms than the older specifications, with outcomes required rather than specifying how certain functions are achieved. In this it is allowing for new developments which had been restricted by the requirements to comply with the older specifications.

In the world of radio telecommunications, the part or the radio frequency spectrum allocated for marine radar has come under increasing political pressure by other sections of the radio community, particularly those involved in cell-phone and related developments. The 3GHz (S-band) section of the spectrum, particularly has been targeted for number of years, but for the moment the marine representation in the ITU has managed to retain the two frequency bands, 3 and 9 GHz.

Ironically, it is developments in the cell-phone and related areas which has helped to develop solid state marine radar to the extent that an affordable radar system can be produced; developments in the 9Ghz (X-band) are not so advanced and are proving extremely expensive.

Comparing new and old technologies

Traditional radar, as we have seen in the earlier chapters, uses a very short pulse, at the frequency of the radar. The ability of the magnetron to transmit this pulse at an absolutely precise frequency is, in modern terms, not very good, and the term 'incoherent' is applied to it. In other words, there is considerable transmission 'overspill' outside the designated frequency and there is small variation in the actual frequency due to physical changes in the magnetron as it heats up or cools and as it ages. To illustrate this, in every-day terms, it can be compared to manually tuning your radio into your favourite station. As you get near the correct frequency, you will start to hear the station transmission faint, muffled or distorted compared to when you are on the frequency when it will be sharp and clear; the faint reception being the frequency 'overspill'. This is one reason why tuning is so important in traditional radar, whether it be manual or automatic; the magnetron transmission frequency can vary slightly and so the received echo signals will return at the same frequency, hence the receiver has to be tuned to this varied frequency.

This 'incoherence' has also caused problems within the radio frequency world with the increasing demand for 'slots' within the spectrum. Marine radar is 'untidy' in this regard and if other radio systems are working on a frequency near the marine radar ones, interference can be caused.

Traditional radar makes use of a strong 'peak power' pulse to achieve the power to get a reasonable echo from targets. New technology techniques use a considerably less 'peak' power, but achieve the power to get a reasonable echo by the comparatively long duration of the pulse. Traditional radar peak power is in the range of 10 to 20 or more kW. NT radar peak power is in the range of 100 to 150 W, about one hundredth of the power of traditional magnetron radars.

Apart from the technical (and political) requirements to tidy up marine radar transmissions, the other advantage which new technology offers is 'solid state' components. The main component missing in NT radar is the magnetron. The need for a much reduced peak power permits 'solid state' components to produce the necessary output, and at the same time produce the precise (coherent) radio frequencies required. In terms of cost, the normal 'life' of a magnetron has been states as about 3000 hours compared to over 50,000 for NT components. Other aspects of replacement and maintenance are also very much reduced and simplified. While the initial cost of NT radar currently is much higher than conventional radar, makers of NT radar claim that the reduced cost of maintenance makes NT radar more cost effective over time.

Pulse compression

In new technology radar, in addition to the pulse, use is made of the information to be gleaned from the variations of the frequency in the returned pulse (echo). Such techniques have been used for many years in other forms of radar, mainly in military use, and in other radar systems where continuous wave (CW) is used which does not have a pulse system.

The pulse used in new technology radars is, in radar terms, extremely long, even in the region of 1000 micro seconds (compared to 1 or even smaller microseconds for conventional radar). This of course makes the pulse in itself useless from the point of view of range discrimination. For this reason, different technology is used to determine range and to distinguish between targets which are close together. This process is called 'pulse compression' and there are two principal different methods of achieving it. One method is 'frequency modulation'. In this the frequency of the transmitted pulse is changes over the duration of the pulse. The Transmitted pulse therefore has a certain characteristic in the nature of the frequency change (the modulation) within the duration of the pulse length. This frequency change is called the 'bandwidth'. The other method is 'phase coding' where the phase of the frequency is changed within the pulse to form a binary series of positives and negatives or '1s' and '0s' (see Figure 18.1).

In the frequency modulation method (called 'chirp') the frequency of the pulse changes over the duration of the pulse, usually increasing in frequency. The pulse therefore has a unique characteristic, in not only its frequency, but also in the change of frequency, the bandwidth. This characteristic is used to distinguish returning echoes from surrounding noise; anything which does not meet the characteristic of the pulse will be rejected.

Pulse compression, as the name suggests, is the process of 'compressing' the relatively long pulse into a short pulse that has the required characteristics for accurate range determination and target discrimination. This is achieved by the use of a pulse compression filter. This uses a mathematical algorithm that combines the change of frequency (the bandwidth) within the pulse, in other words the change of frequency between the start and the end of the pulse, with the pulse length.

By this process a pulse called a 'sinc' pulse is generated. This produces as very narrow (short) pulse with a very big 'amplitude'. This overcomes the low power of the returning echo pulse to distinguish it from the surrounding 'noise', the very sharp 'peak' of the compressed pulse usually stands out sharply above the noise. The extreme 'shortness' of the compressed pulse then meets the requirements for accurate range measurement and target discrimination.

There are two disadvantages to this system, compared to traditional short pulse radar. One is a sidelobe effect produced in the sinc pulse (see the small 'peaks' each side of the main 'peak' in Figure No. 18.2). This is not to be confused with the sidelobe effect produced by the antenna

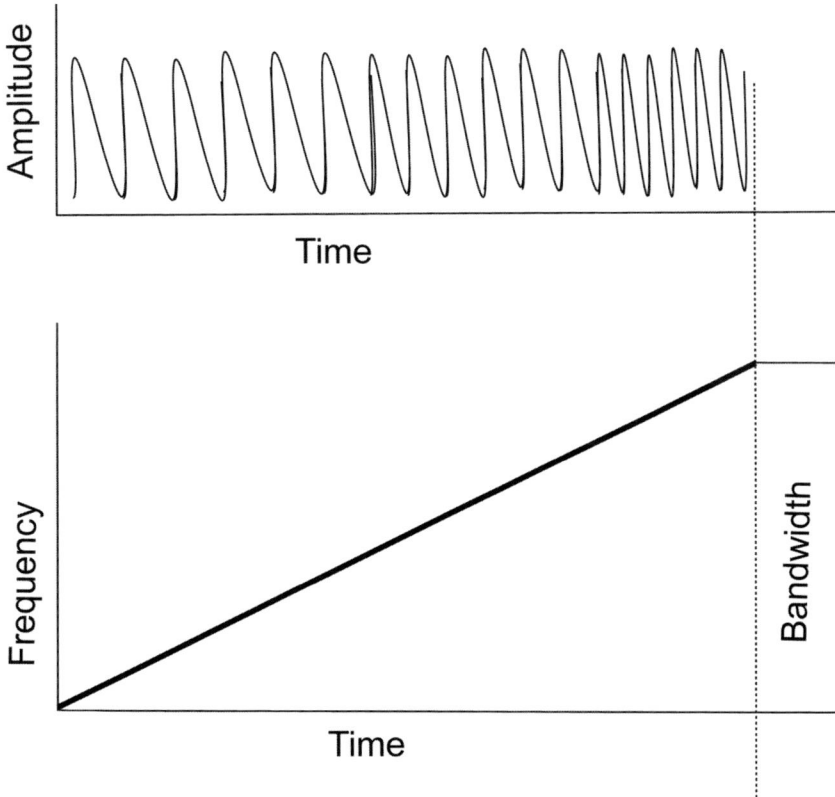

FIGURE 18.1   Change of pulse frequency over time.

(scanner) but it can produce false echoes from very strong targets which can evade the filtering system designed to eliminate such false echoes. This can potentially produce range measurement ambiguity, or actual false target echoes on the same bearing as the real target.

The other disadvantage with such long pulses is the detection of targets close by, the minimum range by which targets can be detected. As both transmitter and receiver are using the same scanner (antenna) the receiver cannot receive an echo until the transmitter has completed its transmission of the pulse. This means that the minimum range, is, in theory, half the length of the uncompressed pulse, in metres. IMO requirements are for a small target to be detected at 40 m horizontally from a 15 m high antenna. To achieve this, a series of pulses of different

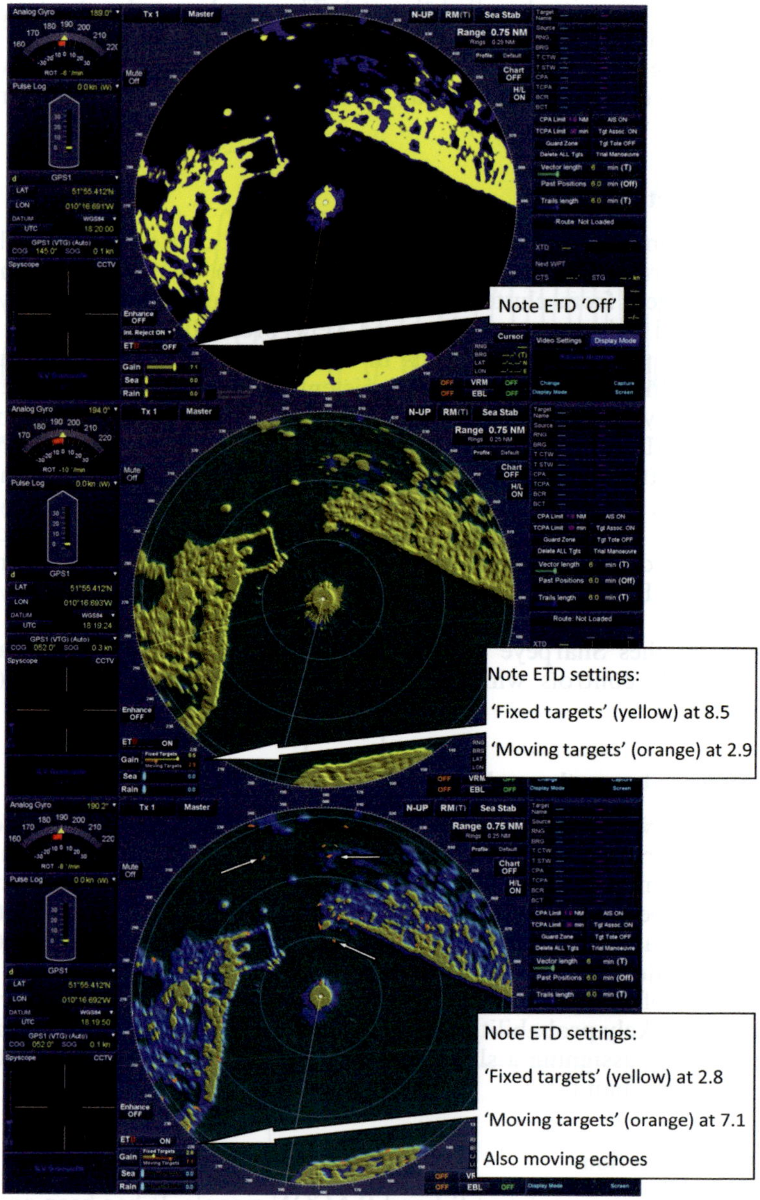

Note ETD 'Off'

Note ETD settings:

'Fixed targets' (yellow) at 8.5

'Moving targets' (orange) at 2.9

Note ETD settings:

'Fixed targets' (yellow) at 2.8

'Moving targets' (orange) at 7.1

Also moving echoes

FIGURE 18.3   Example of enhanced target detection.

S-Band RACONs may be discontinued. This has been influenced by the development of AIS (see Chapter 15 on AIS). For the time being X-Band RACONs will be continued on the presumption that most ships with S-Band radars usually have X-Band as well.

Operator Settings

Certain aspects of setting up and adjusting settings for optimum performance are different for NT radar from that of conventional radar. Two notable adjustments are missing; 'pulse length' and 'tuning'. As discussed above pulse length is a different concept in NT radar compared to conventional. In conventional radar a choice of pulse length is used to either enhance an echo, or to make echoes more distinguishable from each other. NT radar offers no such choice; the target echo response is produced differently, as we have seen. Once an echo is detected, the usual processing is applied to amplify it to produce a usable image on the screen, A large part of this signal processing is amplification as in conventional radar, with the possibility of the user adjusting the 'Gain' as required.

Also missing is 'tuning'. As the equipment is transmitting a 'coherent' pulse, that is its pulse frequency is very precisely defined, the echoes of this pulse, which will be at the same frequency as the transmitted pulse, will match the transmitted pulse. As there is no variation in the pulse and echo frequencies the receiver is constructed to receive that frequency; there is no requirement for receiver adjustment to match the transmitted frequency.

Sea clutter suppression has a similar effect in NT radar to that of conventional radar; the gain is reduced near the origin, depending on the setting of the control. This is coupled with the correlation process for eliminating random echoes. In NT radar the Doppler shift is a further process in detecting 'wanted' targets from surrounding sea clutter.

Rain clutter tends to be less prominent in current NT radar due to it transmitting in the S-band. Pulse Doppler processing is also used for rain clutter suppression. Other than that the rain clutter control behaves in much the same way as conventional radar. The emphasis on the 'leading edge' or change in voltage (see Chapter 4 on the receiver) is applied to the video stage of the incoming echoes. Thus a similar effect to conventional radar is produced that can 'sharpen' the picture and help distinguish wanted targets in rain clutter.

The use of Enhanced Target Detection (ETD) requires use and familiarisation to get most benefit from it. In the Kelvin Hughes system there are two adjustments for the operator to help distinguish 'fixed' and 'moving' targets (see Figure 18.3). This is a new feature and as yet an optimum use of it has not been generally publicised. The main benefit is to distinguish small targets from surrounding clutter. However, as sea clutter is normally moving much of it may show as 'moving' targets, depending on the adjustment set by the operator.

Conclusions

Developments in commercial marine radar have been traditionally driven by market forces. However pressure on the allocation of radio frequencies has prompted new developments. Much of this the technology has been long in use in many non marine applications, but compared to traditional marine radar is extremely costly and non competitive.

NT radar, in the way that it detects targets is fundamentally different from conventional radar. Pulse compression and longer pulses at much lower power, output can be used to achieve the same level of target detection. This also allows solid state components to be used with better reliability and less frequent maintenance.

Clutter rejection is improved over conventional radar with advanced signal processing using correlation and Doppler technology.

Some changes in the way NT radars are set up and adjusted for optimum performance by operators needs to be taken into account. While better clutter rejection is claimed for such radars compared to traditional ones, clutter is still present. OOWs need, as always, to study and test their sets in less critical circumstances so that when in more demanding situations they are fully familiar with their equipment, the controls and settings and the level of confidence which they can give to their radar.

Chapter 19

Notes on Radar Simulation Training

Simulator training in the use of radar is essential. It is almost impossible, in the modern world to achieve a Certificate of Competency without such training. In this chapter the subject of training in the use of radar is discussed, and suggestions made regarding good practice so that mariners at various levels of career development get adequate training. Three levels are normally considered in simulation training:

Operational level. Cadet and trainee, and training and assessment/ examination towards first certificate of competency.

Management level. The additional training required by candidates for examination for senior certificates of competency.

Refresher or upgrading of qualification. This includes additional training that may be required by particular shipping companies and by Bridge Team/Resource Management courses.

Simulators extend from simple software packages to be run on a PC that allow basic skills in plotting and some navigational tasks to be developed, to full mission bridge simulators where a realistic representation of a ship's bridge, its navigational outfit and the environment in which the 'ship' is operating is presented, allowing overall navigational, collision avoidance and decision making skills to be developed. And of course there are various ranges of simulator installations between these two extremes.

In order to meet the requirements of IMO, and the provisions of STCW78/95, there is a minimum outfit of simulation equipment required for simulation training and assessment.

The correct use of simulators complements the experiential learning of time spent on the bridge of ships at sea; it does not replace the requirement for such 'sea time'. It also allows the development of skills which cannot be achieved in actual ships, such as dealing with critical situations. In this section we will examine how various 'skills' in relation to radar, and collision avoidance particularly, can be developed by simulator training.

Radar of course is only one of the 'sensors' feeding information into the bridge team/OOW. And while this work emphasises the use of radar, as stated elsewhere, it cannot be considered in isolation from the other 'sensors', including eyes and ears, which provide information to those on the bridge. In considering simulator training in radar therefore, care has to be taken not to stray into the area of passage planning and making, ship handling and emergencies, bridge teamwork and the whole range of realistic training that modern simulators allow. In the following, the discussion is confined to training in the use of radar and AIS on simulators, mainly for collision avoidance, although other factors must, of necessity, be touched upon.

Operational Level Training

Radar training at the basic or operational level can be divided into three main parts;

1. The radar set. How it works, some theory to understand the information which it presents, the functioning of the different settings and controls.
2. The practical operation of the equipment. This involves the use of the radar in the navigation function, in the traditional sense where information is extracted and plotted on a chart, or increasingly where it is used in conjunction with Electronic Navigation Charts (ENCs).
3. The use of the equipment as an anti-collision aid, with or without AIS.

A live radar set is essential for the first of these tasks. Depending on the simulator fit, simulated radars can, to a greater or lesser degree simulate the different settings for such things as gain, tuning and the clutter controls. Usually these functions do not behave realistically in simulated radar, although there is ongoing development in these areas. For this reason, a live set preferably overlooking a wide expanse of open water and where some marine traffic may be found and other features such as buoys and navigational marks are located. Much of the training in setting up and adjustment of performance can be done on such a set.

For the rest, simulation provides the best medium. Training in the navigation tasks and the anti-collision functions are carried out in situations which can be set up by the instructor to develop the students' particular skills, to make or emphasise particular points relating to a situation, to analyse actions taken by students or case studies and most importantly, to test students' competence.

As a general rule in simulator training, the exercises are divided into three parts:

The briefing.
The exercise 'run'.
The debriefing.

Each of these is as important as the other in the teaching/learning process. The temptation is to concentrate on the 'run' to the detriment of the other two elements, particularly the briefing. If the exercise is to gain its full value then all the elements must have equal emphasis.

The briefing

The students must have a very clear picture of what is expected of them. Often, this may involve a lot of stating the obvious. Instructors should have all the details of the exercise written, and go over each element with the students. If for example, we take a simple plotting exercise, then the course and speed of own ship is obviously the starting point. Then the radar display settings should be dealt with, the scale to be used, the tools (EBL and VRM or cursor), the way the radar person is to call out the range and bearings to the plotter, presuming they are working in teams of two, how the initial plot is to be laid off, the zero speed line (OW) is to be laid off, subsequent plots and completion of the OWA triangle. Then how the report is to be written and stated should be made clear. Obviously as the course progresses the briefing will change, and it will be unnecessary to brief again on many aspects of what has gone before, but care needs to be taken that detail is not overlooked in subsequent exercises, and it is better to repeat certain aspects of the briefing every time.

In the briefing, particularly in more advanced simulation exercises and with more senior and experienced students, the students must be encouraged to 'play'. The reluctance to be seen to make mistakes or wrong decisions is natural. The instructor must take time to encourage the students to participate fully and to explain that this is the only way that they can benefit from the simulation, that it is by making mistakes and wrong decisions that learning is achieved and that the consequence of such errors is harmless in a simulator but might have catastrophic consequences in real life.

The Run

The exercise should be constructed so that initially the students have time to settle in to it before any 'action' starts. Again, taking simple plotting exercises as examples, approaching targets should be far enough away to allow the students to see that their radar is correctly set up and working, that they have the necessary documents and instruments, to identify the targets, to manipulate the radar controls, to

342   RADAR AND AIS FOR WATCHKEEPING OFFICERS

prepare the plotting sheet and gain confidence in what they are doing. The instructor should be in 'radio' contact with the students if possible, but if this is being used then strict radio discipline should be enforced.

Students should be forbidden to exceed the aims of the exercise and fiddling with the radar or other instruments or controls of the simulator. For example in plotting exercises they should not use the automatic plotting aids. These should be disabled in these cases, if possible.

In initial exercises the instructor should spend a lot of time looking over the students' shoulders and ensure the work is being done correctly. A lot of time can be wasted when, at the end of an exercise it is found that some students had not understood the task or otherwise not completed it. Obviously as skill develops there is less need for this, and in fact it encourages confidence if the instructor can leave the students work without this supervision.

The construction and operation of an exercise by the instructor needs to be carefully carried out. It should at all times be as realistic as possible. Nothing can ruin a simulation exercise as much as if the student gets the impression that the instructor is 'chasing' him/her; in other words target ships are behaving unrealistically. The tendency often is to 'challenge' the student, but this needs to be resisted, and it is in the original construction of the exercise that the challenges should be built in. If the student decides not to 'play' or takes some unexpected course of action that avoids the main purpose of the exercise there may be little that can be done about it, but the temptation to put situations to 'catch' such a student should generally be resisted. This does not preclude a well prepared exercise which has situations for most of the foreseeable courses of action which the student might take, but, once again, realistic behaviour of targets is essential.

De-briefing

Possibly this is the most important part of simulation. To gain the most benefit it needs careful and diplomatic handling. The achievement of the aims of an exercise can be completely ruined by poor debriefing. Any sort of 'slagging' or critical comments between students or teams of students should be stopped immediately. A good tactic is to get the students to speak first. This has to be encouraged by leading questions such as 'What did you think of the exercise?'; 'did the action you took achieve the result you anticipated?'. As much as possible the student should comment on his/her own errors such as 'I left the alteration of course too late' or 'I could have altered course more'.

Once the students have had their say, the instructor can now comment. It is best to comment in general terms rather than singling out a particular student or team. For example, attention to the detail of plotting symbols tends to be neglected, and where one or two teams have been lax in this regard, the instructor can say to the whole class 'It is very important to use the correct symbols and terms in the plotting exercise'. If one team has made a serious anti-collision error, once again, the instructor can address the whole class by saying, for example, 'in the case where an alteration to port was made, we can see where this resulted in a dangerous situation ...', this avoids embarrassing the 'offending' team, and all the class benefit from the lessons of the mistake.

While it is natural to focus on the negative aspects, the errors and mistakes, and to correct these, the positives should be emphasised and used where possible. Once again, this should be addressed to the whole class, for example, 'here we can see the benefit of an early and substantial alteration of course to starboard ...', where one or more student/teams have done the right thing.

Finally, and depending on the level and experience of the students, there should be some time for discussion and comparison. In a simple plotting exercise, for example, was the plotting interval long enough? Was there enough time on completion of the plot to allow for decisions to be made? How is it that several teams/students got different results? and so on.

Anti-collision training

Basic plotting

In advance of any time spent on simulation, students should have had adequate time on paper plotting exercises and be fully aware of plotting techniques and terminology. Some of the more advanced aspects of plotting can be instructed in parallel to simulator exercises, but the basic plot, the report and the understanding of the significance of the relative motion plot should be thoroughly understood before simulation exercises start.

The basic plot

This should be carried out in real time with two targets, forward of the beam, and set to pass Own Ship within 2 miles, but not on collision courses. The students should work in teams of two, one at the radar set and one taking down the bearings and ranges called out by the radar operator and plotting them on a paper plotting diagram. Both radar and plot should be in 'North Up' mode, but Own Ship should be on a northerly course.

After about 10 minutes the exercise should be paused and time given

RADAR AND AIS FOR WATCHKEEPING OFFICERS

to allow the students to complete the plot and the report. Once this is done, and the instructor has corrected any errors, the exercise can be run on and the students track the targets to verify the information in the report. It is likely that errors will have been made in the basic plot, the whole process is notoriously error prone, due to students not being familiar with the operation of the controls, with calling out the wrong ranges or bearings, and with the plotter transferring wrong data to the plot. Time and repetition is needed to allow the students develop skill and confidence. After a few exercises errors will dramatically reduce.

A similar exercise, but not the same one, should be repeated to allow the student teams to change roles.

It is important that at this stage particularly that the students' work is supervised and corrected, that correct notations and symbols are used, and that the reports are made out in the correct format.

It would be desirable that a few such exercises, with targets approaching from different quarters would be carried out, but in such courses there is usually a time restraint, and much yet to be covered, so subsequent exercises should build in something new each time. This also helps to hold the students' attention better. The next stage would be situations where targets alter course and/or speed. Once again, but depending on how well the students have developed in handling the basic plot, the exercise could be paused to allow for the basic OWA triangle and report to be done, before continuing to where the targets change course/speed. However, towards the end of these series of exercises, there should be no pause and the emphasis should be increasingly on plotting quickly and accurately and reporting so that decisions can be made in good time.

The next stage is plotting the effect of proposed changes of course and/or speed by own ship. Once again, the exercise can be paused to allow the students to complete the OWA triangle, to report, and to plot the proposed alteration of course/speed. Then, when this has been done, and checked by the instructor, the exercise continues with the students altering course as required. The target/s are plotted to see that they follow the predicted relative tracks. At this stage the alteration of course or speed should be specified by the instructor. Once again, as skill develops, the pause can be dispensed with. In the latter part of this stage, the target/s should alter course or speed, after own ship has altered and the effect this has on the plot studies.

During all these exercises, the target trails should be on (about 6 minute trails if possible) and the attention of the students drawn to the comparison of the trails to the OA relative track. ARPA or other automatic functions should not be available to the students during these series of exercises, but the fact that the OA and WA represent the relative and true vectors of the target should be emphasised, and particularly how these change with the changes in course or speed by own ship, and by the targets.

That most useful plotting aid, the reflection plotter is no longer available on modern sets. However, using delible (erasable) coloured markers, plotting can be done on the face of the radar screen. It is not ideal, as there will be a small amount of parallax, but, like the old reflection plotter, it removes the source of error in transferring data to a paper plotting diagram. It also gives a clearer and more obvious comparison between the plotted OA line and the target trail. The radar should be in North-up stabilised mode and in relative motion. A scale of miles and fractions of a mile will need to be marked off on a flexible plastic rule, or along the edge of a strip of paper.

An interesting variation of this is to do the plot in True Motion Relative Trails mode (or the various names for the same thing by different radar manufacturers). This will result in a relative motion plot on the face of the screen, but the target trails will indicate their true motion. Care should be taken not to confuse students, and demonstrating this should be left until they have a good grasp of the whole relative motion plot principle.

Finally, in the series of manual plotting exercises, the last series of exercises should put the students under a certain amount of pressure. Students should be allowed to make their own decisions about when and by how much to alter course and/or speed. Anti-collision decisions should not be analysed too critically at this stage, although there should be a debrief after each exercise, but the purpose of these final series of manual plotting exercises is to demonstrate the limitations of manual plotting.

ARPA and AIS

In introducing ARPA and ATA functions, care should be taken to deal with each topic in isolation at first and to restrict the students from getting ahead of the instruction programme. This is particularly difficult with students who have been to sea and are reasonably 'up-to-speed' on ARPA, or at least think they are. However, a balance needs to be struck. The instructor needs to ensure that students don't get bored and 'switch off'. This can be particularly difficult where there is a mixed class of students from different backgrounds and experience. It can be useful to pair students, where possible, but again being careful that the more advanced students don't get the impression that they are being used as additional instructors or that they are being held back.

The first series of exercises should concentrate on acquiring targets, on studying the different vectors, particularly in relation to the mode of the display, whether it is relative motion or true, whether the particular set has a 'default' position regarding the vectors vis-a-vis the mode, the pitfalls in misinterpreting the vector, vector time/length. The vectors should be compared to the readout on each target, and care taken to show how targets are identified and related to the readout.

Next, the assessment of the threat posed by the target needs to be examine, i.e. CPA and TCPA determined. This is done by increasing the vector time and causing the relative vectors to lengthen. This will show graphically how close the target will come to own ship. And the vector length will indicate when this will happen. The graphically acquired data should be compared to the readout for each target to ensure consistency, which a. ensures that the correct target is being identified, and b. helps reinforce the data being acquired by the observer. Once this procedure is completed and all targets assessed, the vector length should be restored to 'normal', say 6 or 12 minutes.

The vectors should now be changed to 'True'. This gives a graphical indication of 'Aspect'. This is important in assessing the situation with regard to the collision regulations.

A 'report' can now be completed. This is part of the process of the observer, officer of the watch or bridge team absorbing the information. In dealing with an automated system like ARPA, it can happen that one is looking at the display without really taking in the information being presented. For this reason, a mental or aloud reading of the report is important.

The next step is deciding on a course of action if necessary to avoid a close quarter situation developing.

In this, the 'Trial manoeuvre' is introduced. Not all tracking systems have, or are required to have a trial manoeuvre function, and one make of equipment at least has it only in the True Motion mode. Initial exercises, using trial manoeuvre should have no more than three echoes, one of which is causing a situation where own ship is obliged to alter course. The exercise should take the student/s through the stage of assessing the situation, deciding on an alteration of course and time to alter course, trying this on trial manoeuvre, and then carrying out the manoeuvre. The follow-through should be carefully attended to by continuing to observe the targets to ensure that they have not manoeuvred (history dots) and then own ship should resume course and speed.

One of the difficulties is that in the isolated world of a desktop simulator, the student can become confused between the trial manoeuvre and the 'reality' of the actual manoeuvre. If time and equipment is available, it is worth while doing this piece of instruction on a simulator with visuals and actual steering controls and doing the initial exercises in 'daylight' with clear visibility.

Subsequent exercises can have more complex situations, particularly where targets manoeuvre, but situations with too many targets should be avoided.

In conducting these series of exercises, students must be continuously questioned on their perception of the traffic situation. Once the exercise has been running for a while instructors should frequently ask students to verbally report on targets. The lecturer should hide the readout, and then ask the student to give the report on a particular target, without manipulating any of the controls. This is to emphasise the whole idea of the OOW absorbing what is presented by the equipment.

The use of guard-zones is a debatable subject. The matter is discussed elsewhere in this book, but simulator exercises in the use of guard-zones should be limited to demonstration exercises to show how they are set-up, the alarms which they generate, and their acquisition of targets. The drawbacks of guard-zones should also be shown such as clutter and spurious echoes being acquired, and important targets which appear inside the certain guard-zone configurations and may not be acquired.

Some of the exercise time should be spent on emphasising the impact of AIS in collision avoidance situations. The particular dangers which may exist where erroneous AIS data is presented should be demonstrated. This could be by an exercise where the student is not forewarned that the data is incorrect. Also a demonstration exercise where communication is established between 'ships' but there is delay or confusion resulting in a dangerous situation, should be considered.

Radar and Navigation

Simulation allows for most radar navigational techniques to be practised, and skills developed. This ranges from identifying land, buoys and other echoes of navigational importance, fixing the ship's position, to complex parallel indexing and certain related manoeuvring exercises.

Once again, these exercise should take the student through a carefully controlled series of exercises each one building on the previous to build up the level of knowledge and skill.

In exercising students in recognising land echoes, and possibly other navigational features, much will depend on the simulator software and its representation of radar land images. Some simulators are disappointing in this regard. Often they do not correctly show the shadowing or masking of echoes behind land echoes closer to 'own ship', such as buoys or ship target or other land echoes. However, they all at least can show the outline of the land in own ship's vicinity and will allow radar navigational techniques to be practised.

Initially simple plotting of own ship's position and progress on a paper chart should be done, using EBL and VRM. During these series of exercises, the inherent errors and inaccuracies in radar navigation can be demonstrated. Series of bearings, with and without compass/ gyro error should be plotted, then series of ranges, and then combinations of both.

In subsequent exercises much will depend on what stage the students are at, particularly in coastal navigation. Exercises in coastal navigation with wind and current present can very effectively be done in simulation, and students on the courses preparing for OOW examinations and higher should be well versed in all aspects of coastal navigation, but it should not be the function of the radar and navigational aids course to teach such techniques.

Once the students are 'up to speed' in such position plotting, subsequent exercises can include leeway and tidal stream.

Parallel indexing

Radar, as a navigational tool comes into its own with parallel indexing. Radar simulation likewise comes into its own in allowing extensive practice in the technique and more complex situations, than merely keeping the ship on track.

In the initial series of exercises the students should be exercises in first of all picking off the cross index range, applying it to the radar display and then running the exercise keeping the ship on track. Currents tending to set the ship off the track should be introduced and from the dynamics of the situation the students should learn, by trial and error, how to find the course to steer to keep the ship on track.

The next series of parallel index exercise should be done where there is traffic and own ship is obliged to alter course for it. Using the parallel index line/s the ship should be quickly brought back on track after an alteration of course for another ship.

The third series of exercise should combine current and traffic.

The next series of exercises should involve an alteration of course, in other words going from one parallel index line/s to another. Initially this should be just a simple set-up taking no account of ship's manoeuvring characteristics, but later allowing for wheel-over and distance to new course should be introduced.

Again, care needs to be taken, as the exercise become more complex, that the students can achieve the objects of the exercise. It can happen in alterations of course exercises using parallel index lines that the student loses control of the ship and never gets it onto the new course satisfactorily. This is one area where time should be allocated to allow for repeat exercises to allow students to satisfactorily develop the skill.

Otherwise they will go away demoralised and lacking confidence in parallel index techniques. In this regard, in complex navigational exercises where traffic as well as current are present, the instructor needs to plan and conduct the exercise with great care so that the above is less likely to happen.

Once the student has mastered the basic techniques mentioned above a more complex exercise can be run where the ship is for example passing through a strait, or entering harbour where there is a series of course alteration, and possibly a very few items of traffic.

Radius Turn

The technique of carrying out a radius turn, monitored by the radar should be practised by senior students, it is usually a bit beyond the junior officer level of training and experience. The technique will be normally included in passage planning exercises, but time should be allocated for students to lay off and practice it in isolation.

Maps and Charts

The use of prepared maps for particular passages and of electronic charts, either as radar overlay on ECDIS or chart radars (radar underlay) can be built into exercise at this level. They should however come towards the end of a course when all the more fundamental skills and practices have been achieved. The errors that may occur in using such maps and charts should be heavily emphasised particularly through simulator exercises which demonstrate the consequences of such errors.

Senior Students (management level)

Students who have been officers of the watch, and spent some years away from college will need refresher training in all the basics of radar plotting. Much of this refresher instruction can be done in class-room situations with paper plotting. However, some time needs to be set aside to run basic plotting exercises on the simulator and any course designed for such students should have this in its programme.

Such students should be well practised in the use of automatic plotting aids. Again, however, some refresher instruction will be needed to remind them of best practice and the disciplines of getting the data into their heads. Eradicating bad habits picked up at sea falls into this area of instruction.

Much of what is written above about the conduct of simulator training equally applies to senior students. The danger, in the briefing sessions is the assumption that the student knows or remembers more

Appendix A

Part B – Steering and Sailing Rules

PART B. STEERING AND SAILING RULES

Section I. Conduct of vessels
in any condition of visibility

Rule 4
Application
Rules in this Section apply in any condition of visibility.

Rule 5
Look-out
Every vessel shall at all times maintain a proper look-out by sight and hearing as well as by all available means appropriate in the prevailing circumstances and conditions so as to make a full appraisal of the situation and of the risk of collision.

Rule 6
Safe speed
Every vessel shall at all times proceed at a safe speed so that she can take proper and effective action to avoid collision and be stopped within a distance appropriate to the prevailing circumstances and conditions.

In determining a safe speed the following factors shall be among those taken into account:

(a) By all vessels:
    (i) the state of visibility;
    (ii) the traffic density including concentrations of fishing vessels or any other vessels;
    (iii) the manoeuvrability of the vessel with special reference to stopping distance and turning ability in the prevailing conditions;
    (iv) at night the presence of background light such as from shore lights or from back scatter of her own lights;
    (v) the state of wind, sea and current, and the proximity of navigational hazards;
    (vi) the draught in relation to the available depth of water.
(b) Additionally, by vessels with operational radar:
    (i) the characteristics, efficiency and limitations of the radar equipment;

    (ii) any constraints imposed by the radar range scale in use;

    (iii) the effect on radar detection of the sea state, weather and other sources of interference;

    (iv) the possibility that small vessels, ice and other floating objects may not be detected by radar at an adequate range;

    (v) the number, location and movement of vessels detected by radar;

    (vi) the more exact assessment of the visibility that may be possible when radar is used to determine the range of vessels or other objects in the vicinity.

Rule 7
Risk of collision

(a) Every vessel shall use all available means appropriate to the prevailing circumstances and conditions to determine if risk of collision exists. If there is any doubt such risk shall be deemed to exist.

(b) Proper use shall be made of radar equipment if fitted and operational, including long-range scanning to obtain early warning of risk of collision and radar plotting or equivalent systematic observation of detected objects.

(c) Assumptions shall not be made on the basis of scanty information, especially scanty radar information.

(d) In determining if risk of collision exists the following considerations shall be among those taken into account:

    (i) such risk shall be deemed to exist if the compass bearing of an approaching vessel does not appreciably change;

    (ii) such risk may sometimes exist even when an appreciable bearing change is evident, particularly when approaching a very large vessel or a tow or when approaching a vessel at close range.

Rule 8
Action to avoid collision

(a) Any action to avoid collision shall be taken in accordance with the Rules of this Part and shall, if the circumstances of the case admit, be positive, made in ample time and with due regard to the observance of good seamanship.

(b) Any alteration of course and/or speed to avoid collision shall, if the circumstances of the case admit, be large enough to be readily apparent to another vessel observing visually or by radar; a succession of small alterations of course and/or speed should be avoided.

(c) If there is sufficient sea-room, alteration of course alone may be the most effective action to avoid a close-quarters situation provided that it is made in good time, is substantial and does not result in another close-quarters situation.

(d) Action taken to avoid collision with another vessel shall be such as to result in passing at a safe distance. The effectiveness of the action shall be carefully checked until the other vessel is finally past and clear.

(e) If necessary to avoid collision or allow more time to assess the situation, a vessel shall slacken her speed or take all way off by stopping or reversing her means of propulsion.

(f)   (i) A vessel which, by any of these Rules, is required not to impede the passage or safe passage of another vessel shall, when required by the circumstances of the case, take early action to allow sufficient sea-room for the safe passage of the other vessel.

   (ii) A vessel required not to impede the passage or safe passage of another vessel is not relieved of this obligation if approaching the other vessel so as to involve risk of collision and shall, when taking action, have full regard to the action which may be required by the Rules of this Part.

  (iii) A vessel the passage of which is not to be impeded remains fully obliged to comply with the Rules of this Part when the two vessels are approaching one another so as to involve risk of collision.

Rule 9
Narrow channels

(a) A vessel proceeding along the course of a narrow channel or fairway shall keep as near to the outer limit of the channel or fairway which lies on her starboard side as is safe and practicable.

(b) A vessel of less than 20 metres in length or a sailing vessel shall not impede the passage of a vessel which can safely navigate only within a narrow channel or fairway.

(c) A vessel engaged in fishing shall not impede the passage of any other vessel navigating within a narrow channel or fairway.

(d) A vessel shall not cross a narrow channel or fairway if such crossing impedes the passage of a vessel which can safely navigate only within such channel or fairway. The latter vessel may use the sound signal prescribed in Rule 34(d) if in doubt as to the intention of the crossing vessel.

(e)   (i) In a narrow channel or fairway when overtaking can take place only if the vessel to be overtaken has to take action to permit safe passing, the vessel intending to overtake shall indicate her intention by sounding the appropriate signal prescribed in Rule 34(c)(i). The vessel to be overtaken shall, if in agreement, sound the appropriate signal prescribed in Rule 34(c)(ii) and take steps to permit safe passing. If in doubt she may sound the signals prescribed in Rule 34(d).

   (ii) This Rule does not relieve the overtaking vessel of her obligation under Rule 13.

(f) A vessel nearing a bend or an area of a narrow channel or fairway where other vessels may be obscured by an intervening obstruction shall navigate with particular alertness and caution and shall sound the appropriate signal prescribed in Rule 34(e).

(g) Any vessel shall, if the circumstances of the case admit, avoid anchoring in a narrow channel.

Rule 10
Traffic separation schemes

(a) This Rule applies to traffic separation schemes adopted by the Organization and does not relieve any vessel of her obligation under any other Rule.

(b) A vessel using a traffic separation scheme shall:
   (i) proceed in the appropriate traffic lane in the general direction of traffic flow for that lane;
   (ii) so far as practicable keep clear of a traffic separation line or separation zone;
   (iii) normally join or leave a traffic lane at the termination of the lane, but when joining or leaving from either side shall do so at as small an angle to the general direction of traffic flow as practicable.

(c) A vessel shall, so far as practicable, avoid crossing traffic lanes but if obliged to do so shall cross on a heading as nearly as practicable at right angles to the general direction of traffic flow.

(d) (i) A vessel shall not use an inshore traffic zone when she can safely use the appropriate traffic lane within the adjacent traffic separation scheme. However, vessels of less than 20 metres in length, sailing vessels and vessels engaged in fishing may use the inshore traffic zone.
   (ii) Notwithstanding sub-paragraph (d)(i), a vessel may use an inshore traffic zone when en route to or from a port, offshore installation or structure, pilot station or any other place situated within the inshore traffic zone, or to avoid immediate danger.

(e) A vessel other than a crossing vessel or a vessel joining or leaving a lane shall not normally enter a separation zone or cross a separation line except:
   (i) in cases of emergency to avoid immediate danger;
   (ii) to engage in fishing within a separation zone.

(f) A vessel navigating in areas near the terminations of traffic separation schemes shall do so with particular caution.

(g) A vessel shall so far as practicable avoid anchoring in a traffic separation scheme or in areas near its terminations.

(h) A vessel not using a traffic separation scheme shall avoid it by as wide a margin as is practicable.

(i) A vessel engaged in fishing shall not impede the passage of any vessel following a traffic lane.

(j) A vessel of less than 20 metres in length or a sailing vessel shall not impede the safe passage of a power-driven vessel following a traffic lane.

(k) A vessel restricted in her ability to manoeuvre when engaged in an operation for the maintenance of safety of navigation in a traffic separation scheme is exempted from complying with this Rule to the extent necessary to carry out the operation.

(l) A vessel restricted in her ability to manoeuvre when engaged in an operation for the laying, servicing or picking up of a submarine cable, within a traffic separation scheme, is exempted from complying with this Rule to the extent necessary to carry out the operation.

Section II. Conduct of vessels in sight of one another

Rule 11
Application
Rules in this Section apply to vessels in sight of one another.

Rule 12
Sailing Vessels

(a) When two sailing vessels are approaching one another, so as to involve risk of collision, one of them shall keep out of the way of the other as follows:

   (i) when each has the wind on a different side, the vessel which has the wind on the port side shall keep out of the way of the other;

  (ii) when both have the wind on the same side, the vessel which is to windward shall keep out of the way of the vessel which is to leeward;

 (iii) if the vessel with the wind on the port side sees a vessel to windward and cannot determine with certainty whether the other vessel has the wind on the port or on the starboard side, she shall keep out of the way of the other.

(b) For the purposes of this Rule the windward side shall be deemed to be the side opposite to that on which the mainsail is carried or, in the case of a square-rigged vessel, the side opposite to that on which the largest fore-and-aft sail is carried.

Rule 13
Overtaking

(a) Notwithstanding anything contained in the Rules of Part B. Sections I and II any vessel overtaking any other shall keep out of the way of the vessel being overtaken.

(b) A vessel shall be deemed to be overtaking when coming up with another vessel from a direction more than 22.5 degrees abaft her beam, that is, in such a position with reference to the vessel she is overtaking, that at night she would be able to see only the sternlight of that vessel but neither of her sidelights.

(c) When a vessel is in any doubt as to whether she is overtaking another, she shall assume that this is the case and act accordingly.

(d) Any subsequent alteration of the bearing between the two vessels shall not make the overtaking vessel a crossing vessel within the meaning of these Rules or relieve her of the duty of keeping clear of the overtaken vessel until she is finally past and clear.

Rule 14
Head-on situation

(a) When two power-driven vessels are meeting on reciprocal or nearly reciprocal courses so as to involve risk of collision each shall alter her course to starboard so that each shall pass on the port side of the other.

(b) Such a situation shall be deemed to exist when a vessel sees the other ahead or nearly ahead and by night she would see the mast head lights of the other in a line or nearly in a line and/or both sidelights and by day she observes the corresponding aspect of the other vessel.

(c) When a vessel is in any doubt as to whether such a situation exists she shall assume that it does exist and act accordingly.

Rule 15
Crossing situation

When two power-driven vessels are crossing so as to involve risk of collision, the vessel which has the other on her own starboard side shall keep out of the way and shall, if the circumstances of the case admit, avoid crossing ahead of the other vessel.

Rule 16
Action by give-way vessel

Every vessel which is directed to keep out of the way of another vessel shall, so far as possible, take early and substantial action to keep well clear.

Rule 17
Action by stand-on vessel

(a) (i) Where one of two vessels is to keep out of the way the other shall keep her course and speed.

(ii) The latter vessel may however take action to avoid collision by her manoeuvre alone, as soon as it becomes apparent to her that the vessel required to keep out of the way is not taking appropriate action in compliance with these Rules.

(b) When, from any cause, the vessel required to keep her course and speed finds herself so close that collision cannot be avoided by the action of the give-way vessel alone, she shall take such action as will best aid to avoid collision.

(c) A power-driven vessel which takes action in a crossing situation in accordance with sub-paragraph (a)(ii) of this Rule to avoid collision with another power-driven vessel shall, if the circumstances of the case admit, not alter course to port for a vessel on her own port side.

(d) This Rule does not relieve the give-way vessel of her obligation to keep out of the way.

Rule 18
Responsibilities between vessels
Except where Rules 9, 10 and 13 otherwise require:

(a) A power-driven vessel underway shall keep out of the way of:
    (i) a vessel not under command;
    (ii) a vessel restricted in her ability to manoeuvre;
    (iii) a vessel engaged in fishing;
    (iv) a sailing vessel.

(b) A sailing vessel underway shall keep out of the way of:
    (i) a vessel not under command;
    (ii) a vessel restricted in her ability to manoeuvre;
    (iii) a vessel engaged in fishing.

(c) A vessel engaged in fishing when underway shall, so far as possible, keep out of the way of:
    (i) a vessel not under command;
    (ii) a vessel restricted in her ability to manoeuvre.

(d) (i) Any vessel other than a vessel not under command or a vessel restricted in her ability to manoeuvre shall, if the circumstances of the case admit, avoid impeding the safe passage of a vessel constrained by her draught, exhibiting the signals in Rule 28;
    (ii) A vessel constrained by her draught shall navigate with particular caution having full regard to her special condition.

(e) A seaplane on the water shall, in general, keep well clear of all vessels and avoid impeding the navigation. In circumstances, however, where risk of collision exists, she shall comply with the Rules of this Part.

(f) (i) A WIG craft, when taking off, landing and in flight near the surface, shall keep well clear of all other vessels and avoid impeding their navigation;
    (ii) A WIG craft operating on the water surface shall comply with the Rules of this Part as a power-driven vessel.

Section III. Conduct of vessels
in restricted visibility

Rule 19
Conduct of vessels in
restricted visibility

(a) This Rule applies to vessels not in sight of one another when navigating in or near an area of restricted visibility.

(b) Every vessel shall proceed at a safe speed adapted to the prevailing circumstances and conditions of restricted visibility. A power-driven vessel shall have her engines ready for immediate manoeuvre.

(c) Every vessel shall have due regard to the prevailing circumstances and conditions of restricted visibility when complying with the Rules of Section I of this Part.

(d) A vessel which detects by radar alone the presence of another vessel shall determine if a close-quarters situation is developing and/or risk of collision exists. If so, she shall take avoiding action in ample time, provided that when such action consists of an alteration of course, so far as possible the following shall be avoided:

    (i) an alteration of course to port for a vessel forward of the beam, other than for a vessel being overtaken;

    (ii) an alteration of course towards a vessel abeam or abaft the beam.

(e) Except where it has been determined that a risk of collision does not exist, every vessel which hears apparently forward of her beam the fog signal of another vessel, or which cannot avoid a close-quarters situation with another vessel forward of her beam, shall reduce her speed to the minimum at which she can be kept on her course. She shall if necessary take all her way off and in any event navigate with extreme caution until danger of collision is over.

Appendix B

Radar Plotting Exercises (1)

1. Report on target A, own ship's course 040°, speed 12 knots.

Time	Bearing (degrees true)	Range (Miles)
0000	068	9.2
0006	066	7.4

2. Report on target A, own ship's course 040°, speed 6 knots.

Time	Bearing (degrees true)	Range (Miles)
0010	357	9.5
0020	355	7.6

3. Report on target A, own ship's course 140°, speed 4 knots.

Time	Bearing (degrees true)	Range (Miles)
0805	235	8.0
0820	230	5.7

4. Report on target A, own ship's course 334°, speed 12 knots.

Time	Bearing (degrees true)	Range (Miles)
0805	304	11.0
0810	306	9.6
0815	308	8.1

5. Report on target A, own ship's course 334°, speed 12 knots.

Time	Bearing (degrees true)	Range (Miles)
0805	004	11.0
0810	002	9.6
0815	000	8.1

6. Report on target A, own ship's course 200°, speed 6 knots.

Time	Bearing (degrees true)	Range (Miles)
1000	243	7.0
1010	243	5.6
1020	243	4.2

7. Report on target A, own ship's course 280°, speed 8 knots.

Time	Bearing (degrees true)	Range (Miles)
0800	300	9.1
0807½	298½	8.1
0815	297	7.1

8. Report on target A, own ship's course 350°, speed 10 knots.

Time	Bearing (degrees true)	Range (Miles)
2010	285	8.8
2016	284	8.3
2022	283½	7.8

9. Report on targets A and B, own ship's course 320°, speed 12 knots.

Target	A		B	
Time	Bearing (degrees true)	Range (Miles)	Bearing (degrees true)	Range (Miles)
0800	304	11.0	000	8.0
0803	306	9.6	002	7.8
0806	308	8.1	004	7.6

10. The observations of target A were as follows, own ship's course 110°, speed 12 knots:

Time	Bearing (degrees true)	Range (Miles)
1000	166	8.8
1006	165	7.0
1012	162	5.2

a. Report at 1012.
b. Speed was reduced to 3 knots at 1012. Find the new CPA and TCPA.

11. With own ship on a course of 243° (T) at a speed of 12 knots, observations were made from the radar as follows:

Time	Bearing (degrees true)	Range (Miles)
1105	288	11.0
1115	290	10.0
1125	292	9.0

a. Make a full report on the echo at 1125.

Plotting continues as follows:

Time	Bearing (degrees true)	Range (Miles)
1135	297	7.0
1145	308	5.0

b. Is the echo following the predictions made in your report? If it is not what has happened?

12. Own ship's course 060° (T), speed 18 knots. The following observations were made:

Time	Bearing (degrees true)	Range (Miles)
0900	104	12.0
0905	104	10.8
0910	104	9.6
0915	109	7.8
0920	122	6.0
0925	144	4.7

Explain the movements of the target and predict CPA and TCPA.

13. Own ship's course 250° (T), speed 15 knots. The following observations were made:

Time	Bearing (degrees true)	Range (Miles)
1000	190	11.0
1008	190	8.5
1016	190	6.0
1024	171	5.3
1032	149	5.3

Report on the target at 1016.

Explain the movements of the target subsequent to 1016 and predict CPA and TCPA.

14. Own ship's course 000° (T), speed 12 knots. The following observations were made:

Time	Bearing (degrees true)	Range (Miles)
1000	008	12.3
1006	008	10.3
1012	008	8.2

Find the CPA and TCPA, speed and aspect of the target at 1012.

At 1018 own ship reduced speed to 3 knots, and at 1030 the engines were put full astern and all way taken off. Assuming these manoeuvres to have instantaneous effect, at what time approximately can it be expected to hear the fog signal of the other vessel, assuming that such a signal can be heard at a distance of 2 miles?

15. Own ship's course 240° (T), speed 8 knots. The following observations were made:

Time	Bearing (degrees true)	Range (Miles)
1000	220	7.0
1007½	221	6.5
1015	222	6.0

Find the CPA and TCPA, speed and aspect of the target at 1015.

At 1040 speed was reduced and at 1050 speed had settled at 4 knots. Estimate the least time after 1050 that speed may be resumed to maintain a 3 mile clearance of the target, assuming the target maintains course and speed.

16. Own ship's course 260° (T), speed 10 knots. The following observations were made:

Time	Bearing (degrees true)	Range (Miles)
0400	233	10.3
0406	234	9.1
0412	235	8.0

Find the CPA and TCPA, speed and aspect of the target at 0412.

At 0418 speed was reduced and at 0430 speed had settled at 3 knots. Estimate the least time after 0430 that speed may be resumed to maintain a 4 mile clearance of the target, assuming the target maintains course and speed.

17. Own ship's course 308° (T), speed 13 knots. The following observations were made:

Time	Bearing (degrees true)	Range (Miles)
1000	273	10.0
1005	273½	8.9
1010	274½	7.8

With a head-reach of 2 miles in 20 minutes, find the time to stop engines so that the other vessel will pass 3 miles ahead.

18. Own ship's course 230° (T), speed 9 knots. The following observations were made:

Time	Bearing (degrees true)	Range (Miles)
0710	272	9.7
0716	271	8.9
0722	270	8.1

At 0722 course is altered 60° to starboard. Assuming the alteration to have immediate effect, find the new CPA and TCPA.

19. Own ship's course 160° (T), speed 9 knots. The following observations were made:

Time	Bearing (degrees true)	Range (Miles)
0800	220	9.2
0805	221	7.9
0810	222	6.6

Find the CPA and TCPA, speed and aspect of the target at 0810.

At 0813 course was altered 55° to starboard and at 0817 the ship was steady on her new course. At 0820 the target bore 173° (T) range 4.0 miles. What is the new CPA and TCPA?

20. Own ship course 007° (T) speed 13.7 knots target was observed at:

Time	Bearing (degrees true)	Range (Miles)
0945	008	11.5
0950	007½°	9.0
0952	007	8.0

At 0954 course was altered to 067° (T) being steadied on 2 minutes later (0956). Forecast report at 1000.

21. Own ship is steering 209° (T) at a speed of 12 knots. From the following observations:

Time	Bearing (degrees true)	Range (Miles)
1003	170° (T)	10.9 miles
1013	172° (T)	8.6 miles
1023	175° (T)	6.2 miles

Find the true course and speed of the target and also its CPA and TCPA.

At 1023 the observing ship alters course 45° to port and at 1033 when settled on the new course the bearing and range of the target is as follows: 192° (T) 3.7 miles.

Find the new nearest point of approach and time of nearest point of approach of the target. What assumption is made throughout this problem?

22. With own ship steering course 193° (T) 7½ knots the following observations were taken of a target, which was known to be a light vessel:

Time	Bearing (degrees true)	Range (Miles)
1553	263	9.0
1601	267½	8.6
1609	272	8.2

Find the set, drift and rate of the tide.

23. With own ship steering 030° (T), speed 10 knots, the range and bearing of a lightvessel was 346° (T), range 8 miles at 0912 hours. Assuming that there is a tide setting 270° (T) at 3 knots, predict the light vessel's CPA and TCPA.

24. With own ship steering 000° (T) at a speed of 15 knots, the following observations are made on the radar:

Target	A		B	
Time	Bearing (degrees true)	Range (Miles)	Bearing (degrees true)	Range (Miles)
1200	320°	9.6	045°	11.7
1206	316°	8.3	046°	9.8
1212	311°	7.1	048°	8.0

Target 'A' is known to be a light vessel. Find set, drift and rate of the tide also the CPA of the light vessel. Give a full report on Echo B at 1212.

At 1212, own speed is reduced to 7½ knots. Find the alteration of own course which is necessary to enable own ship to pass the light vessel at the same distance as originally predicted.

What effect will the reduction of speed and the intended alteration of course have on the CPA and TCPA of target B?

Answers

1. Brg. 066°, drawing left; Rng.7.4 miles, closing; CPA 1 mile; TCPA 24 mins; Co. 294°; spd. 11 kts; ASP R46°.

2. Brg. 355°, drawing left; Rng. 7.6 miles, closing; CPA 1.5 miles; TCPA 37 mins; Co. 160°; spd. 7.5 kts; ASP G15°.

3. Brg. 230°, drawing left; Rng. 5.7 miles, closing; CPA 1.5 miles; TCPA 35 mins; Co. 085°; spd. 7.5 kts; ASP R35°.

4. Brg. 308°, drawing right;, range 8.1 miles, closing; CPA 2 miles; TCPA 26 mins; Co. 305°; spd. 20 kts; ASP Green 55°.

5. Brg. 000°, drawing left; range 8.1 miles, closing; CPA 2 n.m.; TCPA 28 mins; Co. 070°; spd. 12 kts; ASP R55°.

6. Brg. 243°, steady; range 4.1 miles closing; CPA coll.; TCPA 29 mins; Tgt. Co. 108°; Tgt. spd. 6.0 kts; ASP R44°.

7. Brg. 297°, drawing left; Rng. 7.1 miles, closing; CPA 1.7 miles; TCPA 50 mins; Co. 201°; spd. 4.5 kts; ASP R85°.

8. Brg. 283.5°, drawing left; Rng. 7.8 miles, closing; CPA 1. miles; TCPA 93 mins; Co. 020°; spd. 9.5 kts; ASP R84°.

9. A. Brg 308°, drawing right; range 8.1 miles, closing; CPA 2 miles; TCPA 17 mins; Co. 098°; spd. 18kts; ASP G30°; B. Brg. 004°, drawing right; range 7.6 miles, closing; CPA 6.3 miles; TCPA 30 mins; Co. 330°; spd. 7kts; ASP R150°.

10. Brg. 162°, drawing left; Rng. 5.2 miles, closing; CPA 1. miles; TCPA 17 mins; Co. 032°; spd. 16.5 kts; ASP R50°; New CPA 3.3 miles; New TCPA 16 mins.

11. Brg. 292°, drawing right; Range 9.0 miles, closing; Aspect Red 106° CPA 3.4 miles; TCPA 78 mins. No. Vessel altered 60 to port at 1125.

12. Target has altered course 90° to port, maintaining speed New CPA 4.4 miles; TCPA 3 mins, i.e. at 0928.

13. True course and speed of target = 321° T 17 Kn Target stopped at 10:16.

14. CPA 0; TCPA 1038; 8 kts; R11°; 1047.

15. CPA 1.7 miles; TCPA at 1146; ASP G136°; approx. 30 mins. After 1050.

16. CPA 0.7 miles; TCPA at 0452; ASP G70°; 7 kts; 36 mins.

17. 1027.

18. CPA 5.8 miles; TCPA at 0747.

19. CPA 0.8′; TCPA at 0834; tgt. Co. 047°; Tgt. spd. 9 kts; ASP R41°; new CPA 3.0′; new TCPA 0828.

20. Forecast report at 1000: Target bearing 350°, drawing left; Range 4.7 miles, closing; ASP R26°; CPA = 3.6 miles; TCPA in 6 mins.

21. True course and speed of target = 288° T at 10.2 Kn.
CPA 1.2 miles; TCPA at 10.48.
New CPA 2.9 miles; new TCPA at 10:39.
It is assumed that the target maintains course and speed.

22. Set 336°; rate 2.5 kts; drift 6 cables.

23. CPA 4 miles; TCPA at 09.57.

24. Set 248°; rate 4 kts; drift 0.8 n.m.; CPA of lightvessel = 3.8 n.m.; Full report of 'B' 1212: Bearing 048° Drawing right; Range 8 n.m. closing; Aspect R44°; CPA = 1.3 miles; TCPA in 26 mins; Alter course 18° to starboard; New CPA 2.2 n.m.; new TCPA in 30 mins.

Radar Plotting Exercises (2)

1. Own ship course 149° (T), speed 12 knots. Observations were obtained as follows:

0700 Target Brg. 199° (T)	Range 11.2 miles.
0710 Target Brg. 197° (T)	Range 9.2 miles.
0720 Target Brg. 196° (T)	Range 7.2 miles.

   At 0725 own ship alters course 45° to starboard and reduces speed to 8 knots. Predict the earliest time to resume course and speed if the target is not to come within 3 miles. Predict the aspect and relative bearing at that time.
   (Ans. Time to resume 0740, Rel. Brg. R34°, Aspect R 105°.)

   Plotting continues as follows:

0727 Target Brg. 193° (T)	Range 6.0 miles.
0735 Target Brg. 181° (T)	Range 5.6 miles.
0743 Target Brg. 167° (T)	Range 5.3 miles.

   a. What action, if any, has been taken by the target?
   b. Own ship resumes course and speed at 0743, what is new CPA?
   c. Predict relative bearing and aspect at 0831.

   (Ans. a. Target has altered course 40° to starboard and reduced to 9.5 kts.
   b. CPA = 2.7 n.m.
   c. Rel. Brg. = R 26° Aspect = G 176°.)

2. Own ship course 311° (T) speed 12 knots. Observations were made as follows:

1000 Target Brg. 001°	Range 11.0′.
1010 Target Brg. 001°	Range 9.0′.
1020 Target Brg. 001°	Range 7.0′.

   a. Report at 1020.
   b. At 1021½ own ship alters course 50° to starboard and reduces to 9 knots. The manoeuvre is complete at 1027 with the target

bearing 351°, range 5.8′. Predict the relative bearing and aspect of the target at 1037.
c. When the target bears Red 301°, course only is resumed. Predict nearest point of approach.
d. To what speed should own ship also increase if target is to pass 3 miles off?
e. If own ship does increase to the above speed, at what time might you see her masts in line (Aspect 0°).

(Ans. a. At 10:20 Target bearing 001° steady; Range 7.0 n.m. decreasing; Aspect R65°; CPA collision in 35 mins.
 b. At 10:37. Brg. 325°, Aspect = R101°.
 c. If course only is resumed CPA 3.7, TCPA 10:50.
 d. Increase to 13 kts.
 e. Aspect 0 at 1100.)

3. Own ship's course 163° (T) at speed 15 knots. Observations of a lightvessel were made as follows:

Time
1200   Target Brg. 193° (T)   Range 12.0 n.m.
1210   Target Brg. 196° (T)   Range 9.5 n.m.
1220   Target Brg. 200° (T)   Range 7.0 n.m.

a. Find the set and rate of the current.
b. Find the alteration of course necessary at 1220 to ensure that the ship leaves the light vessel to starboard by not less than 3 miles.

At 1230 when on new course, own ship reduces to 7.5 knots due to restricted visibility.

c. What further alteration is necessary to maintain the passing distance?
d. Forecast the TCPA.

(Ans. a. Set and rate 263° (T) at 5.5 kts.
 b. Alteration of course necessary, 8° to port.
 c. Further alteration of 22° to port.
 d. TCPA 13:10.)

4. Own ship course 095° (T) at a speed of 15 knots. Observations were made as follows:

Time
0731   Target Brg. 162° (T)   Range 7.4 n.m.
0739   Target Brg. 162° (T)   Range 6.3 n.m.
0747   Target Brg. 162° (T)   Range 5.2 n.m.

a. Make a full report at 0747.
b. At 0747 speed is reduced to 5 knots. Find the earliest time that speed may be resumed so that the target will not pass within 3 miles of own ship.
c. If at 0747 speed had been maintained and course altered 70° to port when might course have been resumed so that target would not have come within 2 miles of own ship?

(Ans. a. Report at 07:47: Bearing G67° steady; Range 5.2 n.m. closing; Aspect R80°; Tgt. Co. 063°; Tgt. Speed 4.25; CPA collision; TCPA in 36 mins.
b. Time to resume 08:09.
c. Time to resume 08:19.)

5. Own ship's course 000° (T), speed 12 knots. Observations were taken as follows:

Time
1000   Target Brg. 310°   Range 11.0 n.m.
1010   Target Brg. 311°   Range 9.0 n.m.
1020   Target Brg. 313°   Range 7.0 n.m.

At 1025 own ship alters 45° to starboard.

a. Find the earliest time to resume so that target will not pass closer than 2 n.m.

Plotting continues as follows:

Time
1031   Target Brg. 308°   Range 5 n.m.
1041   Target Brg. 285°   Range 3.9 n.m.
1051   Target Brg. 253°   Range 3.8 n.m.

b. Find what action has been taken by the target.

(Ans. a. Earliest time to resume 11.15.
b. Target has altered 48° to starboard.)

6. Own ship course 100° (T) at speed 15 knots. Observations were made as follows:

Time
1000   Target Brg. 120°   Range 12.0 n.m.
1010   Target Brg. 119½°   Range 10.0 n.m.
1020   Target Brg. 119°   Range 8.0 n.m.

At 1028 own ship altered course to 160° (T). She was steady on the new course at 1030 at which time the target bore 119° at range 6 n.m. Predict earliest time to resume such that target will not come within 3 n.m. (*Ans. Earliest time to resume 10.41.*)

Plotting continues:

Time
1035   Target Brg. 110°   Range 5.2 n.m.
1040   Target Brg. 097°   Range 4.5 n.m.

a. What action has been taken by the target?
b. In the light of the targets alteration, at what time may own ship now resume course such that the target will pass not closer than 3 miles. (*Ans. a. Target has stopped; b. 10.53.*)

7. Own ship's course 067° (T) at a speed of 10 knots:

Time
0817   Target Brg. 047°   Range 5.0 n.m.
0820   Target Brg. 050°   Range 3.8 n.m.
0823   Target Brg. 057°   Range 2.6 n.m.

Give a full report at 0823. What would be the vessels aspect at 0826?

(*Ans. Full report at 08:23: Bearing 057° drawing right; Range 2.6 n.m. closing; CPA = 0.8 miles; TCPA in 6 mins; Aspect = G40°; Aspect at 08:26 = G55°.*)

8. Own ship's course 157° (T), speed 10 knots, the following observations were made:

Time
0205   Target Brg. 097°   Range 6.5 n.m.
0211   Target Brg. 100°   Range 5.2 n.m.
0217   Target Brg. 105°   Range 3.9 n.m.

Find 9 minutes after the last observation the relative bearing,

range, aspect, true course and speed of the target. On what relative
bearing would you expect the masts of the other vessel to be in line?
(i.e. Aspect 180° or zero.)

*(Ans. At 02:26: Rel. Brg. = R34°; Range 2.1 n.m.; Aspect G80°;
Course 222° (T) at 14 kts; Masts in line when target bears G65°.)*

9. Own ship steering 125° (T) at 15 knots observations are made as
   follows:

   Time
2140	Target Brg. 082° (T)	Range 8.0 n.m.
2146	Target Brg. 082° (T)	Range 6.5 n.m.
2152	Target Brg. 082° (T)	Range 5.0 n.m.

   If at 2152 course is altered

a.	40° to starboard	Find CPA and time of CPA.
b.	40° to port	Find CPA and time of CPA.
c.	90° to port	Find CPA and time of CPA.

   *(Ans. (a) CPA = 3 n.m. at 22:23.
   (b) CPA = 2.3 n.m. at 22:04.
   (c) CPA = 4.2 n.m. at 21:58.)*

10. A vessel steering 010° (T) at 10 knots observations of a target are
    taken as follows:

    Time
    | 1030 | Target Brg. 050° (T) | Range 9.5 n.m. |
    | 1036 | Target Brg. 050° (T) | Range 8.3 n.m. |
    | 1042 | Target Brg. 050° (T) | Range 7.1 n.m. |

    At 10.42, course is altered (instantaneously) 70° to starboard.

    Find: (i) New CPA.
    (ii) New CPA if alteration is deferred until 1054.

    *(Ans. (i) New CPA = 4.7 n.m.
    (ii) New CPA = 3 n.m.)*

11. Own ship steering 200° (T) at 12 knots a target bore:

Time
0700   Target Brg. 244°   Range 7.0 n.m.
0705   Target Brg. 244°   Range 5.8 n.m.
0710   Target Brg. 244°   Range 4.6 n.m.

At 0710 speed was reduced to 6 knots which was effective at 0720.

a. Find the target's CPA and TCPA.
b. Forecast the target's position at 0740.

(Ans. a. 1.7' at 0733, b. bearing 148° by 2.0'.)

12. Own vessel steering 030° (T) and a reduced speed of 10 knots a target was observed as follows:

Time
1427   Target Brg. 352°   Range 5.5 n.m.
1432   Target Brg. 350½°   Range 4.0 n.m.
1436   Target Brg. 349°   Range 3.0 n.m.

a. Find the effect on the nearest approach, caused by reducing speed to 5 knots at 1436 if the new speed is reached at 1444.
b. Comment on this proposed manoeuvre and suggest alternatives, speed available in 8 mins 14 knots. Show the result of your proposed manoeuvre on the plot.

(Ans. a. original CPA 0.3' at 1448, new CPA less than 0.2' at 145.
       b. Not advisable. An immediate bold alteration to starboard with or without an increase in speed.)

13. Own ship steering 215° (T) speed 15 knots. The following observations were made by radar:

Time
0731   Target Brg. 282° (T)   Range 7.4 miles.
0739   Target Brg. 282° (T)   Range 6.3 miles.
0747   Target Brg. 282° (T)   Range 5.2 miles.

a. Make a full report at 0747.
b. At 0747 speed reduced to 5 knots. Find time that speed may be resumed so that target does not pass within 3 miles of own ship.
c. If at 0747 speed had been maintained and course altered 70° to port when might course have been resumed so that target does not come within 2 miles of own ship?

(Ans. a. Tgt. Brg. 282°, closing; CPA Collision; TCPA 0821; Tgt.
   Co. 183° tgt. Speed 13.5 kts; Aspect R 85°;
   b. 0808;
   c. 0820.)

14. Own ship steering 160° (T) speed 12 knots the following obser-
    vations were made by radar:

    Time
    1613   Target Brg. 159°   Range 10.5 n.m.
    1619   Target Brg. 161°   Range 8.3 n.m.
    1625   Target Brg. 163°   Range 6.0 n.m.

    At 1628 own course was altered to 250° (T) and speed reduced to
    9 knots owing to reduced visibility.

    At 1629 the manoeuvre was completed and the target bore 163°
    range 4.7 n.m.

    Predict range and bearing of target at 1647. Subsequently the
    positions of the target were:

    Time
    1641   Target Brg. 158°   Range 4.0 n.m.
    1653   Target Brg. 145°   Range 3.1 n.m.
    1705   Target Brg. 127°   Range 2.5 n.m.

    a. Has prediction been accurate? If it has not why not?

    (Ans. a. Brg. 109° range 2.5';
        b. Target has altered course to port, from 328° to 288° and
        reduced speed from 11 knots to 7 knots.)

15. Own ship steering 030° (T), at a reduced speed of 9.3 knots, the
    following ranges and bearings of a target were taken:

    Time
    0311   Target Brg. 048°   Range 11.0 n.m.
    0318   Target Brg. 047°   Range 8.8 n.m.
    0325   Target Brg. 046°   Range 6.4 n.m.

    a. Assuming both vessels keep courses and speeds, at what time
        will target be expected at 4.0 n.m. range.
    b. When it reaches 4.0 n.m. would a reduction of speed from the
        present 9.3 knots to 3 knots give a satisfactory approach of 2
        miles or more?

   c. If not to what speed should your ship reduce to get a clearance of 2 miles? (Assume alterations have immediate effect.)

(Ans. a. 0332;
     b. No. CPA would be 1.4' TCPA at 0350;
     c. When the target reached 4.0' it would be necessary to take all way off, and even at that the CPA may be less than 2.0'.)

16. Own ship steering 030° (T) at 12 knots

Time	Target A True Brg.	Range	Target B (Lt v/l) True Brg.	Range
1010	065	11.0	025.0	12
1015	065	9.5	024.5	11
1020	065	8.0	024.0	10

   a. Find the targets' course, speed and TCPA.
     (Ans. 287, 10.8 kts, 1047.)

   b. Find alteration of course required at 1025 so that the target A will pass 1.5 n.m. to port. *(Ans. 052°.)*
   c. At the time of target A's CPA find the alteration necessary to pass 2 n.m. due east of the light vessel. *(Ans. 1044, 022°.)*

17. The following echoes are observed on the radar, ship steering 300° (T) at 15 knots.

Time	Rel. Brg.	Range
1212	332	11.0
1218	332	9.5
1224	332	8.0
1224	Course altered 40° to Starboard	
1230	329	6.9
1236	319	5.8
1242	305	4.9

Find
   a. True Course and speed of the target prior to the alteration.
      (Ans. 235°, 9 knots.)
   b. The time and relative bearing at CPA subsequent to the
      alteration of course. (Ans. 1248. 300° Rel.)
   c. Discuss the development of the plot subsequent to 1224.
      (Ans. Target has A/C to Starbd, to 264°.)

18. Own ship steering 000° at 12 knots. Visibility 0.5 miles.

Time	Target A Bearing (T)	Target A Range (miles)	Target B Bearing (T)	Target B Range (miles)	Target C Bearing (T)	Target C Range (miles)
0000	090	6	270	5.0	000	6.0
0006	092	6	270	4.7	000	5.5
0012	094	6	270	4.4	000	5.0

Determine the courses and speeds of each target.

(Ans. Tgt. A Co. 002°, spd. 9 kts; Tgt. B Co. 015°, spd. 13 kts;
Tgt. Co. 000°, spd. 8 kts.)

Discuss fully the possible courses of action to be taken by own
ship.

If own ship reduces speed to 6 knots at 0018 predict the CPAs of
each target and explain the outcome of the action. Assume
immediate effect of the speed reduction.

(Ans. Tgt. A, CPA 5.8 miles; Tgt. B CPA 3.6 miles, Tgt. C CPA 0,
but range increasing very slowly.)

19. Own ship steering 090° at 12 knots, visibility about 0.5 miles. In
    the attached diagram, target A is a buoy. Targets B and C are other
    ships. The diagram represents a relative motion display on the 12′
    scale with true vectors, vectors 12 minutes, at time 0430.

   a. From the diagram, calculate the CPA and TCPA of each target.
      (Ans. A:0.4 miles in 48 mins, B: 2.8 miles in 25 mins, C: 2.0 miles
      in 32 mins.)
   b. What is the direction and rate of the tidal stream?
      (Ans. 012° by 4 knots.)

At 0436 course is altered to pass 2 miles south of the buoy (Target A).

c. What course is now being steered? *(Ans. 106°.)*
d. At what time will the buoy be abeam? *(Ans. 0513.)*
e. What are now the CPAs and TCPAs of Targets B and C? What possible courses of action can be taken to avoid a close quarter situation with any target?
(Ans. Tgt. B 1.3' at 0506; Tgt. C 1.0' at 0457; About the only course of action is a reduction in speed. This will change the course to steer to pass 2' south of the buoy Tgt. A.)

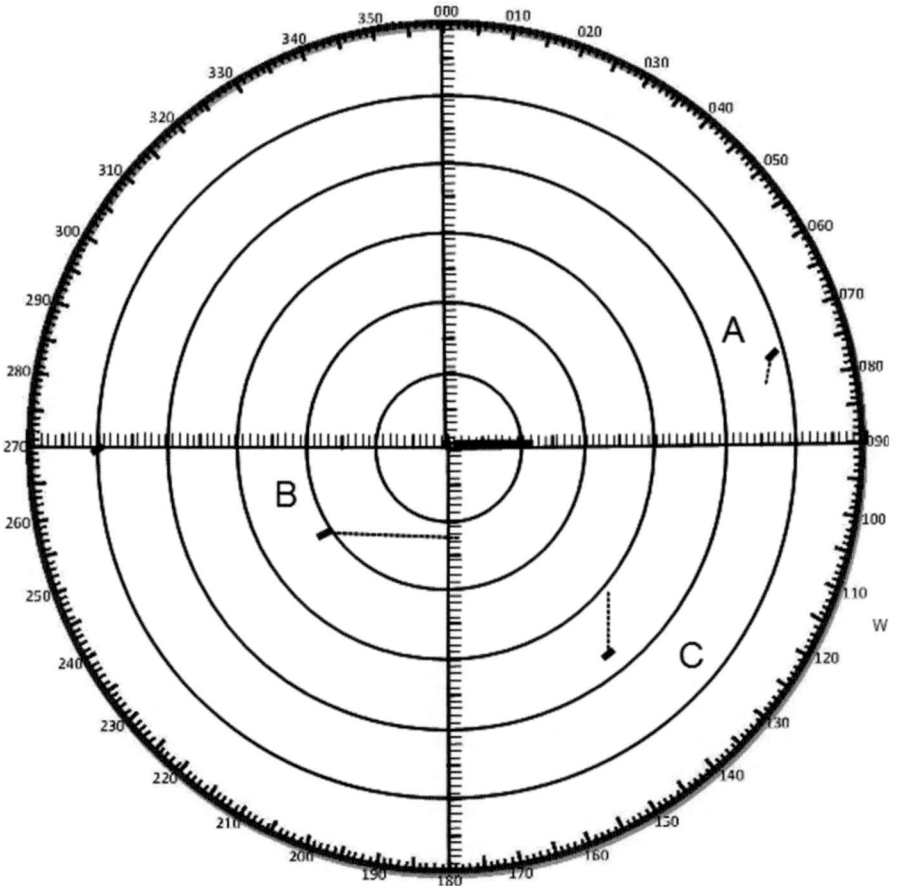

FIGURE FOR Q19.

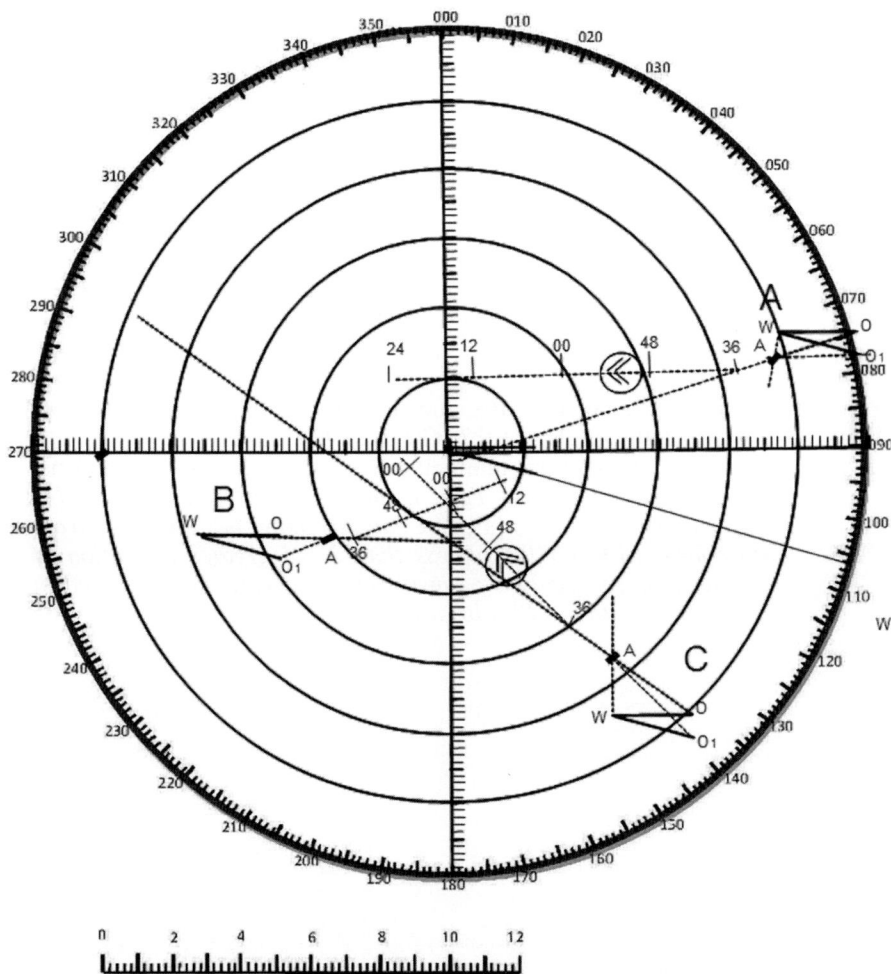

FIGURE FOR Q19 (answer).

20. Own ship steering 090° at 15 knots, visibility about 0.5 miles. In the attached diagram, targets A, B and C are presumably other ships. The diagram represents a relative motion display on the 12′ scale with relative vectors, vectors 12 minutes at time 1200.

   a. From the diagram, calculate the true courses and speeds of each target. (*Ans. A: 022°, 10 knots; B: 155°, 9 knots; C: 095°, 19 knots.*)

   At 1212 it is proposed to alter course 40° to starboard.

   b. Indicate where each target might be expected at 1224 (presuming all maintain their courses and speeds and that own manoeuvre has immediate effect). (*Ans. A: brg. 093° × 3.6′; B: brg. 190° × 3.0′; C: brg. 180° × 2.3′.*)
   c. What would be the CPA and TCPA of each target? (*Ans. A: 3.4′ at 1230; B:3.2′ at 1227; C:1.5′ at 1233.*)
   d. Is this a satisfactory manoeuvre? Discuss alternative actions. (*Ans. It is obvious that Tgts. A and B will go well clear by this manoeuvre. However, Tgt. C would now be crossing from starboard, with a CPA of less than 2′. A greater alteration of course than 40° might be considered, or a reduction in speed in conjunction with the 40° alteration.*)

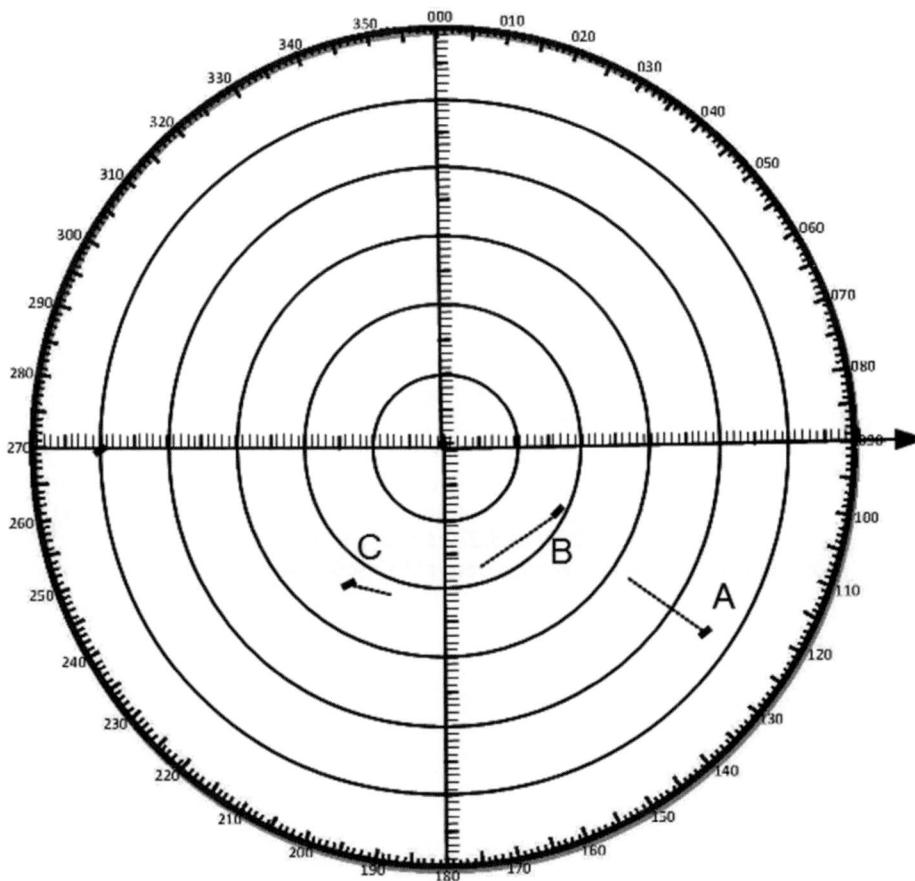

FIGURE FOR Q20.

21. Own ship steering 210° at 15 knots, visibility about 1.0 miles. In the attached diagram, targets A, B and C are presumably other ships. The diagram represents a relative motion display on the 12′ scale with true vectors, vectors 12 minutes at time 2300.

Target D is a buoy and when it is abeam it is required to alter course to 150°.

a. From the diagram, calculate the CPA and TCPA of each target at 2300.
   (Ans. A:1.2′ in 18 mins; B: 2.9′ in 30 mins; C: 3.3′ in 25 mins.)
b. At what time will the buoy (target D) be abeam?
   (Ans. 26 mins.)
c. Indicate where each target might be expected in relation to own ship at the time the buoy will be abeam (presuming all maintain their courses and speeds).
   (Ans. A: brg. 010° × 1.6′; B: brg. 067° × 3.0′; C: brg. 160° × 3.0.)
d. What would be the CPA of each target after the alteration of course (presuming the manoeuvre to have immediate effect)?
   (Ans. A: 3.4′ at 1230; B:3.2′ at 1227; C:1.5′ at 1233.)
e. Is this a satisfactory manoeuvre? Discuss alternative actions.

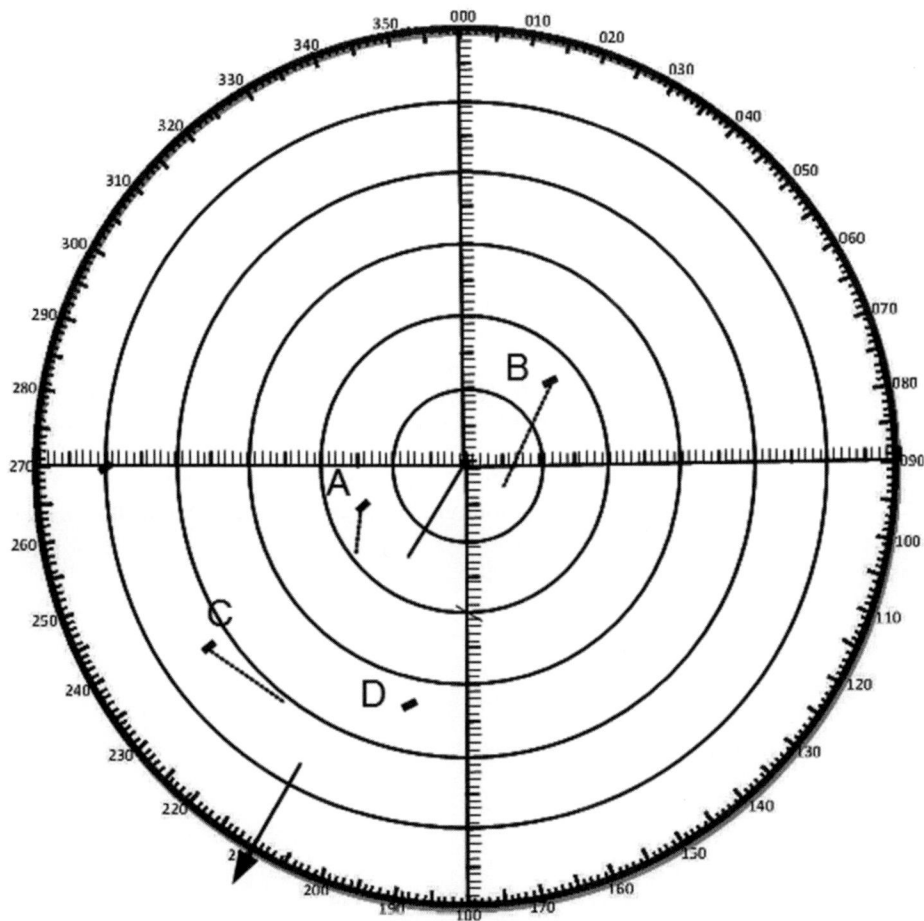

FIGURE FOR Q21.

22. Own ship steering 028° at 20 knots, visibility is variable from 0.3 miles to about 1.0 mile. In the attached diagram, target A is a buoy, targets B, C and D are presumably other ships. The diagram represents a relative motion display on the 12′ scale with true vectors, vectors 12 minutes at time 0400.

One minute after the buoy is abeam it is required to alter course to 050°.

   a. At what time will the buoy (target A) be abeam, and at what range? (*Ans. At 0411, 2.4′.*)
   b. Indicate where each target might be expected in relation to own ship at the time of the alteration of course (presuming all maintain their courses and speeds).
   (*Ans. B: brg. 062° × 2.0′; C: brg. 034° × 6.3′; D: brg. 063° 4.0′.*)
   c. What would be the CPA of each target after the alteration of course (presuming the manoeuvre to have immediate effect)? (*Ans. B: 1.6′ at 0415½; C: 2.3′ at 0418; D: 0.9′ at 0419½.*)
   d. Is this a satisfactory manoeuvre? Discuss alternative actions.

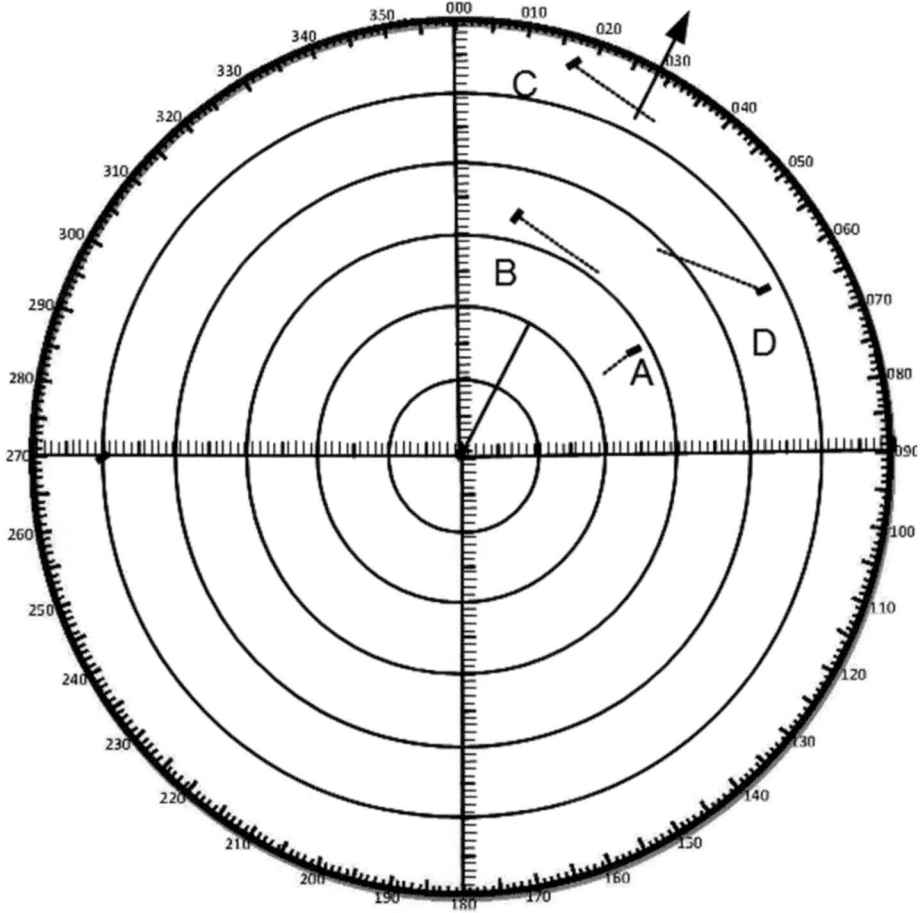

FIGURE FOR Q22.

23. Own ship steering 235° at 20 knots, visibility is about 0.3 miles. In the attached diagram, targets A, B, C and D are presumably other ships. The diagram represents a relative motion display on the 12′ scale with relative vectors, vectors 12 minutes at time 0815.

When Target C is 8 miles from own ship it is intended to alter course so that it passes ahead with a CPA of 2 miles to port.

   a. What course must be steered to achieve this and at what time must the alteration be made (presuming the manoeuvre to have immediate effect)? *(Ans. 264° at 0818½.)*
   b. What would be the CPAs and TCPAs of each target after the alteration of course (presuming all maintain their courses and speeds). *(Ans. A: 2′ 14 mins after a/c; B: 0.9′ 34 mins after a/ c;C: 0.6′ 21 mins after a/c; D: 1.0′ 19 mins after a/c.)*
   c. Is this a satisfactory manoeuvre? Discuss alternative actions.

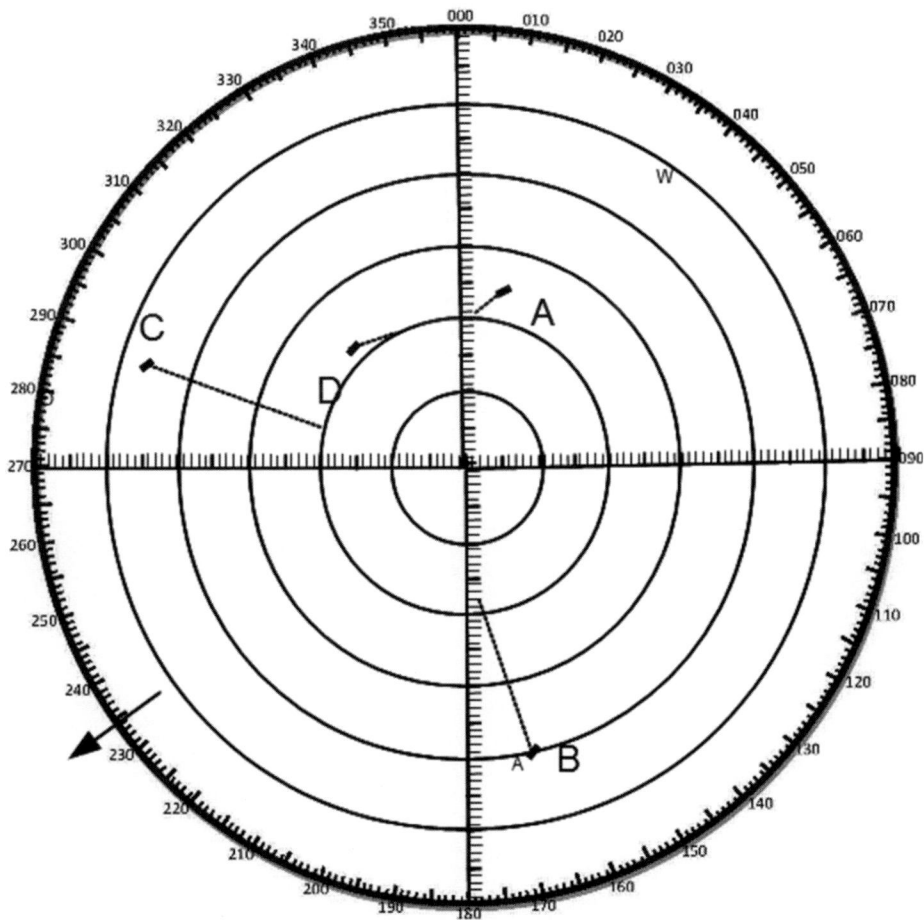

FIGURE FOR Q23.

24. Own ship steering 298° at 14 knots, visibility is about 0.5 miles. In the attached diagram, targets A, B, C and D are presumably other ships. The diagram represents a 6 miles relative motion display on the 12′ scale with true vectors, vectors 6 minutes at time 1010.

At 1013 it is intended to alter course 30° to starboard.

a. What would be the CPAs and TCPAs of each target after the alteration of course (presuming immediate effect of the manoeuvre and all targets maintain their courses and speeds). *(Ans.: A: 0.7′. 16 mins after a/c; B: 0.9′. 11 mins after a/c; C:1.6′. 11 mins after a/c; D: N/A.)*

b. Is this a satisfactory manoeuvre? Discuss alternative actions.

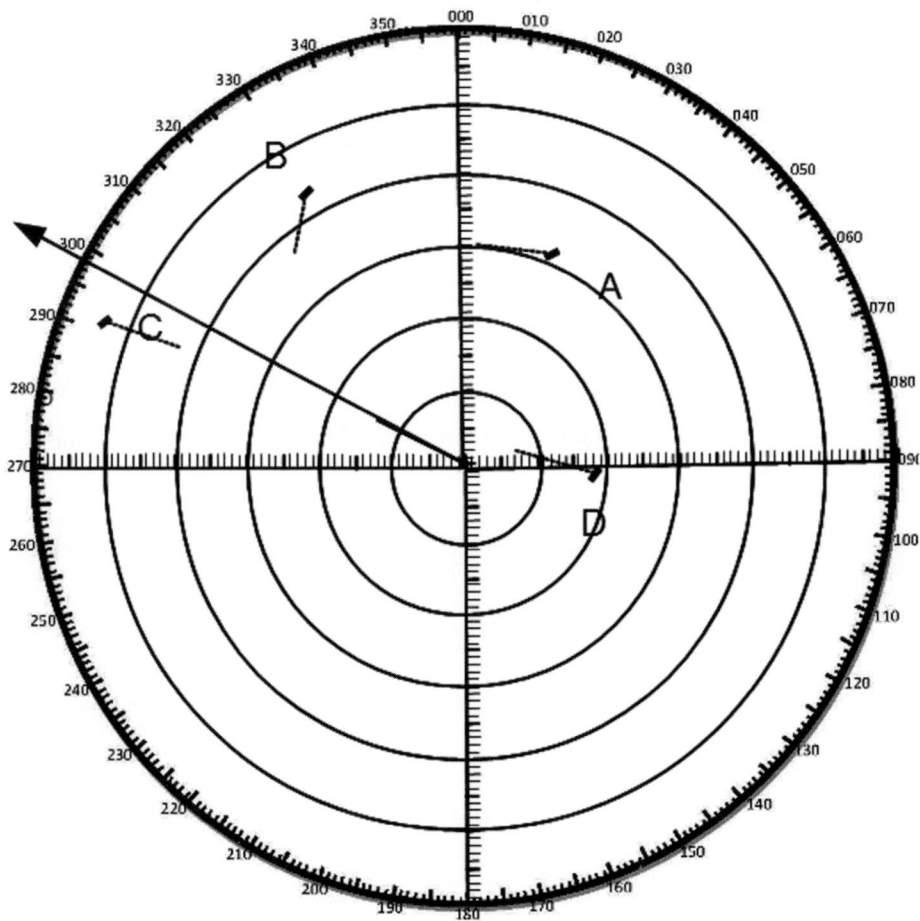

FIGURE FOR Q24.

25. Own ship steering 090° at 12 knots, visibility about 0.5 miles. In the attached diagram, targets A, B and C are other ships. The diagram represents a relative motion display on the 12′ scale with true vectors, vectors 6 minutes at time 0430.

    a. From the diagram, calculate the CPA and TCPA of each target. *(Ans. A:1.3 miles in 20 mins, B: 1.4 miles in 12 mins, C: 0.2 miles in 18 mins.)*

At 0436 Speed is reduced, it taking 6 minutes to reduce to 6 knots.

    b. What will be the CPA and TCPA of each target when speed is reduced? *(Ans. A: 0.4′ at 0454, B: 1.4′ at 0442, C: 1.0 at 0451.)*

    c. Show on the diagram the anticipated position, with true vector of each target at 0442.

    d. Is this a satisfactory manoeuvre? Discuss alternative manoeuvres.

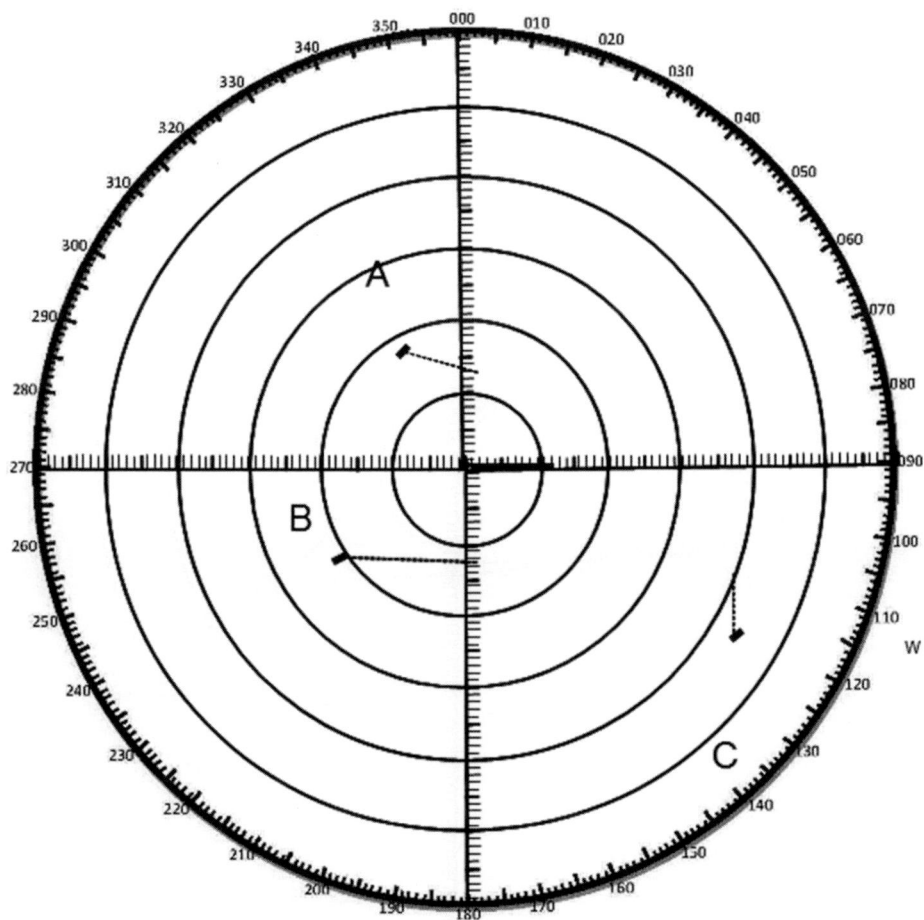

FIGURE FOR Q25.

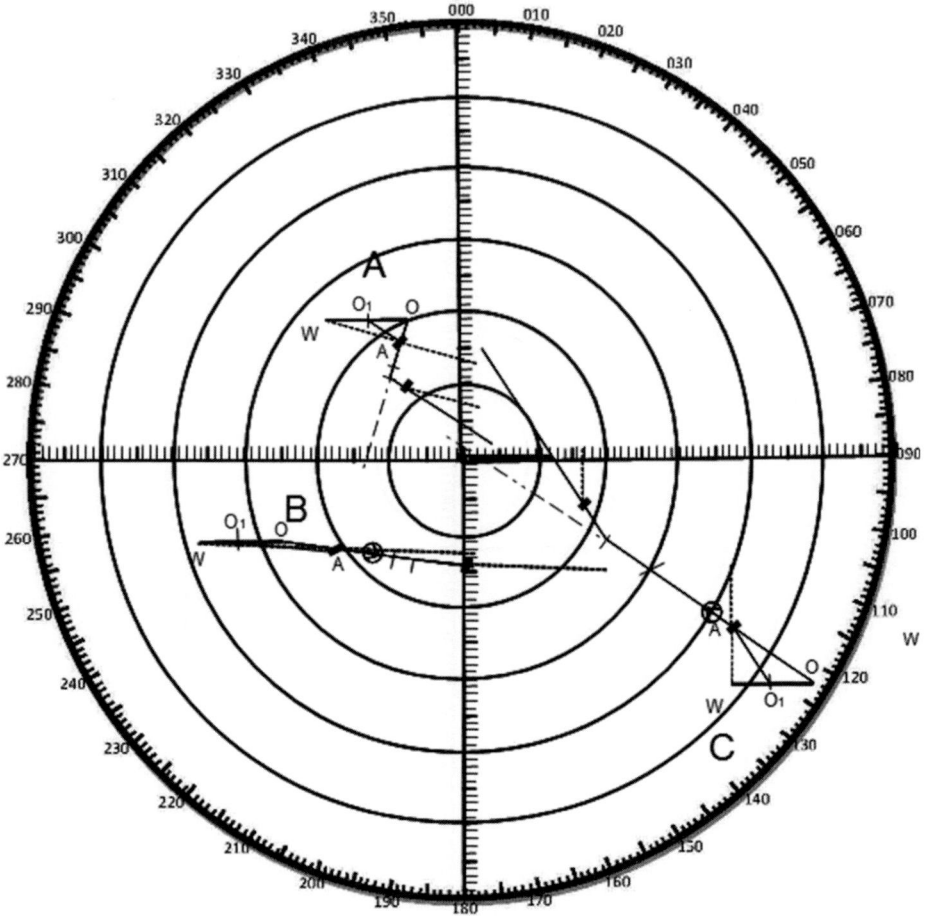

FIGURE FOR Q25 (answer).

Appendix C

IMO Symbols and Abbreviations

INTERNATIONAL MARITIME ORGANIZATION
4 ALBERT EMBANKMENT
LONDON SE1 7SR

Telephone: 020 7735 7611
Fax: 020 7587 3210

IMO

E

Ref. T2-OSS/2.7.1

SN/Circ.243
15 December 2004

**GUIDELINES FOR THE PRESENTATION OF
NAVIGATION-RELATED SYMBOLS, TERMS AND ABBREVIATIONS**

1   The Sub-Committee on Safety of Navigation (NAV), at its fiftieth session (5 to 9 July 2004), agreed on Guidelines for the presentation of navigation-related symbols, given in annex 1, and terms and abbreviations, given in annex 2, and that they should be used for the display of navigation-related information on all shipborne navigational equipment and systems in consistent and uniform manner.

2   The Maritime Safety Committee, at its seventy-ninth session (1 to 10 December 2004), concurred with the Sub-Committee's views, approved the annexed Guidelines and encouraged their use for all shipborne navigational systems and equipment.

3   Member Governments are invited to bring the annexed Guidelines to the attention of all concerned.

***

ANNEX 1

Guidelines for the Presentation of Navigation-related Symbols

1   Purpose

The purpose of these annexed guidelines is to provide guidance on the appropriate use of navigation-related symbols to achieve a harmonized and consistent presentation.

2   Scope

The use of these guidelines will insure that the symbols used for the display of navigation-related information on all shipborne navigational systems and equipment are presented in a consistent and uniform manner.

3   Application

These guidelines apply to all shipborne navigational systems and equipment. The symbols listed in the appendix should be used for the display of navigation-related information to promote consistency in the symbol presentation on navigational equipment.

The symbols listed in the Appendix should replace symbols which are currently contained in existing performance standards. Where a standard symbol is not available, another symbol may be used, but this symbol should not conflict with the symbols listed in the appendix.

APPENDIX

Navigation-related Symbols

Table 1: Own Ship Symbols

Topic	Symbol	Description
Own ship		Double circle, located at own ship's reference position. Use of this symbol is optional, if own ship position is shown by the combination of Heading Line and Beam Line.
Own Ship True scale outline		True scale outline located relative to own ship's reference position, oriented along own ship's heading. Used on small ranges/large scales.
Own Ship Radar Antenna Position		Cross, located on a true scale outline of the ship at the physical location of the radar antenna that is the current source of displayed radar video.
Own Ship Heading line		Solid line thinner than the speed vector line style, drawn to the bearing ring or of fixed length, if the bearing ring is not displayed. Origin is at own ship's reference point.
Own Ship Beam line		Solid line of fixed length; optionally length variable by operator. Midpoint at own ship's reference point.
Own Ship Speed vector		Dashed line – short dashes with spaces approximately twice the line width of heading line. Time increments between the origin and endpoint may optionally be marked along the vector using short intersecting lines. To indicate Water/Ground stabilization optionally one arrowhead for water stabilization and two arrowheads for ground stabilization may be added.
Own Ship Path prediction		A curved vector may be provided as a path predictor.
Own Ship Past Track		Thick line for primary source. Thin line for secondary source. Optional time marks are allowed.

Table 2: Tracked Radar Target Symbols

Topic	Symbol	Description
Tracked Target including Dangerous Target		Solid filled or unfilled circle located at target position. The course and speed vector should be displayed as dashed line, with short dashes with spaces approximately twice the line width. Optionally, time increments, may be marked along the vector. For a **"Dangerous Target"**, bold, red (on colour display) solid circle with course and speed vector, flashing until acknowledged.
Target in Acquisition State		Circle segments in the acquired target state. For automatic acquisition, bold circle segments, flashing and red (on colour display) until acknowledged.
Lost Target		Bold lines across the circle, flashing until acknowledged.
Selected Target		A square indicated by its corners centred around the target symbol.
Target Past Positions		Dots, equally spaced by time.
Tracked Reference Target	R	Large R adjacent to designated tracked target. Multiple reference targets should be marked as R1, R2, R3, etc.

Table 3: AIS Target Symbols

Topic	Symbol	Description
AIS Target (sleeping)		An isosceles, acute-angled triangle should be used. The triangle should be oriented by heading, or COG if heading missing. The reported position should be located at centre and half the height of the triangle. The symbol of the sleeping target should be smaller than that of the activated target.
Activated AIS Target Including Dangerous Target		An isosceles, acute-angled triangle should be used. The triangle should be oriented by heading, or COG if heading missing. The reported position should be located at centre and half the height of the triangle. The COG/SOG vector should be displayed as a dashed line with short dashes with spaces approximately twice the line width. Optionally, time increments may be marked along the vector. The heading should be displayed as a solid line thinner than speed vector line style, length twice of the length of the triangle symbol. Origin of the heading line is the apex of the triangle. The turn should be indicated by a flag of fixed length added to the heading line. A path predictor may be provided as curved vector. For a **"Dangerous AIS Target"**, bold, red (on colour display) solid triangle with course and speed vector, flashing until acknowledged.
AIS Target – True Scale Outline		A true scale outline may be added to the triangle symbol. It should be: Located relative to reported position and according to reported position offsets, beam and length. Oriented along target's heading. Used on low ranges/large scales.
Selected target		A square indicated by its corners should be drawn around the activated target symbol.
Lost target		Triangle with bold solid cross. The triangle should be oriented per last known value. The cross should have a fixed orientation. The symbol should flash until acknowledged. The target should be displayed without vector, heading and rate of turn indication.
Target Past Positions		Dots, equally spaced by time.

Table 4: Other Symbols

Topic	Symbol	Description
AIS Based AtoN **Real Position of Charted Object**		Diamond with crosshair centred at reported position. (Shown with chart symbol. Chart symbol not required for radar.)
AIS Based AtoN **Virtual position**		Diamond with crosshair centred at reported position.
Monitored Route		Dashed bold line, waypoints (WPT) as circles.
Planned or Alternate Route		Dotted line, WPT as circles.
Trial Manoeuvre	T	Large T on screen.
Simulation Mode	S	Large S on screen.
Cursor		Crosshair (two alternatives, one with open centre).
Range Rings		Solid circles.
Variable Range Markers (VRM)		Circle. Additional VRM should be distinguishable from the primary VRM.
Electronic Bearing Lines (EBL)		Dashed line. Additional EBL should be distinguishable from the primary EBL.

Topic	Symbol	Description
Acquisition/ Activation Area		Solid line boundary for an area.
Event Mark		Rectangle with diagonal line, clarified by added text (e.g. "MOB" for man overboard cases).

***

ANNEX 2

Guidelines for the Presentation of Navigation-related Terms and Abbreviations

1   Purpose

The purpose of these guidelines is to provide guidance on the use of appropriate navigation-related terminology and abbreviations intended for presentation on shipborne navigational displays. These are based on terms and abbreviations used in existing navigation references.

2   Scope

These guidelines are issued to ensure that the terms and abbreviations used for the display of navigation-related information on all shipborne navigation equipment and systems are consistent and uniform.

3   Application

These guidelines apply to all shipborne navigational systems and equipment including, radar, ECDIS, AIS, INS and IBS. When navigation-related information is displayed as text, the standard terms or abbreviations listed in the Appendix should be used, instead of using terms and abbreviations which are currently contained in existing performance standards.

Where a standard term and abbreviation is not available, another term or abbreviation may be used. This term or abbreviation should not conflict with the standard terms or abbreviations listed in the Appendix and provide a clear meaning. Standard marine terminology should be used for this purpose. When the meaning is not clear from its context, the term should not be abbreviated.

Unless otherwise specified, standard terms should be shown in lower case while abbreviations should be presented using upper case.

APPENDIX

List of Standard Terms and Abbreviations

Term	Abbreviation	Abbreviation	Term
Acknowledge	ACK	ACK	Acknowledge
Acquire, Acquisition	ACQ	ACQ	Acquire, Acquisition
Acquisition Zone	AZ	ADJ	Adjust, Adjustment
Adjust, Adjustment	ADJ	AFC	Automatic Frequency Control
Aft	AFT	AFT	Aft
Alarm	ALARM	AGC	Automatic Gain Control
Altitude	ALT	AIS	Automatic Identification System
Amplitude Modulation	AM	ALARM	Alarm
Anchor Watch	ANCH	ALT	Altitude
Antenna	ANT	AM	Amplitude Modulation
Anti Clutter Rain	RAIN	ANCH	Anchor Watch
Anti Clutter Sea	SEA	ANCH	Vessel at Anchor (applies to AIS)
April	APR	ANT	Antenna
Audible	AUD	APR	April
August	AUG	AUD	Audible
Automatic	AUTO	AUG	August
Automatic Frequency Control	AFC	AUTO	Automatic
Automatic Gain Control	AGC	AUX	Auxiliary System/Function
Automatic Identification System	AIS	AVAIL	Available
Auxiliary System/Function	AUX	AZ	Acquisition Zone
Available	AVAIL	BITE	Built in Test Equipment
Background	BKGND	BKGND	Background
Bearing	BRG	BRG	Bearing
Bearing Waypoint To Waypoint	BWW	BRILL	Brilliance
Brilliance	BRILL	BWW	Bearing Waypoint To Waypoint
Built in Test Equipment	BITE	C	Carried (e.g. carried EBL origin)
Calibrate	CAL	C UP (See note 2)	Course Up
Cancel	CNCL	CAL	Calibrate
Carried (e.g. carried EBL origin)	C	CCRP	Consistent Common Reference Point
Centre	CENT	CCRS	Consistent Common Reference System
Change	CHG	CENT	Centre
Circular Polarised	CP	CHG	Change
Clear	CLR	CLR	Clear
Closest Point of Approach	CPA	CNCL	Cancel
Consistent Common Reference Point	CCRP	COG	Course Over the Ground
Consistent Common Reference System	CCRS	CONT	Contrast
Contrast	CONT	CORR	Correction
Correction	CORR	CP	Circular Polarised
Course	CRS	CPA	Closest Point of Approach
Course Over the Ground	COG	CRS	Course
Course Through the Water	CTW	CTS	Course To Steer
Course To Steer	CTS	CTW	Course Through the Water
Course Up	C UP (See note 2)	CURS	Cursor
Cross Track Distance	XTD	D	Dropped (e.g. dropped EBL origin)
Cursor	CURS	DATE	Date
Dangerous Goods	DG	DAY/NT	Day/Night
Date	DATE	DEC	December

Term	Abbreviation	Abbreviation	Term
Day/Night	DAY/NT	DECR	Decrease
Dead Reckoning, Dead Reckoned Position	DR	DEL	Delete
December	DEC	DELAY	Delay
Decrease	DECR	DEP	Departure
Delay	DELAY	DEST	Destination
Delete	DEL	DEV	Deviation
Departure	DEP	DG	Dangerous Goods
Depth	DPTH	DGAL (See note 2)	Differential Galilleo
Destination	DEST	DGLONASS (See note 2)	Differential GLONASS
Deviation	DEV	DGNSS (See note 2)	Differential GNSS
Differential Galilleo	DGAL (See note 2)	DGPS (See note 2)	Differential GPS
Differential GLONASS	DGLONASS (See note 2)	DISP	Display
Differential GNSS	DGNSS (See note 2)	DIST	Distance
Differential GPS	DGPS (See note 2)	DIVE	Vessel Engaged in Diving Operations (applies to AIS)
Digital Selective Calling	DSC	DPTH	Depth
Display	DISP	DR	Dead Reckoning, Dead Reckoned Position
Distance	DIST	DRG	Vessel Engaged in Dredging or Underwater Operations (applies to AIS)
Distance Root Mean Square	DRMS (See note 2)	DRIFT	Drift
Distance To Go	DTG	DRMS (See note 2)	Distance Root Mean Square
Drift	DRIFT	DSC	Digital Selective Calling
Dropped (e.g. dropped EBL origin)	D	DTG	Distance To Go
East	E	E	East
Electronic Bearing Line	EBL	EBL	Electronic Bearing Line
Electronic Chart Display and Information System	ECDIS	ECDIS	Electronic Chart Display and Information System
Electronic Navigational Chart	ENC	ENC	Electronic Navigational Chart
Electronic Position Fixing System	EPFS	ENH	Enhance
Electronic Range and Bearing Line	ERBL	ENT	Enter
Enhance	ENH	EP	Estimated Position
Enter	ENT	EPFS	Electronic Position Fixing System
Equipment	EQUIP	EQUIP	Equipment
Error	ERR	ERBL	Electronic Range and Bearing Line
Estimated Position	EP	ERR	Error
Estimated Time of Arrival	ETA	ETA	Estimated Time of Arrival
Estimated Time of Departure	ETD	ETD	Estimated Time of Departure
Event	EVENT	EVENT	Event
Exclusion Zone	EZ	EXT	External
External	EXT	EZ	Exclusion Zone
February	FEB	FEB	February
Fishing Vessel	FISH	FISH	Fishing Vessel
Fix	FIX	FIX	Fix
Forward	FWD	FM	Frequency Modulation
Frequency	FREQ	FREQ	Frequency
Frequency Modulation	FM	FULL	Full
Full	FULL	FWD	Forward
Gain	GAIN	GAIN	Gain
Galilleo	GAL	GAL	Galilleo
Geometric Dilution Of Precision	GDOP	GC	Great Circle

Term	Abbreviation	Abbreviation	Term
Global Maritime Distress and Safety System	GMDSS	GDOP	Geometric Dilution Of Precision
Global Navigation Satellite System	GNSS	GLONASS	Global Orbiting Navigation Satellite System
Global Orbiting Navigation Satellite System	GLONASS	GMDSS	Global Maritime Distress and Safety System
Global Positioning System	GPS	GND	Ground
Great Circle	GC	GNSS	Global Navigation Satellite System
Grid	GRID	GPS	Global Positioning System
Ground	GND	GRI	Group Repetition Interval
Group Repetition Interval	GRI	GRID	Grid
Guard Zone	GZ	GRND	Vessel Aground (applies to AIS)
Gyro	GYRO	GYRO	Gyro
Harmful Substances (applies to AIS)	HS	GZ	Guard Zone
Head Up	H UP (See note 2)	H UP (See note 2)	Head Up
Heading	HDG	HCS	Heading Control System
Heading Control System	HCS	HDG	Heading
Heading Line	HL	HDOP	Horizontal Dilution Of Precision
High Frequency	HF	HF	High Frequency
High Speed Craft (applies to AIS)	HSC	HL	Heading Line
Horizontal Dilution Of Precision	HDOP	HS	Harmful Substances (applies to AIS)
Identification	ID	HSC	High Speed Craft (applies to AIS)
In	IN	I/O	Input/Output
Increase	INCR	ID	Identification
Indication	IND	IN	In
Information	INFO	INCR	Increase
Infrared	INF RED	IND	Indication
Initialisation	INIT	INF RED	Infrared
Input	INP	INFO	Information
Input/Output	I/O	INIT	Initialisation
Integrated Radio Communication System	IRCS	INP	Input
Interference Rejection	IR	INT	Interval
Interswitch	ISW	IR	Interference Rejection
Interval	INT	IRCS	Integrated Radio Communication System
January	JAN	ISW	Interswitch
July	JUL	JAN	January
June	JUN	JUL	July
Latitude	LAT	JUN	June
Limit	LIM	LAT	Latitude
Line Of Position	LOP	LF	Low Frequency
Log	LOG	LIM	Limit
Long Pulse	LP	LOG	Log
Long Range	LR	LON	Longitude
Longitude	LON	LOP	Line Of Position
Loran	LORAN	LORAN	Loran
Lost Target	LOST TGT	LOST TGT	Lost Target
Low Frequency	LF	LP	Long Pulse
Magnetic	MAG	LR	Long Range
Manoeuvre	MVR	MAG	Magnetic
Manual	MAN	MAN	Manual
Map(s)	MAP	MAP	Map(s)
March	MAR	MAR	March

Term	Abbreviation
Maritime Mobile Services Identity number	MMSI
Maritime Pollutant (applies to AIS)	MP
Maritime Safety Information	MSI
Marker	MKR
Master	MSTR
Maximum	MAX
May	MAY
Medium Frequency	MF
Medium Pulse	MP
Menu	MENU
Minimum	MIN
Missing	MISSING
Mute	MUTE
Navigation	NAV
Normal	NORM
North	N
North Up	N UP (See note 2)
November	NOV
October	OCT
Off	OFF
Officer of the Watch	OOW
Offset	OFFSET
On	ON
Out/Output	OUT
Own Ship	OS
Panel Illumination	PANEL
Parallel Index Line	PI
Passenger Vessel (applies to AIS)	PASSV
Performance Monitor	MON
Permanent	PERM
Person Overboard	POB
Personal Identification Number	PIN
Pilot Vessel (applies to AIS)	PILOT
Port/Portside	PORT
Position	POSN
Positional Dilution Of Precision	PDOP
Power	PWR
Predicted	PRED
Predicted Area of Danger	PAD
Predicted Point of Collision	PPC
Pulse Length	PL
Pulse Modulation	PM
Pulse Repetition Frequency	PRF
Pulse Repetition Rate	PRR
Pulses Per Revolution	PPR
Racon	RACON
Radar	RADAR
Radius	RAD
Rain	RAIN
Range	RNG
Range Rings	RR
Raster Chart Display System	RCDS

Abbreviation	Term
MAX	Maximum
MAY	May
MENU	Menu
MF	Medium Frequency
MIN	Minimum
MISSING	Missing
MKR	Marker
MMSI	Maritime Mobile Services Identity number
MON	Performance Monitor
MP	Maritime Pollutant (applies to AIS)
MP	Medium Pulse
MSI	Maritime Safety Information
MSTR	Master
MUTE	Mute
MVR	Manoeuvre
N	North
N UP (See note 2)	North Up
NAV	Navigation
NORM	Normal
NOV	November
NUC	Vessel Not Under Command (applies to AIS)
OCT	October
OFF	Off
OFFSET	Offset
ON	On
OOW	Officer of the Watch
OS	Own Ship
OUT	Out/Output
PAD	Predicted Area of Danger
PANEL	Panel Illumination
PASSV	Passenger Vessel (applies to AIS)
PDOP	Positional Dilution Of Precision
PERM	Permanent
PI	Parallel Index Line
PILOT	Pilot Vessel (applies to AIS)
PIN	Personal Identification Number
PL	Pulse Length
PM	Pulse Modulation
POB	Person Overboard
PORT	Port/Portside
POSN	Position
PPC	Predicted Point of Collision
PPR	Pulses Per Revolution
PRED	Predicted
PRF	Pulse Repetition Frequency
PRR	Pulse Repetition Rate
PWR	Power
RACON	Racon
RAD	Radius
RADAR	Radar
RAIM	Receiver Autonomous Integrity Monitoring
RAIN	Anti Clutter Rain

Term	Abbreviation
Raster Navigational Chart	RNC
Rate Of Turn	ROT
Real-time Kinemetic	RTK
Receiver	RX (See note 2)
Receiver Autonomous Integrity Monitoring	RAIM
Reference	REF
Relative	REL (See note 3)
Relative Motion	RM
Revolutions per Minute	RPM
Roll On/Roll Off Vessel (applies to AIS)	RoRo
Root Mean Square	RMS
Route	ROUTE
Safety Contour	SF CNT
Sailing Vessel (applies to AIS)	SAIL
Satellite	SAT
S-Band (applies to Radar)	S-BAND
Scan to Scan	SC/SC
Search And Rescue Transponder	SART
Search And Rescue Vessel (applies to AIS)	SARV
Select	SEL
September	SEP
Sequence	SEQ
Set (i.e., set and drift, or setting a value)	SET
Ship's Time	TIME
Short Pulse	SP
Signal to Noise Ratio	SNR
Simulation	SIM (See note 4)
Slave	SLAVE
South	S
Speed	SPD
Speed and Distance Measuring Equipment	SDME
Speed Over the Ground	SOG
Speed Through the Water	STW
Stabilized	STAB
Standby	STBY
Starboard/Starboard Side	STBD
Station	STN
Symbol(s)	SYM
Synchronisation	SYNC
Target	TGT
Target Tracking	TT
Test	TEST
Time	TIME
Time Difference	TD
Time Dilution Of Precision	TDOP
Time Of Arrival	TOA
Time Of Departure	TOD
Time to CPA	TCPA
Time To Go	TTG
Time to Wheel Over Line	TWOL

Abbreviation	Term
RAIN	Rain
RCDS	Raster Chart Display System
REF	Reference
REL (See note 3)	Relative
RIM	Vessel Restricted in Manoeuvrability) (applies to AIS)
RM	Relative Motion
RMS	Root Mean Square
RNC	Raster Navigational Chart
RNG	Range
RoRo	Roll On/Roll Off Vessel (applies to AIS)
ROT	Rate Of Turn
ROUTE	Route
RPM	Revolutions per Minute
RR	Range Rings
RTK	Real-time Kinemetic
RX (See note 2)	Receiver
S	South
SAIL	Sailing Vessel (applies to AIS)
SART	Search And Rescue Transponder
SARV	Search And Rescue Vessel (applies to AIS)
SAT	Satellite
S-BAND	S-Band (applies to Radar)
SC/SC	Scan to Scan
SDME	Speed and Distance Measuring Equipment
SEA	Anti Clutter Sea
SEL	Select
SEP	September
SEQ	Sequence
SET	Set (i.e., set and drift, or setting a value)
SF CNT	Safety Contour
SIM (See note 4)	Simulation
SLAVE	Slave
SNR	Signal to Noise Ratio
SOG	Speed Over the Ground
SP	Short Pulse
SPD	Speed
STAB	Stabilized
STBD	Starboard/Starboard Side
STBY	Standby
STN	Station
STW	Speed Through the Water
SYM	Symbol(s)
SYNC	Synchronisation
T	True
TCPA	Time to CPA
TCS	Track Control System
TD	Time Difference
TDOP	Time Dilution Of Precision
TEST	Test
TGT	Target

Term	Abbreviation
Track	TRK
Track Control System	TCS
Track Made Good	TMG (See note 5)
Trail(s)	TRAIL
Transceiver	TXRX (See note 2)
Transferred Line Of Position	TPL
Transmitter	TX (See note 2)
Transmitting Heading Device	THD
Trial	TRIAL (See note 4)
Trigger Pulse	TRIG
True	T
True Motion	TM
Tune	TUNE
Ultrahigh Frequency	UHF
Universal Time, Co-ordinated	UTC
Unstabilised	UNSTAB
Variable Range Marker	VRM
Variation	VAR
Vector	VECT
Very High Frequency	VHF
Very Low Frequency	VLF
Vessel Aground (applies to AIS)	GRND
Vessel at Anchor (applies to AIS)	ANCH
Vessel Constrained by Draught (applies to AIS)	VCD
Vessel Engaged in Diving Operations (applies to AIS)	DIVE
Vessel Engaged in Dredging or Underwater Operations (applies to AIS)	DRG
Vessel Engaged in Towing Operations (applies to AIS)	TOW
Vessel Not Under Command (applies to AIS)	NUC
Vessel Restricted in Manoeuvrability) (applies to AIS)	RIM
Vessel Traffic Service	VTS
Vessel Underway Using Engine (applies to AIS)	UWE
Video	VID
Voyage	VOY
Voyage Data Recorder	VDR
Warning	WARNING
Water	WAT
Waypoint	WPT
West	W
Wheel Over Line	WOL
Wheel Over Time	WOT
X-Band (applies to Radar)	X-BAND

Abbreviation	Term
THD	Transmitting Heading Device
TIME	Ship's Time
TIME	Time
TM	True Motion
TMG (See note 5)	Track Made Good
TOA	Time Of Arrival
TOD	Time Of Departure
TOW	Vessel Engaged in Towing Operations (applies to AIS)
TPL	Transferred Line Of Position
TRAIL	Trail(s)
TRIAL (See note 4)	Trial
TRIG	Trigger Pulse
TRK	Track
TT	Target Tracking
TTG	Time To Go
TUNE	Tune
TWOL	Time to Wheel Over Line
TX (See note 2)	Transmitter
TXRX (See note 2)	Transceiver
UHF	Ultrahigh Frequency
UNSTAB	Unstabilised
UTC	Universal Time, Co-ordinated
UWE	Vessel Underway Using Engine (applies to AIS)
VAR	Variation
VCD	Vessel Constrained by Draught (applies to AIS)
VDR	Voyage Data Recorder
VECT	Vector
VHF	Very High Frequency
VID	Video
VLF	Very Low Frequency
VOY	Voyage
VRM	Variable Range Marker
VTS	Vessel Traffic Service
W	West
WARNING	Warning
WAT	Water
WOL	Wheel Over Line
WOT	Wheel Over Time
WPT	Waypoint
X-BAND	X-Band (applies to Radar)
XTD	Cross Track Distance

List of Standard Units of Measurement and Abbreviations

Unit	Abbreviation	Abbreviation	Unit
cable length	cbl	cbl	cable length
cycles per second	cps	cps	cycles per second
degree(s)	deg	deg	degree(s)
fathom(s)	fm	fm	fathom(s)
feet/foot	ft	ft	feet/foot
gigaHertz	GHz	GHz	gigaHertz
hectoPascal	hPa	hPa	hectoPascal
Hertz	Hz	Hz	Hertz
hour(s)	hr(s)	hr(s)	hour(s)
kiloHertz	kHz	kHz	kiloHertz
kilometre	km	km	kilometre
kiloPascal	kPa	kPa	kiloPascal
knot(s)	kn	kn	knot(s)
megaHertz	MHz	MHz	megaHertz
minute(s)	min	min	minute(s)
Nautical Mile(s)	NM	NM	Nautical Mile(s)

Notes:

1.   Terms and abbreviations used in nautical charts are published in relevant IHO publications and are not listed here.

2.   In general, terms should be presented using lower case text and abbreviations should be presented using upper case text. Those abbreviations that may be presented using lower case text are identified in the list, e.g. "dGNSS" or "Rx".

3.   Abbreviations may be combined, e.g. "CPA LIM" or "T CRS". When the abbreviation for the standard term "Relative" is combined with another abbreviation, the abbreviation "R" should be used instead of "REL", e.g. "R CRS".

4.   The use of the abbreviations "SIM" and "TRIAL" are not intended to replace the appropriate symbols listed in annex 1.

5.   The term "Course Made Good" has been used in the past to describe "Track Made Good". This is a misnomer in that "courses" are directions steered or intended to be steered with respect to a reference meridian. "Track Made Good" is preferred over the use of "Course Made Good".

6.   Where information is presented using SI units, the respective abbreviations should be used.

MCA Notes on Use of VHF and VIS

Extracts from

MARINE GUIDANCE NOTE

MCA The Maritime and Coastguard Agency   MGN 324 (M + F)

RADIO: OPERATIONAL GUIDANCE ON THE USE OF VHF RADIO AND AUTOMATIC IDENTIFICATION SYSTEMS (AIS) AT SEA

Notice to all Owners, Masters, Officers and Pilots of Merchant Ships, Owners and Skippers of Fishing Vessels and Owners of Yachts and Pleasure Craft

This notice replaces Marine Guidance Notes MGN 22, 167 & 277

Summary

Given the continuing number of casualties where the misuse of VHF radio has been established as a contributory factor it has been decided to re-issue the MCA Operational Guidance Notes on the use of VHF Radio. It has also been decided to include operational guidance notes for AIS equipment on board ship formerly contained in Marine Guidance Notice 277.

Key Points
- The use of marine VHF equipment must be in accordance with the International Telecommunications Union (ITU) Radio Regulations.
- Although the use of VHF radio may be justified on occasion as a collision avoidance aid, the provisions of the Collision Regulations should remain uppermost.
- There is no provision in the Collision Regulations for the use of AIS information therefore decisions should be taken based primarily on visual and/or radar information.
- IMO Guidelines on VHF Communication Techniques are given in Appendix I.

- Typical VHF ranges and a Table of Transmitting frequencies in the Band 156 – 174 MHz for Stations in the Maritime Mobile Service is shown at Appendix II.
- IMO Guidelines for the Onboard Operational Use of Shipborne Automatic Identification Systems (AIS) is shown in Appendix III.
- MCA Guidance on the use of AIS in Navigation together with a list of MCA AIS base stations is shown in Appendix IV.

1.   The International Maritime Organisation (IMO) has noted with concern the widespread misuse of VHF channels at sea especially the distress, safety and calling Channels 16 (156.8 MHz) and 70 (156.525 MHz), and channels used for port operations, ship movement services and reporting systems. Although VHF at sea makes an important contribution to navigation safety, its misuse causes serious interference and, in itself, becomes a danger to safety at sea. IMO has asked Member Governments to ensure that VHF channels are used correctly.

2.   All users of marine VHF on United Kingdom vessels, and all other vessels in United Kingdom territorial waters and harbours, are therefore reminded, in conformance with international and national legislation, marine VHF apparatus may only be used in accordance with the International Telecommunications Union's (ITU) Radio Regulations. These Regulations specifically prescribe that:

(a)   Channel 16 may only be used for distress, urgency and very brief safety communications and for calling to establish other communications which should then be concluded on a suitable working channel;
(b)   Channel 70 may only be used for Digital Selective Calling not oral communication;
(c)   On VHF channels allocated to port operations or ship movement services such as VTS, the only messages permitted are restricted to those relating to operational handling, the movement and the safety of ships and to the safety of persons;
(d)   All signals must be preceded by an identification, for example the vessel's name or callsign;
(e)   The service of every VHF radio telephone station must be controlled by an operator holding a certificate issued or recognised by the station's controlling administration. This is usually the country of registration, if the vessel is registered. Providing the Station is so controlled, other persons besides the holder of the certificate may use the equipment.

3.   Appendix I to this notice contains the IMO Guidance on the use

of VHF at sea. Masters, Skippers and Owners must ensure that VHF channels are used in accordance with this guidance.

4.   Appendix II to this notice illustrates typical VHF ranges and a table of transmitting Frequencies in the Band 156 – 174 MHz for Stations in the Maritime Mobile Service, incorporating changes agreed by the 1997 World Radio Conference.

5.   Channels 6, 8, 72 and 77 have been made available, in UK waters, for routine ship-to-ship communications, Masters, Skippers and Owners are urged to ensure that all ship-to-ship communications working in these waters is confined to these channels, selecting the channel most appropriate in the local conditions at the time.

6.   Channel 13 is designated for use on a worldwide basis as a navigation safety communication channel, primarily for intership navigation safety communications. It may also be used for the ship movement and port services.

Use of VHF as Collision Avoidance Aid

7.   There have been a significant number of collisions where subsequent investigation has found that at some stage before impact, one or both parties were using VHF radio in an attempt to avoid collision. The use of VHF radio in these circumstances is not always helpful and may even prove to be dangerous.

8.   At night, in restricted visibility or when there are more than two vessels in the vicinity, the need for positive identification is essential but this can rarely be guaranteed. Uncertainties can arise over the identification of vessels and the interpretation of messages received. Even where positive identification has been achieved there is still the possibility of a misunderstanding due to language difficulties however fluent the parties concerned might be in the language being used. An imprecise or ambiguously expressed message could have serious consequences.

9.   Valuable time can be wasted whilst mariners on vessels approaching each other try to make contact on VHF radio instead of complying with the Collision Regulations. There is the further danger that even if contact and identification is achieved and no difficulties over the language of communication or message content arise, a course of action might still be chosen that does not comply with the Collision

Regulations. This may lead to the collision it was intended to prevent.

10.   In 1995, the judge in a collision case said "It is very probable that the use of VHF radio for conversation between these ships was a contributory cause of this collision, if only because it distracted the officers on watch from paying careful attention to their radar. I must repeat, in the hope that it will achieve some publicity, what I have said on previous occasions that any attempt to use VHF to agree the manner of passing is fraught with the danger of misunderstanding. Marine Superintendents would be well advised to prohibit such use of VHF radio and to instruct their officers to comply with the Collision Regulations."

11.   In a case published in 2002 one of two vessels, approaching each other in fog, used the VHF radio to call for a red to red (port to port) passing. The call was acknowledged by the other vessel but unfortunately, due to the command of English on the calling vessel, what the caller intended was a green to green (starboard to starboard) passing. The actions were not effectively monitored by either of the vessels and collision followed.

12.   Again in a case published in 2006 one of two vessels, approaching one another to involve a close quarter's situation, agreed to a starboard to starboard passing arrangement with a person on board another, unidentified ship, but not the approaching vessel. Furthermore, the passing agreement required one of the vessels to make an alteration of course, contrary to the requirements of the applicable Rule in the COLREGS. Had the vessel agreed to a passing arrangement requiring her to manoeuvre in compliance with the COLREGS, the ships would have passed clear, despite the misidentification of ships on the VHF radio. Unfortunately by the time both vessels realised that the ships had turned towards each other the distance between them had further reduced to the extent that the last minute avoiding action taken by both ships was unable to prevent a collision.

13.   Although the practice of using VHF radio as a collision avoidance aid may be resorted to on occasion, for example in pilotage waters, the risks described in this note should be clearly understood and the Collision Regulations complied with.

Use of VHF Automatic Identification Systems (AIS)

14.   AIS operates primarily on two dedicated VHF channels (AIS1 – 161.975 MHz and AIS2 – 162.025 MHz). Where these channels are not available regionally, the AIS is capable of automatically switching to alternate designated channels. AIS has now been installed on the majority of commercial vessels, and has the potential to make a significant contribution to safety. However the mariner should treat the AIS information with caution, noting the following important points:

15.   Mariners on craft fitted with AIS should be aware that the AIS will be transmitting own-ship data to other vessels and shore stations.
  To this end they are advised to:

> 15.1 initiate action to correct improper installation;
>
> 15.2 ensure the correct information on the vessel's identity, position, and movements (including voyage-specific, see Annex IV) is transmitted; and
>
> 15.3 ensure that the AIS is turned on, at least within 100 nautical miles of the coastline of the United Kingdom.

16.   The simplest means of checking whether own-ship is transmitting correct information on identity, position and movements is by contacting other vessels or shore stations. Increasingly, UK Coastguard and port authorities are being equipped as AIS shore base stations. As more shore base stations are established, AIS may be used to provide a monitoring system in conjunction with Vessel Traffic Services and Ship Reporting (SOLAS Chapter V, Regulations 11 and 12 refer).

17.   Many ship owners have opted for the least-cost AIS installation to meet the mandatory carriage requirement. By doing so, many of the benefits offered by graphic display (especially AIS on radar) are not realised with the 3-line 'Minimum Keyboard Display' (MKD).

18.   The Pilot Connector Socket and suitable power outlet should be located somewhere of practical use to a marine pilot who may carry compatible AIS equipment. This should be somewhere close to the wheelhouse main conning position. Less accessible locations in chart rooms, at the after end of the wheelhouse are not recommended.

19.   The routine updating of data into the AIS, at the start of the voyage and whenever changes occur, should be included in the navigating officer's checklist and include:

– ship's draught;
– hazardous cargo;
– destination and ETA;
– route plan (way points);
– correct navigational status;
– short safety-related messages.

20.   The quality and reliability of position data obtained from targets
will vary depending on the accuracy of the transmitting vessel's GNSS
equipment. It should be noted that older GNSS equipment may not
produce Course Over Ground and Speed Over Ground (COG/SOG)
data to the same accuracy as newer equipment.

21.   Operational guidance for Automatic Identification Systems (AIS)
on board ships can be found in the MCA Guidance on the Safety of
Navigation – Implementing SOLAS Chapter V (accessible from the
MCA website at www.mcga.gov.uk) and reproduced in Appendix IV of
this notice.

APPENDIX I

GUIDANCE ON THE USE OF VHF AT SEA

(Extract from: IMO Resolution A.954 (23). Proper use of VHF Channels at Sea (Adopted on 5th December 2003))

1.   VHF COMMUNICATION TECHNIQUE

1.1 Preparation
Before transmitting, think about the subjects which have to be communicated and, if necessary, prepare written notes to avoid unnecessary interruptions and ensure that no valuable time is wasted on a busy channel.

1.2 Listening
Listen before commencing to transmit to make certain that the channel is not already in use. This will avoid unnecessary and irritating interference.

1.3 Discipline
    (a) VHF equipment should be used correctly and in accordance with the Radio Regulations. The following in particular should be avoided:

    (b) calling on channel 16 for purposes other than distress, and very brief safety communications, when another calling channel is available;

    (c) non-essential transmissions, e.g. needless and superfluous signals and correspondence;

    (d) communications not related to safety and navigation on port operation channels; communication on channel 70 other than for Digital Selective Calling;

    (e) occupation of one particular channel under poor conditions;

    (f) transmitting without correct identification;

    (g) use of offensive language.

1.4 Repetition
Repetition of words and phrases should be avoided unless specifically requested by the receiving station.

1.5 Power reduction
When possible, the lowest transmitter power necessary for satisfactory communication should be used.

1.6 Automatic identification system (AIS)

AIS is used for the exchange of data in ship-to-ship communications and also in communication with shore facilities. The purpose of AIS is to help identify vessels, assist in target tracking, simplify information exchange and provide additional information to assist situational awareness. AIS may be used together with VHF voice communications.

AIS should be operated in accordance with Resolution A.917 (22) as amended by Resolution A.956 (23) on Guidelines for the onboard operation use of shipborne automatic identification systems.

1.7 Communications with coast stations

On VHF channels allocated to port operations service, the only messages permitted are restricted to those relating to the operational handling, the movement and safety of ships and, in emergency, to the safety of persons, as the use of these channels for ship-to-ship communications may cause serious interference to communications related to the movement and safety of shipping in port areas.

Instructions given on communication matters by shore stations should be obeyed.

Communications should be carried out on the channel indicated by the shore station. When a change of channel is requested, this should be acknowledged by the ship.

On receiving instructions from a shore station to stop transmitting, no further communications should be made until otherwise notified (the shore station may be receiving distress or safety messages and any other transmissions could cause interference).

1.8 Communications with other ships

VHF Channel 13 is designated by the Radio Regulations for bridge to bridge communications. The ship called may indicate another working channel on which further transmissions should take place. The calling ship should acknowledge acceptance before changing channels.

The listening procedure outlined above should be followed before communications are commenced on the chosen channel.

1.9 Distress communications

Distress calls/messages have absolute priority over all other communications. When heard, all other transmissions should cease and a listening watch should be kept.

Any distress call/message should be recorded in the ship's log and passed to the master.

On receipt of a distress message, if in the vicinity, immediately acknowledge receipt. If not in the vicinity, allow a short interval of time to elapse before acknowledging receipt of the message in order to permit ships nearer to the distress to do so.

1.10 Calling

In accordance with the radio regulations Channel 16 may only be used for distress, urgency and very brief safety communications and for calling to establish other communications which should then be conducted on a suitable working channel.

Whenever possible, a working frequency should be used for calling. If a working frequency is not available, Channel 16 may be used, provided it is not occupied by a distress call/message.

In case of difficulty to establish contact with a ship or shore station, allow adequate time before repeating the call. Do not occupy the channel unnecessarily and try another channel.

1.11 Changing channels

If communications on a channel are unsatisfactory, indicate change of channel and await confirmation.

1.12 Spelling

If spelling becomes necessary use the spelling table contained in the International Code of Signals and the radio regulations and the IMO Standard Marine Communication Phrases (SMCP)

1.13 Addressing

The words "I" and "You" should be used prudently. Indicate to whom they refer. Example of good practice: "Seaship, this is Port Radar, Port Radar, do you have a pilot?" "Port Radar, this is Seaship, I do have a pilot."

1.14 Watchkeeping

Every ship, while at sea, is required to maintain watches. Continuous watch keeping is required on VHF DSC Channel 70 and also when practicable, a continuous listening watch on VHF Channel 16.

In certain cases Governments may require ships to keep a watch on other channels.

APPENDIX II

VHF RANGES

It should be noted that the fact that a transmitter and receiver are within radio sight does not automatically guarantee that an acceptable signal will be received at that point. This will depend, amongst other things on the power of transmission, the sensitivity of the receiver and the quality and position of the transmitting and receiving aerials. The range may also be affected to some degree by the pressure, temperature and humidity of the air between the transmitter and receiver.

OPERATION OF AIS ON BOARD

(Extract from IMO Resolution A.917. (22). Guidelines for the onboard operational use of shipborne Automatic Identification Systems (AIS) (Adopted on 29th November 2001). As amended by Resolution A.956. (23). (Adopted 5th December 2003).

INHERENT LIMITATIONS OF AIS

31.   The officer of the watch (OOW) should always be aware that other ships, in particular leisure craft, fishing boats and warships, and some coastal shore stations including Vessel Traffic Service (VTS) centres, might not be fitted with AIS.

32.   The OOW should always be aware that other ships fitted with AIS as a mandatory carriage requirement might switch off AIS under certain circumstances by professional judgement of the master.

33.   In other words, the information given by the AIS may not be a complete picture of the situation around the ship.

34.   The users must be aware that transmission of erroneous information implies a risk to other ships as well as their own. The users remain responsible for all information entered into the system and the information added by the sensors.

35.   The accuracy of the information received is only as good as the accuracy of the AIS information transmitted.

36.   The OOW should be aware that poorly configured or calibrated ship sensors (position, speed and heading sensors) might lead to incorrect information being transmitted. Incorrect information about one ship displayed on the bridge of another could be dangerously confusing.

37.   If no sensor is installed or if the sensor (e.g. the gyro) fails to provide data, the AIS automatically transmits the 'not available' data value. However the built in integrity check cannot validate the contents of the data processed by the AIS.

38.   It would not be prudent for the OOW to assume that the information

received from the other ship is of a comparable quality and accuracy to that which might be available on own ship.

USE OF AIS IN COLLISION AVOIDANCE SITUATIONS

39. The potential of AIS as an anti collision device is recognised and AIS may be recommended as such a device in due time.

40. Nevertheless, AIS information may be used to assist collision avoidance decision making. When using the AIS in the ship to ship mode for anti collision purposes, the following precautionary points should be borne in mind:

    (a) AIS is an additional source of navigational information. It does not replace, but supports, navigational systems such as radar target tracking and VTS; and

    (b) The use of AIS does not negate the responsibility of the OOW to comply at all times with the Collision Regulations

41. The user should not rely on AIS as the sole information system, but should make use of all safety relevant information available

42. The use of AIS on board ship is not intended to have any special impact on the composition of the navigational watch, which should be determined in accordance with the STCW Convention.

43. Once a ship has been detected, AIS can assist tracking it as a target. By monitoring the information broadcast by that target, its actions can also be monitored. Changes in heading and course are, for example, immediately apparent, and many of the problems common to tracking targets by radar, namely clutter, target swap as ships pass close by and target loss following a fast manoeuvre, do not affect AIS. AIS can also assist in the identification of targets, by name or call sign and by ship type and navigational status.

USE OF AIS IN NAVIGATION

*(Extract from MCA Guidance on the Safety of Navigation –
Implementing SOLAS Chapter V)*

1.   AIS is designed to be able to provide additional information to existing Radar or ECDIS displays. Until the optimum display modes have been fully evaluated and decided upon internationally, AIS will comprise "stand alone" units without integration to other displays.

2.   AIS will provide identification of targets together with the static and dynamic information listed in the IMO Guidelines paragraph.12. Mariners should, however, use this information with caution noting the following important points:

(a)   Collision avoidance must be carried out in strict compliance with the COLREGs. There is no provision in the COLREGs for use of AIS information therefore decisions should be taken based primarily on visual and / or radar information.

(b)   The use of VHF to discuss actions to take between approaching ships is fraught with danger and still discouraged. (See above). The MCA's view is that identification of a target by AIS does not remove the danger. Decisions on collision avoidance should be made strictly according to the COLREGs.

(c)   Not all ships will be fitted with AIS, particularly small craft and fishing boats. Other floating objects which may give a radar echo will not be detected by AIS.

(d)   AIS positions are derived from the target's GNSS position. (GNSS=Global Navigation Satellite System, usually GPS). This may not coincide exactly with the target.

(e)   Faulty data input to AIS could lead to incorrect or misleading information being displayed on other vessels. Mariners should remember that information derived from radar plots relies solely upon data measured by the own-ship's radar and provides an accurate measurement of the target's relative course and speed, which is the most important factor in deciding upon action to avoid collision. Existing ships of less than 500 gt. which are not required to fit a gyro compass are unlikely to transmit heading information.

(f)   A future development of AIS is the ability to provide synthetic AIS targets and virtual navigation marks enabling coastal authorities to provide an AIS symbol on the display in any position. Mariners should bear in mind that this ability could lead to the appearance of "virtual" AIS targets and therefore take particular care when an AIS target is not complemented by a radar target. AIS will sometimes be able to detect targets which are in a radar shadow area.

Index